Advances in

ASTRONOMY AND ASTROPHYSICS

VOLUME 8

CONTRIBUTORS TO THIS VOLUME

John V. Evans
Tor Hagfors
E. J. Öpik

ADVANCES IN
ASTRONOMY and
ASTROPHYSICS

edited by

ZDENĚK KOPAL

Department of Astronomy
University of Manchester
Manchester, England

Volume 8

 ACADEMIC PRESS New York and London • 1971

ACADEMIC PRESS, INC.
111 Fifth Avenue, New York, New York 10003

United Kingdom Edition published by
ACADEMIC PRESS, INC. (LONDON) LTD.
Berkeley Square House, London W1X 6BA

LIBRARY OF CONGRESS CATALOG CARD NUMBER: 61-18299

PRINTED IN THE UNITED STATES OF AMERICA

CONTENTS

Microwave Studies of Thermal Emission from the Moon

TOR HAGFORS

Radar Studies of the Moon

JOHN V. EVANS AND TOR HAGFORS

Cratering and the Moon's Surface

E. J. Öpik

CONTRIBUTORS TO VOLUME 8

JOHN V. EVANS, *Lincoln Laboratory, Massachusetts Institute of Technology, Lexington, Massachusetts*

TOR HAGFORS,* *Institute for Telecommunication Sciences and Aeronomy, ESSA, Boulder, Colorado and Radio Observatorio de Jicamarca, Instituto Geofisico del Peru, Lima, Peru*

E. J. ÖPIK, *Armagh Observatory, Armagh, Northern Ireland and Department of Physics and Astronomy, University of Maryland, College Park, Maryland*

* Present address: Lincoln Laboratory, Massachusetts Institute of Technology, Lexington, Massachusetts.

CONTENTS OF PREVIOUS VOLUMES

Microwave Studies of Thermal Emission from the Moon

TOR HAGFORS*

Institute for Telecommunication Sciences and Aeronomy
ESSA, Boulder, Colorado
and
Radio Observatorio de Jicamarca
Instituto Geofisico del Peru
Lima, Peru

I

INTRODUCTION

The first observations of microwave emission from the Moon were reported by Dicke and Beringer as early as 1946. These initial exploratory studies were followed by extensive observations by Piddington and Minnett (1949) at a wavelength of 1.25 cm. It was found from these studies that the emission temperature of the Moon as a whole varies as an approximately sinusoidal function of lunar phase, but with an amplitude very considerably less than that at infrared frequencies, as measured by Pettit and Nicholson (1930) and by Pettit (1935). A quantitative understanding of this behavior of the thermal emission was gained through the pioneering theoretical work of Troitsky (1954).

* Present address: Lincoln Laboratory, Massachusetts Institute of Technology, Lexington, Massachusetts.

With the advent of new radio telescopes and new measurement techniques, a large number of observations of various properties of the lunar thermal emission have been made, over wavelengths ranging from 0.13 cm (Fedoseev, 1963) to 168 cm (Baldwin, 1961). These measurements have either given mean temperature over the whole lunar disk or the average temperature over a limited region of the lunar surface. The emission temperature has also been studied as a function of lunar phase, as a function of time during lunar eclipses, as a function of position on the lunar surface, and as a function of the polarization of the emitted radiation. Many of these characteristics have been studied extensively at a number of different wavelengths.

As a result of these studies, comparisons with other types of observations, and with measurements of electrical properties of both man-made and naturally occurring materials, many scientists have drawn quite detailed, far-reaching, and often controversial conclusions about the likely nature of the lunar surface layer. The recent landings of instrumented packages and of man on the Moon have provided data which contain much more accurate information about the nature of the lunar soil than one can hope to obtain from ground-based studies. The *in-situ* measurements of the properties of the lunar surface soil will, however, only provide accurate information on a certain number of very limited areas on the Moon. Ground based observations, including microwave emission measurements, may thus in the future assume the role of interpolating between or extrapolating from the various spots on the Moon where the properties of the surface are well known. One should also keep in mind that many of the microwave observation techniques which have found application to the study of the Moon, in the future certainly will be applied to planetary studies and as such are of considerable value in themselves.

In what follows we first review methods of observation and various limitations imposed in practice by the equipment, by noise, etc. We then proceed to discuss the origin of the thermal emission of the moon at microwave frequencies and to evaluate the effect of various specific surface attributes on the characteristics of the microwave emission. The extensive observational material which is now available is briefly reviewed and discussed in the light of the theoretical predictions based on various lunar surface models. In the final section, a summary is given of the various conclusions which may be drawn with reasonable confidence from the microwave observations about some of the physical properties of the lunar surface material.

II

METHODS OF OBSERVATIONS AND DEFINITIONS

In this section we shall define a number of quantities used in the subsequent descriptions of the observation of microwave emission from the Moon. We introduce these quantities in order to facilitate our discussion of observational methods and sources of error. For greater detail the reader is referred to a number of excellent standard texts such as Pawsey and Bracewell (1955), Bracewell (1962), or Steinberg and Lequeux (1963).

The monochromatic brightness of an emitting black (perfectly absorbing) body is defined by

$$B(\nu) = \frac{2KT}{c^2}\, \nu^2. \tag{1}$$

Here $B(\nu)$ is the amount of power emitted per unit projected area per steradian and per Hertz. This approximate formula is applicable with very high accuracy for all microwave observations. Here K is Boltzmann's constant (1.38×10^{-23} joules/°K), c is the velocity of light (2.9979×10^8 m/sec), T is the absolute temperature, and ν is the frequency of the radiation. The power given by (1) is the total power in the two orthogonal polarizations of the emitted radiation.

A body which is not perfectly absorbing is often referred to as a gray body. Let the absorption coefficient near the frequency ν be $e(\nu)$. For a given temperature T, the amount of emission from a gray body is reduced by a factor of $e(\nu)$ with respect to the emission from a perfectly black body of the same temperature. The apparent temperature, or brightness temperature, is given by $T_b(\nu) = e(\nu)T$. The brightness temperature of a body is, therefore, equal to the temperature of a blackbody with the monochromatic temperature $T_b(\nu)$.

The most important element in a radio telescope is the antenna. The purpose of the antenna is to transform the flux from the source into an electrical signal which can be made available to a radio receiver. The antenna can be thought of as an aperture which absorbs the radiation incident upon it. This ability to absorb radiation will be strongly dependent on the direction of the incoming flux with respect to the axis of the antenna beam. Let the direction of the antenna beam be represented by the vector $\mathbf{n_0}$ and the direction of the incoming flux by $\mathbf{n}$. If the incoming flux from this direction is $S(\mathbf{n})$, the power available at the antenna terminals will be

$$P = S(\mathbf{n}) \cdot A(\mathbf{n} - \mathbf{n_0}) \tag{2}$$

where $A(\mathbf{n} - \mathbf{n_0})$ is the effective collecting area of the antenna in a direction offset from the antenna beam axis by $\mathbf{n} - \mathbf{n_0}$. With an extended source with a brightness distribution $B(\mathbf{n})$, the differential flux from the direction $\mathbf{n}$ will be

$$dS(\mathbf{n}) = B(\mathbf{n}) \cdot d(\mathbf{n}) \tag{3}$$

where $d(\mathbf{n})$ signifies a solid angle element. The total available power at the antenna terminals will hence be

$$P(\mathbf{n_0}) = \int B(\mathbf{n})\, A(\mathbf{n} - \mathbf{n_0})\, d(\mathbf{n}). \tag{4}$$

The collecting aperture $A(\mathbf{n} - \mathbf{n}_0)$ will usually be a sharply peaked function about the direction $\mathbf{n} = \mathbf{n}_0$. The quality of the antenna is often specified in terms of the angular separation of the peak at $\mathbf{n}_0$ and some direction in which the function has dropped to half the maximum value. Twice this angle is referred to below as half power beam width. Not all the power will enter the antenna through the main beam. The fraction of the power entering through the main beam if the antenna were placed in an enclosure of uniform brightness is referred to as the beam efficiency. It rarely exceeds 60% and is often considerably smaller. In practice there will be losses in the antenna feedlines. This causes the available power at the antenna terminals to be divided between signals generated by thermal effects in the equivalent antenna loss resistance and the signals actually entering the antenna aperture. The evaluation of the various antenna parameters, such as beam efficiency and ohmic losses, is often very difficult and the uncertainty in some of these parameters may be the chief source of error in many of the observations of thermal emission from the Moon. In addition, for some experiments relying on accurate pointing of the antenna, the pointing accuracy may become a limiting factor. Even the best antennas will have pointing uncertainties of some $20''$ of arc and atmospheric refraction effects can become quite serious, particularly at low angles of elevation, i.e., less than $15°$ from the horizon.

In addition to the *intensity* of the flux from a source, one is often interested in knowing the polarization of the received radiation. This is achieved by splitting the received flux into two orthogonally polarized modes by an antenna feed system which is sensitive to two orthogonally polarized waves, either linearly or circularly polarized. The particular set of orthogonal polarizations is not important since any set of orthogonal polarizations may be synthesized by linear combination of any other orthogonal set. In many systems in use in lunar observations, only one linear polarization is observed at one time. The orthogonal polarization is obtained by rotating the feed element, and such a rotation may be accompanied by small changes in the collecting aperture due to asymmetries in feed supports, etc.

The receivers used in radio astronomy can be thought of as bandpass amplifiers with power detectors at the output. A certain amount of noise is added by the amplifier. This noise power is added on to the thermal noise available at the antenna terminals. The total equivalent temperature will, therefore, be the sum of the equivalent temperature of the signal available at the antenna terminals and the noise temperature of the receiver. In most cases the noise temperature of the receiver is less than $1000°K$ and in some of the more recent masers and parametric amplifiers, the noise contribution from the amplifier may only amount to a few tens a degrees. The precision with which the antenna temperature can be measured is determined by the number of independent samples N involved in a measurement. If the width of the passband of the amplifier is $\Delta\nu$, one can make $\Delta\nu$ independent power measurements per second. With an integration time t_0, the number of samples becomes $t_0\Delta\nu$. The relative uncertainty in the temperature determination hence becomes

$$\frac{\Delta T}{T} = \frac{1}{(t_0 \cdot \Delta\nu)^{1/2}}. \tag{5}$$

Bandwidths used in radio observations of the Moon are frequently on the order of 100 to 1000 MHz and typical mean temperatures T are on the order of 300 to 1000°K, including receiver noise. This means that the statistical uncertainty in the temperature determinations can be made better than 0.1°K with an integration time of 1 sec. In many cases, gain and zero point drift are much more important contributors to the precision of the measurements. Elaborate calibration schemes have been worked out and are currently in use whereby the theorectical precision dictated by statistical fluctuations alone can be approached.

III

THE ORIGIN OF THE RADIO BRIGHTNESS OF THE MOON

The radio brightness of the Moon may arise as a result of several different mechanisms. The Moon may appear bright at radio wavelengths because of reflected solar radiation. This, of course, is the primary reason for the Moon to appear bright to the eye. The radiation from the Moon could result from thermal emission from the interior. This is, as we shall see, the primary source of radiation from the Moon. A third cause of radiation might be the reflection of galactic radio noise at the lunar surface.

Consider first the reflection of solar radiation from the Moon. The solar flux density at the Moon is $B_s(\nu)\Omega_s$, where $B_s(\nu)$ is the mean disk brightness of the Sun and where Ω_s is the solid angle subtended by the sun at the Moon. A fraction R of this flux is reflected and let us assume that this flux is uniformly distributed in all directions. The resulting flux at the Earth of reflected solar radiation hence becomes

$$S = \Omega_m\{R\Omega_s B_s(\nu)/4\pi\} \tag{6}$$

where Ω_m is the solid angle subtended by the Moon at the Earth. It follows that the mean equivalent brightness temperature of the Moon due to reflection from the Sun is

$$T_{ms} = T_s\{R\Omega_s/4\pi\}. \tag{7}$$

The fraction R is quite close to 0.07 (see Evans and Hagfors chapter, p. 29) at all microwave frequencies and the expression within curly brackets in (7) is approximately equal to 0.35×10^{-6}. A solar brightness temperature of 10^6 °K will hence only contribute 0.35°K to the disk brightness temperature. Figure 1 shows the brightness temperature of the Sun referred to the size of the visible disk for quiet

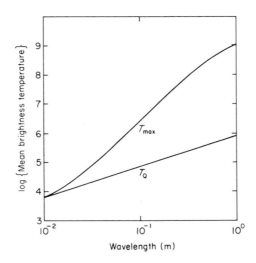

FIG. 1.   Extremes of solar brightness plotted against wavelength.

and for disturbed conditions for wavelengths between 1 cm and 1 m. It will be seen that the contribution from the quiet Sun is always less than 10^6 °K and hence will contribute less than 1°K to the lunar brightness. For wavelengths in excess of 10 cm, however, the contribution to the lunar brightness from the disturbed Sun may well become appreciable. Data therefore have to be collected with this in mind. In what follows we shall neglect the contribution to the brightness due to reflection of radiation from the Sun.

The cosmic noise temperature at 1000 MHz varies from about 2° to about 20°K depending on the direction. Since the reflectivity is on the order of 0.07 this means that the cosmic radiation reflected in the Moon will contribute less than 2°K to the brightness of the Moon. Since the thermal emission temperature of the Moon is about 230°K, this is rather inappreciable. However, the cosmic noise temperature depends on frequency as $\nu^{-2.8}$ so that at some frequency slightly below 100 MHz the cosmic noise reflected in the Moon will contribute more than the thermal emission. This, incidentally, may offer some interesting possibilities for measuring the spherical albedo of the Moon at frequencies below 100 MHz.

Emission of thermal origin is the only source of radio brightness we shall be concerned with henceforth. The monochromatic brightness associated with *one* polarization as a result of thermal emission was shown by Piddington and Minnett (1949) and by Troitsky (1954) to be

$$T_b(\nu) = [1 - R(\nu, \theta_{\text{out}})] \int_{-\infty}^{0} T(z)\, k(\nu)\, \frac{dz}{\cos \theta_{\text{i}}}\, e^{-|z|k(\nu)/\cos \theta_{\text{i}}}\,. \tag{8}$$

The significance of the quantities not previously defined is as follows:

$$k(\nu) = \text{power absorption coefficient at frequency } \nu,$$
$$T(z) = \text{temperature at depth } z,$$
$$R(\nu, \theta_{\text{out}}) = \text{power reflection coefficient for the mode of}$$
$$\text{polarization observed.}$$

The remaining geometrical quantities are explained in Fig. 2. For the particu-

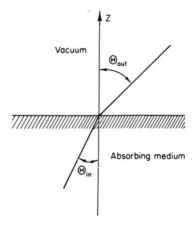

F<small>IG</small>. 2.   Geometry of thermal radio waves emitted from a dielectric surface.

lar linear polarization with E-field *in* the plane of incidence (in the plane of the paper in Fig. 2) or E-field *normal* to this plane, the reflection coefficients are usual taken as

$$R_{\parallel}(\nu, \theta) = \left| \frac{\epsilon \cos \theta - (\epsilon - \sin^2 \theta)^{1/2}}{\epsilon \cos \theta + (\epsilon - \sin^2 \theta)^{1/2}} \right|^2,$$

$$R_{\perp}(\nu, \theta) = \left| \frac{\cos \theta - (\epsilon - \sin^2 \theta)^{1/2}}{\cos \theta + (\epsilon - \sin^2 \theta)} \right|^2 \tag{9}$$

where ϵ is the complex relative dielectric constant of the lossy dielectric. Before proceeding we should enumerate the various assumptions which are implicit in Eq. (8) for the surface brightness. As it stands, it assumes that the real and imaginary parts of the dielectric constant are independent of depth causing the rays in the medium to follow straight lines. The possibility of scattering of the waves due to inhomogenieties embedded in the medium has also been excluded. The choice of the Fresnel reflection coefficients to evaluate the emissivity of the surface is, furthermore, based on the assumption of a smooth boundary. All of these assumptions, it should be kept in mind, are sure to be violated to a greater or lesser degree when we are dealing with the lunar surface. In each experiment discussed it is therefore necessary somehow to assess the importance of the various assumptions.

The assumption about independence of electrical properties with depth may be removed by raytracing through a slowly varying medium. If the variation is too rapid, a full wave solution must be invoked to determine the coupling of thermal emission from a volume element to the waves emanating on the outside of the medium. In either of these cases, the calculations may be carried out for specific models, but the results are likely to be very involved. The assumption about absence of scattering within the medium can probably not be removed from the theory in practice. The third assumption regarding the smoothness of the surface can be taken into account for some idealized models as we shall see below.

IV

VARIATIONS IN THE BRIGHTNESS WITH TIME

The temperature $T(z)$ at a given point in the surface layer is in part determined by solar irradiation and also, perhaps, by internal heat sources of radioactive origin. The equations required to determine $T(z)$ as a function of time, i.e., $T(z, t)$ are (Wesselink, 1948; Jaeger, 1953)

$$\frac{\partial T(z, t)}{\partial t} = \frac{\kappa}{\rho C_\mathrm{m}} \frac{\partial^2 T(z, t)}{\partial z^2}$$

$$(1 - R_\mathrm{i})\sigma\, T(0, t)^4 - (1 - R_\mathrm{v})\, S_0 f(t) = - \left.\frac{\partial T(z, t)}{\partial z}\right|_{z=0}.$$

(10)

Here

κ = thermal conductivity,

ρ = density of the material,

C_m = specific heat (referred to unit mass),

σ = 5.679×10^{-8} W m^{-2} · deg^{-4} (Stefan Boltzmann's constant),

S_0 = solar constant = 2 cal/cm^2 · min = 1.4 kW/m^2.

The reflectivities R_i and R_v refer to the wavelengths where the principal emission (infrared) and the principal absorption (visible) take place, respectively.

The combination $a^2 = \kappa/\rho C_m$ is also known as the diffusivity. The insolation function is denoted by $f(t)$ and will, of course, be a function of position on the Moon. This insolation function has two different forms, one arising as a result of the rotation of the Moon on its axis and the other as a result of a lunar eclipse.

A. Lunation Variations

Jaeger (1953) and recently Krotikov and Shchuko (1963) solved these equations for lunation variations assuming no lateral heat conduction and no net mean heat flux from the Moon. It was also assumed that the Moon rotates on an axis which is perpendicular to the ecliptic (it actually deviates from this by only $1°32'$). The period of the insolation function $f(t)$ is a synodic month. These solutions of Eq. (10) take the thermal conductivity as well as the specific heat of the surface material as constants independent of the temperature. In view of the extreme surface temperature variations on the Moon, this may not be an accurate assumption. In fact, laboratory experiments on materials which one might expect to encounter on the Moon have shown a marked temperature dependence of the thermal conductivity κ, possibly because of radiative heat transfer inside the material (Bernett et al., 1963; Buettner, 1963; Watson, 1964). The variation of specific heat C_m with temperature does not, however, seem to be fast enough to be of importance (Buettner, 1963). Troitsky et al. (1968) have investigated the effect of temperature dependent thermal parameters and have concluded that a temperature dependent κ will primarily affect the mean surface temperature and only at wavelengths shorter than 1.5 cm.

The common approach to the problem, therefore, is to represent the surface temperature $T_s(\phi, \psi, t)$ as a Fourier series:

$$T_s(\phi, \psi, t) = T_{s\,0}(\psi) + \sum_{l=1}^{\infty} T_{s\,l}(\psi) \cos[(2\pi t/t_0 - \phi)l - \phi_{s\,l}] \qquad (11)$$

where

ψ = selenographic latitude,

ϕ = selenographic longitude,

t_0 = synodic month = 29.53 days,

$T_{s\,l}(\psi)$ = latitude dependent lth harmonic amplitude of the surface temperature variation,

$\phi_{s\,l}$ = phase of lth harmonic of the surface temperature variation.

The various harmonic components may be determined from observational data. Infrared data unfortunately are not very abundant at present, and the scant data available are somewhat contradictory. Sinton (1962) has made high resolution maps on the daylight side of the Moon, but could not obtain good data for the nighttime temperatures due to lack of sensitivity. From his daytime observations, however, the latitude dependence of the temperature is found to follow a law of the form

$$T_{s\,i}(\psi) = T_{s\,i}(0) \cos^n \psi,$$

where $n = 0.25 \pm 0.1$. This is in fair agreement with the theoretical predictions of Krotikov and Shchuko (1963). The lunar midnight temperature has been variously quoted as $120° \pm 2°$ by Pettit and Nicholson (1930), as $106°$ by Murray and Wildey (1964), and as $100°$ by Saari (1964). Low (1965) measured the mean temperature of the cold limb of the Moon and found $90°$. Regarding the midday temperature, there is reasonably good agreement among various workers that it is about $400° \pm 10°$ (Pettit and Nicholson, 1930; Geoffrion et al., 1960; Sinton, 1962; Moroz, 1965). It appears that the values in Table I for the amplitudes and phase of the first few Fourier coefficients are in agreement with the best available data (Troitsky, 1965).

TABLE I
Surface Temperature Parameters

$(\kappa\rho C_m)^{-1/2}$ cgs units	$T_{s\,0}$	$T_{s\,1}$	$T_{s\,2}$	$\phi_{s\,1}$	$\phi_{s\,2}$	$T_{s\,m}$	$T_{s\,n}$
1000	219°	170°	36°	2°	—6°	400°	100°

Substitution of this form of surface temperature variation into the heat conduction equation and subsequent substitution of the resulting $T(z, t)$ into Eq. (8) for the brightness temperature give (Troitsky, 1954)

$$T_b(\nu, \theta_{\text{out}}) = [1 - R(\nu, \theta_{\text{out}})] \left\{ T_{s\,0}(\psi) + AL_E \cos \theta_i \right.$$
$$\left. + \sum_{l=1}^{\infty} \frac{T_{s\,l}(\psi) \cos[(2\pi t/t_0 - \phi)l - \phi_{s\,l} - \gamma_l]}{(1 + 2\delta_l \cos \theta_i + 2\delta_l^2 \cos^2 \theta_i)^{1/2}} \right\} \tag{12}$$

where

$$\delta_l = (l\pi\rho C_m/t_0\kappa)^{1/2}/k(\nu) = \beta_1 \sqrt{l}L_E = \sqrt{l}L_E/L_T = (l\pi/t_0)^{1/2} L_E/a,$$

$L_E = \lambda/2\pi \sqrt{\epsilon} \tan \Delta = $ depth of penetration of electromagnetic wave,

$\tan \Delta = $ loss tangent of dielectric,

$\lambda = $ free space wavelength of the emitted radiation,

$L_T = (t_0\kappa)^{1/2}/\pi\rho C_m = (\sqrt{t_0}/\pi)(\gamma\rho C_m)^{-1} = $ depth of penetration,

$\gamma = (\kappa\rho C_m)^{-1/2} = $ characteristic parameter of material,

$\gamma_l = \tan^{-1}[l \cos \theta_i/(1 + \delta_l \cos \theta_i)]$.

Troitsky (1961) suggested introducing the specific loss tangent $b = \tan \Delta/\rho$ as a parameter, since this quantity is practically independent of density for a given material. With the specific loss tangent introduced one obtained for δ_1

$$\delta_1 = \lambda C_m/2b(\pi t_0 \epsilon)^{1/2}.$$

Note that both θ_{out} and θ_i are functions of ψ and ϕ. The term $AL_E \cos \theta_i$ was inserted by Troitsky (1965) to account for a possible mean thermal flux out of the lunar surface, A being the mean temperature gradient near the surface caused by internal heat sources. We therefore see that the amount of variation of lunar brightness through a lunation depends on the ratio of the depths of penetration of the electromagnetic and the thermal waves. Since the former is frequency dependent, the temporal variation in brightness will be a function of the frequency of observation. In fact, the depth of penetration should be proportional to wavelength. A demand for more sophisticated models involving variations both of thermal and electrical properties with depth is suggested by some of the observational results to be presented. The accuracy of these, however, does not seem to warrant such complications at present.

B. Eclipse Variations

Troitsky (1965) has discussed the variation in brightness of the radio emission of the Moon during an eclipse in terms of Eq. (10), assuming the surface temperature to be known from infrared measurements. The surface temperature variation is divided into three separate intervals. In each of these, the temperature is assumed to vary linearly with time. Let the times of first, second, third, and fourth contact during the eclipse be denoted by t_1, t_2, t_3, and t_4, respectively, and let the associated temperatures be T_1, T_2, T_3, and T_4. The slopes of the three straight lines of the surface temperature variation are then determined by

$$\alpha_n = \frac{T_{n+1} - T_n}{t_{n+1} - t_n}, \qquad n = 1, 2, 3.$$

The minimum temperature in the microwave region occurs at t_3 and the drop in brightness at the center of the lunar disk becomes

$$\Delta T(\nu) = \left[-(1 - R_0) \frac{4k(\nu)\,a}{3\sqrt{\pi}} \{\alpha_1(t_3 - t_1)^{3/2} + (\alpha_2 - \alpha_1)(t_3 - t_2)^{3/2}\} \right.$$
$$\left. + (1 + R_0) \frac{k(\nu)^2\,a^2}{2} \{\alpha_1(t_3 - t_1)^2 + (\alpha_2 - \alpha_1)(t_3 - t_2)^2\} \right]. \qquad (13)$$

Hence the drop in microwave brightness depends on the diffusivity a and the power absorption coefficient $k(\nu)$ in addition to the quantities derived from observations of the infrared surface temperature. R_0 is the power reflection coefficient at normal incidence. The microwave brightness variation thus depends on the ratio of the square root of the diffusivity and the depth of penetration of the electromagnetic wave L_E, both for the eclipse and for the lunation variation of the insolation.

V

VARIATIONS IN BRIGHTNESS
AND POLARIZATION ACROSS THE MOON

Equation (12) predicts a variation in brightness across the disk of the Moon for two primary reasons. For one, the emissivity represented by the first parenthetical factor varies with angle of incidence θ_{out} on the Moon. For the other, the brightness temperature represented by the second parenthetical factor in Eq. (12) is changing partly because of the change in the phase of the insolation over the disk, and partly because of the variation in the depth of penetration of the electromagnetic waves with angle of incidence. We shall primarily be concerned with the former cause of brightness variation because this offers a means of deducing a single parameter, viz. the dielectric constant of the medium.

Figure 3a shows an example of the brightness distribution across a lunar diameter for a pencil beam antenna on the assumption of uniform lunar temperature. The upper curve represents a linearly polarized component in the direction of the diameter of the disk (i.e., E-field in the local plane of incidence). The lower curve represents the orthogonally polarized linear polarization. Both curves assume a perfectly smooth, slightly lossy dielectric sphere. The ratio of the two components, assuming a dielectric medium, is given by

$$\frac{T_{\parallel}(\nu, \theta_{out})}{T_{\perp}(\nu, \theta_{out})} = \frac{\epsilon(\nu)[\cos\theta_{out} + (\epsilon(\nu) - \sin^2\theta_{out})^{1/2}]^2}{[\epsilon(\nu)\cos\theta_{out} + (\epsilon(\nu) - \sin^2\theta_{out})^{1/2}]^2}. \tag{14}$$

Figure 3b shows a plot of this ratio as a function of distance from the center of the lunar disk. It can be seen that a measurement of this ratio as a function of angle of incidence provides a convenient method for determining the dielectric constant of the lunar surface material.

Unfortunately the situation is not quite as simple as one might at first imagine. One major difficulty stems from the finite beamwidth of the antenna. This leads to a convolution of the distribution of emitted radiation with the antenna beam pattern as expressed in Eq. (4). The effect of such convolution is to reduce the apparent degree of polarization quite drastically, and the amount of reduction must be evaluated from an accurate knowledge of the shape of the antenna polar diagram.

It is everyday experience that the Moon is not a smooth, spherical body and

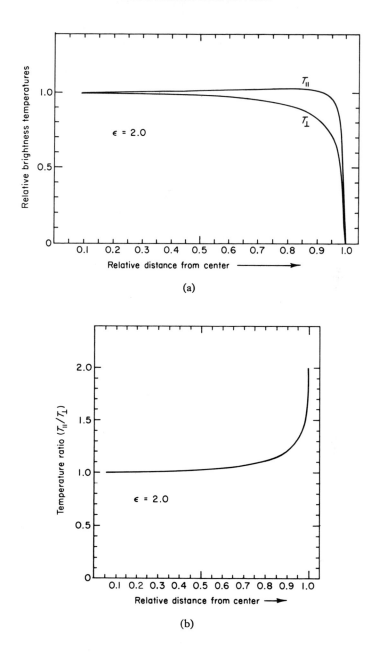

FIG. 3. (a) Brightness distribution across lunar disk for the two principal linear polarizations for uniform surface temperature; (b) ratio of emission temperatures plotted against distance from the center of the Moon.

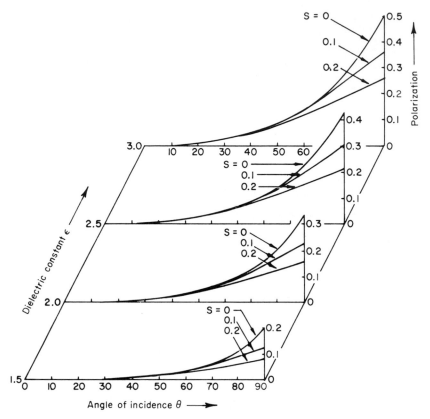

Fɪɢ. 4 (a).   Polarization of thermal emission as modified by surface roughness. Gaussian surface correlation function: S = roughness parameter, S = 0 = smooth.

one might suspect that the surface roughness could play an important role in modifying the emissivities. One way to study the effect of deviations from a smooth surface on the emissivity is to imagine the surface to be locally smooth so that Fresnel's laws of reflection or refraction may be employed, but with parameters deviating locally from those of a smooth sphere in a statistical manner. Several authors have attempted to do this (Soboleva, 1963; Moran, 1965; Davies and Gardner, 1966; Alekseev *et al.*, 1968; Rea *et al.*, 1968; Hagfors and Moriello, 1965). In the approach used in the last of these references, the surface is assumed to deviate from the mean surface by a random amount $Z(x, y)$ depending on the position on the surface. This random variable is taken

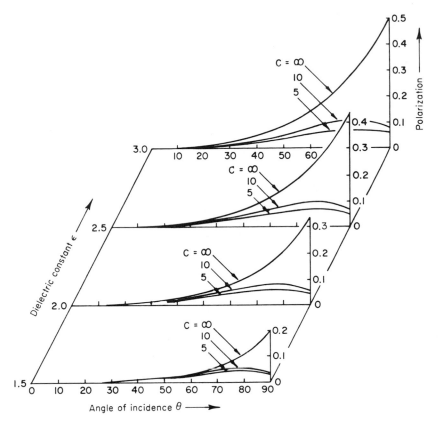

Fig. 4 (b). Polarization of thermal emission as modified by surface roughness. Exponential surface correlation function: C = roughness parameter, $C = \infty$ = smooth.

to be Gaussian with zero mean and with root-mean-square value h_0. The covariance of Z is defined by

$$\langle Z(x, y)\, Z(x + \Delta x, y + \Delta y)\rangle = h_0^2 \rho(\Delta r)$$

where

$$\Delta r = (\Delta x^2 + \Delta y^2)^{1/2}$$

(the autocorrelation function $\rho(\Delta r)$ not to be confused with density ρ). When the wave vector of the wave arriving at the observer is $\mathbf{k}$ and the wave vector of the wave inside the medium is $\mathbf{K}$, it turns out that the dominant contributions to the observed power at the observer come from surface elements with a normal given by

$$\mathbf{n} = \frac{\mathbf{K} - \mathbf{k}}{|\mathbf{K} - \mathbf{k}|}. \tag{15}$$

This contribution becomes relatively more important as the wavelength of the radiation is made small compared with the radius of curvature of the surface. Under the further assumption that

$$h_0[\mathbf{e}_z \cdot (\mathbf{K} - \mathbf{k})] > 1$$

which corresponds to "deep" phase variations across the boundary, the integration was carried out numerically over an omnidirectional angular power spectrum (i.e., over $\mathbf{K}$) of unpolarized radiation to give the amount of power polarized in the plane of incidence ($T_{\parallel}$) and across the plane of incidence ($T_{\perp}$) at the observer. The polarization of the radiation at the observer is defined by

$$P = \frac{T_{\parallel} - T_{\perp}}{T_{\parallel} + T_{\perp}} = \frac{(T_{\parallel}/T_{\perp}) - 1}{(T_{\parallel}/T_{\perp}) + 1}. \tag{16}$$

In Fig. 4 (a), the computations were carried out for

$$\rho(\varDelta r) = \exp(-\varDelta r^2/2L^2) \tag{17}$$

with the parameter S of the diagram given by

$$S = h_0/L. \tag{17a}$$

In Fig. 4 (b), similar computations were carried out for

$$\rho(\varDelta r) = \exp(-\varDelta r/l) \tag{18}$$

and the parameter C of the diagram is defined by

$$C^2 = \frac{l^2\lambda^2}{h_0{}^4\pi^2(\sqrt{\epsilon} - 1)}. \tag{18a}$$

In both cases, there is a pronounced effect of the roughness on the polarization, particularly when the distribution of surface slopes is wide. Note in particular the slight decrease in polarization near grazing incidence for the wide angular distribution of surface slopes.

The polarization could also be reduced with respect to the prediction of Eq. (14) if there were a transition layer of intermediate dielectric constant on surface of the Moon. One model which has been worked out consists of a layer of dielectric constant ϵ_1 resting on top of another layer of dielectric constant ϵ_2.

Assuming that the loss tangent in the top layer is small and that the depth of the top layer is more than a wavelength and of random depth, one obtains for the ratio of the two polarizations

$$\frac{T_{\parallel}(\nu, \theta_{\text{out}})}{T_{\perp}(\nu, \theta_{\text{out}})} \qquad (19)$$

$$= \frac{\epsilon_2[\cos\theta + (\epsilon_2 - \sin^2\theta)^{1/2}][\cos\theta(\epsilon_2 - \sin^2\theta)^{1/2} + \epsilon_1 - \sin^2\theta]}{[\epsilon_2\cos\theta + (\epsilon_2 - \sin^2\theta)^{1/2}][\epsilon_1\cos\theta(\epsilon_2 - \sin^2\theta)^{1/2} + \epsilon_2(\epsilon_1 - \sin^2\theta)/\epsilon_1]}$$

where we have put $\theta_{\text{out}} = \theta$ for ease of writing. A wavelength dependence could arise with such a model if the surface layer becomes thin with respect to the wavelength at some wavelength of observation. Matveev (1967) has considered a model where the refractive index $n(z) = (\epsilon(z))^{1/2}$ depends on depth z as

$$n(z) = 1 + n_0[1 - (1 - \zeta)\, e^{z/z_0}] \qquad (z < 0) \qquad (20)$$

and has found that the simple Fresnel equation (14) for the power ratio can be used if an effective dielectric constant $\epsilon_{\text{eff}}(\nu)$ is substituted. Figure 5 shows the

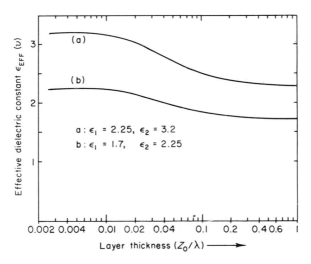

FIG. 5. Diagram showing the effective dielectric constant as a function of layer thickness. (a) $\epsilon_1 = 2.25$, $\epsilon_2 = 3.2$; and (b) $\epsilon_1 = 1.7$, $\epsilon_2 = 2.25$.

value of $\epsilon_{\text{eff}}(\nu)$ plotted against z_0/λ, i.e., the transition layer thickness in wavelengths, for two different combinations of $\epsilon_1 = (1 + n_0\zeta)^2$ and $\epsilon_2 = (1 + n_0)^2$, viz. a: $\epsilon_1 = 2.25$, $\epsilon_2 = 3.2$; and b: $\epsilon_1 = 1.7$, $\epsilon_2 = 2.25$. It may be appreciated that a marked wavelength dependence of the polarization of the emission could arise with such a model.

It has also been suggested that the radiation from the Moon must be split up into a part which is unpolarized due to diffusive scattering on the surface and another part which appears to arise as if from a smooth Moon (Losovsky, 1967). However, there does not at present exist any good method to determine which fraction to assign to the two parts.

The polarization measurement, which at first thought seems to offer such a simple way for determining the dielectric constant of the surface material, thus is influenced by a number of complicating factors which make the interpretation of the results of necessity ambiguous.

VI

OBSERVATIONAL RESULTS

In this section, we shall briefly review the observational results in the light of the theories outlined in the previous sections. We shall attempt to summarize the conclusions which may be drawn from these relating to properties of the material of the lunar surface layer as well as to variations with depth of some of these properties. Results concerning the statistics of the topography of the lunar surface are also obtained.

A. *Lunation Variations of Emission Temperature*

A summary of the observations of lunation variations of apparent emission temperature as a function of wavelength is best given in tabular form. The observational results are listed in Table II in order of increasing wavelength of observation. The mean temperature, in some references referred to the subearth point and in others referred to a mean over the disk as noted, is given in the second column; the third column shows the amplitude of the first harmonic of the lunation variation of the temperature; and the fourth column gives the phase lag of the first harmonic with respect to the phase of the surface temperature variation. The following columns give some details on the observational procedure and the reference to the publication of the data. The data pertaining to the temperature variation at the center of the Moon and to the whole disk of the Moon are not directly comparable. The mean disk temperature is related to the mean temperature at the center of the Moon by a relation (Krotikov and Shchuko, 1963)

$$T_{d0} = b_0 T_{c0} \tag{21}$$

where $0.91 \leqslant b_0 \leqslant 0.96$, depending on the dielectric constant ϵ. The amplitude of the variations with the period of a lunation are related by

$$T_{d1} = b_1 T_{c1} \tag{22}$$

TABLE II

LIST OF RESULTS OF MEASUREMENTS OF LUNATION VARIATION OF
THERMAL EMISSION FROM THE MOON

Wavelength (cm)	T_0 (°K)	T_1 (°K)	ϕ_1 (deg)	Half power beamwidth	Reference
0.10[c] *	229	115	18	3.9′	Low and Davidson (1965)
0.13[c]	216	120	16	10′	Fedoseev (1963)
0.15[d] †	265	145	—	5′	Sinton (1955)
0.18[c]	240	115	14	6′	Naumov (1963)
0.32[c]	210	65	10	9′	Tolbert and Coates (1963)
0.33[c]	196	70	27	2.9′	Gary et al. (1965)
0.40[d]	230	73	24	25′	Kislyakov (1961)
0.40[c]	228	85	27	1.6′	Kislyakov and Salomonovich (1963)
0.40[d]	204	56	23	36′	Kislyakov and Plechkov (1964)
0.80[c]	197	32	40	18′	Salomonovich (1958)
0.80[c]	211	40	30	2′	Salomonovich and Losovsky (1962)
0.86[c]	225	45	40	6′	Gibson (1958)
1.18[c]	220	29	48	3.5′	Moran (1965)
1.25[c]	215	35	45	23′	Piddington and Minnett (1949)
1.37[c]	220	24	43	4.0′	Moran (1965)
1.6[d]	208	37	30	44′	Kamenskaya et al. (1962)
1.63[d]	224	36	34	26′	Zelinskaya et al. (1959)
1.63[d]	207	32	10	44′	Dmitrenko et al. (1964)
2.0[c]	190	20	40	4′	Salomonovich and Koshchenko (1961)
3.15[c]	195	12	44	9′	Mayer (1961)
3.2[c]	223	17	45	6′	Koshchenko et al. (1961)
3.2[d]	216	16	15	40′	Bondar et al. (1962)
9.4[d]	220	5.5	5	2°20′	Medd and Broten (1961)
9.6[d]	230	—	—	19′	Koshchenko et al. (1961)
11[d]	214	—	—	17′	Mezger and Strassl (1959)
14.2[d]	221	—	—	2°	Krotikov and Shchuko (1965)
20.8[d]	205	5	49	36′	Waak (1961)
21.0[d]	232			10	Heiles and Drake (1963)
21.0[d]	250	5		35	Mezger and Strassl (1959)
21.0[d]	230			90	Razin and Fedorev (1963)
22	270			15′	Davies and Jenisson (1960)
25.0[d]	226			1°17′	Alekseev et al. (1967)
30.2[d]	227			1°18′	Alekseev et al. (1967)
31.25[d]	227				Troitsky et al. (1967)
32.3	233			3°00′	Razin and Fedorev (1963)
35.2[d]	225			—	Troitsky et al. (1967)
40[d]	224			—	Troitsky et al. (1967)
54[d]	218.5			—	Troitsky et al. (1967)
60.24[d]	216.5			—	Troitsky et al. (1967)
70.16[d]	217			1°30′	Krotikov et al. (1964)

* Superscript c refers to center of disk temperature.
† Superscript d refers to average disk temperature.

where $0.75 \leqslant b_1 \leqslant 1$, depending on ϵ and on δ_1. We shall here convert the average disk results to center of disk results by assuming

$$b_0 = 0.94,$$
$$b_1 = 0.78. \tag{23}$$

This will introduce a systematic error, but one which is appreciably smaller than the random error in the observational results. The ratio $T_{c\,0}/T_{c\,1} = M(\lambda)$ is plotted as a function of wavelength in Fig. 6. The points which were derived by

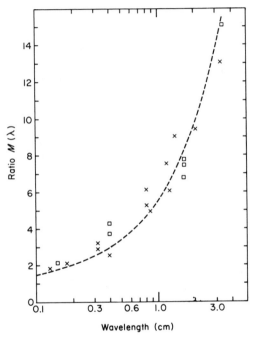

FIG. 6. Plot of $M(\lambda) = T_{c\,0}/T_{c\,1}$ against wavelength. Dotted curve corresponds to $\delta_1 = 2.5\lambda$.

measurements of $T_{c\,0}$ and $T_{c\,1}$ directly are indicated as crosses, the ones obtained by converting from disk brightness variation are shown as small squares. For comparison with theory, a curve is plotted for the surface temperature as summarized by the data in Table I for

$$\delta_1 = L_E/L_T = 2.5\lambda \tag{24}$$

where λ is expressed in centimeters. As can be seen, this relationship gives a satisfactory agreement between observations and the simple theory presented above. The constant of proportionality is somewhat higher than that derived

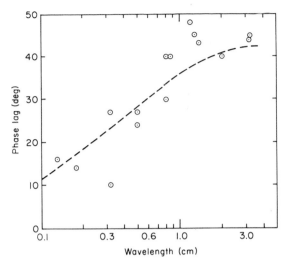

FIG. 7. Phase angle of the lunation variation of emission temperature plotted against wavelength. Dotted curve corresponds to $\delta_1 = 2.5\lambda$.

by Troitsky (1965) but in agreement with Linsky's conclusion (1966). Figure 7 shows the phase angles of the emission temperature variation pertaining to the subearth point also plotted against wavelength. Shown dotted is the expected relationship for $\delta_1 = 2.5\lambda$. The spread in the data is quite large, but subject to this uncertainty agreement does exist between the δ_1 derived from amplitude and from phase observations on the basis of a simple homogeneous surface model. The relationship (24) does indeed classify the lunar surface material as a slightly lossy dielectric, as was assumed in the survey of the theories. From the lunation data, we therefore conclude that the parameter $\gamma C_m / b\sqrt{\epsilon}$ has the following value:

$$\frac{\gamma C_m}{b\sqrt{\epsilon}} = 1.4 \cdot 10^4 \text{ cm}^{-1} \text{ sec}^{1/2}. \tag{25}$$

For the lunation variations of the temperature, Troitsky (1965) estimates that $L_T = (10 \pm 5)$ cm and the above parameter therefore must be an average of the parameter over a depth of some 10 cm. Since γ is decreasing with depth, ϵ increasing, and C_m remaining approximately constant, one would expect the above surface parameter to decrease with depth in the absence of appreciable changes in b possibly caused by composition changes.

B. Eclipse Variations of Emission Temperature

Variations in emission temperature during a lunar eclipse have been observed in the microwave region between wavelengths of 0.8 mm (Baldock et al., 1965) and 2.2 cm (Plechkov and Porfiryev, 1965). Due to extreme observational difficulties primarily caused by atmospheric effects, the results of the observations are conflicting and so noisy that it is in most cases not possible to derive physically meaningful parameters from them. Troitsky (1965) has, however, ventured to select some of the observational data for analysis and has derived the values for the parameter $\gamma C_m/b\sqrt{\epsilon}$ for comparison with the lunation data. The particulars of the data summarized by Troitsky are listed in Table III. As can be seen, the value of the surface parameter varies rather widely from one observation to the other. The mean value of the data gives

$$\frac{\gamma C_m}{b\sqrt{\epsilon}} = 1.4 \cdot 10^4 \text{ cm}^{-1} \text{ sec}^{1/2},$$

which is in surprisingly good agreement with the value derived from lunation data. Troitsky, selecting only a few of the data in Table III, derives a slightly lower value for this parameter and argues that the absence of decrease in $\gamma C_m/b\sqrt{\epsilon}$ with depth as evidenced by the comparison of lunation and eclipse data, must be taken as an indication of a *decrease* in specific loss tangent b with depth. This he attributes to an accumulation of meteoric material of relativity high conductivity in the topmost few millimeters of the lunar surface. However this may be, it does appear to the present author that this is a bold conclusion to draw from data encumbered with such large experimental uncertainties.

TABLE III

Observational Data Used in Eclipse Analysis

Wavelength (cm)	T_e/T_{em} (%)	T_{em}	$(\gamma C_m/b\sqrt{\epsilon})10^{-4}$	Reference
0.12	22.5	315	0.9	Kamenskaya et al. (1965)
0.32	6	277	2.1	Jacobs et al. (1964)
0.4	8	270	1.6	Kamenskaya et al. (1965)
0.4	12	270	0.9	Kamenskaya et al. (1965)
0.4	10	270	0.9	Pleckhov and Porfiryev (1965)
0.43	4	268	2.6	Tolbert et al. (1962)
0.6	8	267	0.9	Kamenskaya et al. (1965)
0.75	5	253	1.5	Kamenskaya et al. (1965)
0.86	7	247	0.8	Gibson (1961)
1.6	3	235	1.3	Kamenskaya et al. (1965)
1.6	6	235	0.7	Plechkov (1965)
2.2	1	228	3.1	Plechkov and Porfiryev (1965)

C. Variation of Mean Temperature with Wavelength

It will be remembered that a term was inserted into Eq. (12) to account for the possibility of a mean vertical temperature gradient in the lunar surface material. A series of experiments to detect a possible variation in brightness temperature with wavelength was carried out at the Research Institute of Radio Physics of the U.S.S.R. during 1961 to 1964. The observations were carried out at the wavelengths 0.4, 1.6, 9.6, 14.0, 32.5, 35.0, and 50.0 cm. The results of these observations are shown in Fig. 8 together with a "mean"

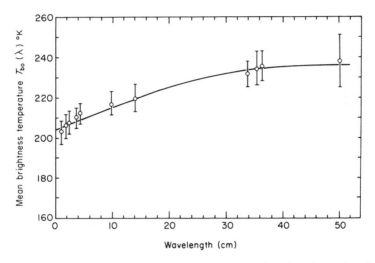

FIG. 8. Plots of the mean emission temperature as a function of wavelength.

curve drawn through the points by Troitsky who has analyzed the data (1965). Assuming L_E to be 20λ, he determines that

$$A = \text{grad } T = 2\text{–}4 \text{ deg/m}$$

and with a reasonable guess for the parameter γ (≈ 600) determines that the mean heat flux to the surface of the moon from within is

$$Q \approx 10^{-6} \text{ cal/cm}^2 \text{ sec.}$$

This is several times as high as the corresponding heat flux from the Earth and might mean that the concentration of radioactive elements in the lunar interior is higher than on Earth. The conclusion has been challenged by Linsky (1966) on the grounds that a temperature-dependent thermal conductivity could lead to the observed wavelength dependence of the mean brightness temperature.

However, Troitsky *et al.* (1968), reexamining the problem with temperature dependent parameters, find that their previous conclusions about thermal heat flux essentially remain unaltered.

D. Observations of Polarization of the Emission

Observations of the polarization of the thermal emission from the Moon have not been carried out as extensively as observations of lunation variations. This is due to the requirement for an antenna beam which has an angular extent small compared with that of the Moon. Table IV summarizes some of the results

TABLE IV

RESULTS OF POLARIZATION OBSERVATIONS

Wave-length (cm)	Smooth ϵ	Corrected ϵ	Roughness model	Reference
0.8	1.5 ± 0.1	1.7	15° Gaussian	Moran (1965)
		2.3	30° Gaussian	Moran (1965)
2.1	1.8	1.50	15° cone	Baars *et al.* (1965)
3.2	1.55	1.65	20° cone	Soboleva (1963)
3.7	1.7 ± 0.1	1.8	S = 0.2 (Eq. 17a)	Hagfors and Hull (1966)
6.	2.0 ± 0.05	2.2 ± 0.1	8–16°	Davies and Gardner (1966)
6.3	1.9	2.0	20° cone	Golnev and Soboleva (1964)
11	2.05 ± 0.05	2.25 ± 0.05	8–16°	Davies and Gardner (1966)
21	2.3 ± 0.15	2.50 ± 0.15	8–16°	Davies and Gardner (1966)
21	2.1 ± 0.3			Heiles and Drake (1963)

obtained. The column referred to as "roughness" specifies the type of model used for the surface in order to derive an estimate of the dielectric constant. The "cone" model means that the surface normal is assumed to be distributed with constant probability within a certain cone. The 8–16° models mean that the surface normals are distributed with constant probability between the cones with apex angle 8 and 16°. The Gaussian model means that the surface normal is distributed around the mean normal with a Gaussian probability distribution with the specified opening angle. This model is closely related to the one specified by Eq. (17) of the present chapter. According to Table IV, there appears to be a gradual increase in the dielectric constant with increasing wavelength from between 1.5 and 1.8 near 1 cm to near 2.3 near 10 cm. This could be an indication that there is an increase in dielectric constant with depth such as suggested by Matveev's model (1967). It would only take a transition layer of about 1 cm to produce the observed increase (see Fig. 5). Another possibility for the explanation of the variation of ϵ with wavelength, which cannot be discarded, is that small scale surface roughness plays a more important and unpredictable role at the shorter wavelengths (see Losovsky, 1967).

VII

SUMMARY

The present account of radio observations of the Moon will no doubt have shown the reader that the observational results allow for a considerable latitude of interpretations. For this reason, we shall not attempt to construct a comprehensive model of the lunar surface based on these results, but rather produce a short summary of the conlusions for use by those who may want to arrive at a model which is in agreement with *all* available data.

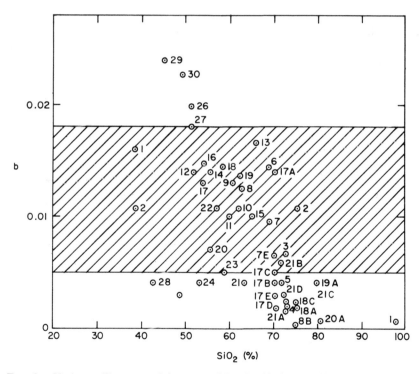

FIG. 9. Various silicon-containing materials classified according to specific loss tangent and silicon content. Key: 1, quartz, sand; 2, obsidian; 3, ignimbrid; 4, liparite; 5, granite; 6, 7, pumice stone; 8–10, tuffes; 11, trachytic lava; 12–19, volcanic ash; 20, 21, simtes; 22, andesitobasalt; 23, diorite; 24, 25, gabbro; 26, 27, basalts; 28, iolite; 29, dunite; 30, dolerite; 17A–17F, australites; 18A–18C, indochinites; 19A, moldavites; 20A, silica glass; 21A–21D, philippinites.

Subject to the difficulties of interpretation, the radio observations of the Moon show that:

(a) The mean emission temperature of the Moon is near 230°K.

(b) There appears to be a mean thermal flux out of the Moon corresponding to $Q = 10^{-6}$ cal/cm² sec.

(c) The upper surface layer of the Moon behaves as a slightly lossy dielectric with ratio of penetration depths of electromagnetic and monthly thermal waves given by

$$\delta_1 = L_E/L_T = 2.5\lambda,$$

where the wavelength λ is expressed in centimeters.

(d) The dielectric constant of the lunar surface material lies in the range $\epsilon = 1.5$–2.5, the lower values being obtained at short wavelengths where the depth of penetration of the electromagnetic wave is small. This may mean that there is a gradual increase of dielectric constant with depth. Such low values for the dielectric constant are indicative of a tenuous surface layer.

(e) The value of $\gamma C_m/b \sqrt{\epsilon}$ is 1.4×10^4 cm⁻¹ sec$^{1/2}$. Here $\gamma = (\kappa \rho C_m)^{-1/2}$, C_m = specific heat, ρ = density, and κ = thermal conductivity and b is the specific loss tangent $b = \Delta/\rho$. This indicates that b lies somewhere in the region 5×10^{-3}–2×10^{-2} cm³ g⁻¹. For comparison, Fig. 9 shows various SiO_2-containing materials classified according to their values of specific loss tangent.

References

Alekseev, V. A., Aleshina, T. N., and Krotikov, V. D. (1967). *Radiofizika* **10**, 603.
Alekseev, V. A., Aleshina, T. N., Krotikov, V. D., and Troitsky, V. S. (1968). *Sov. Astron. AJ (English Transl.)* **11**, 860.
Baars, J. W., Mezger, P. G., Savin, N., and Wendker, H. (1965). *Astron. J.* **70**, 132.
Baldock, R. V., Bastin, J. A., Clegg, P. E., Emery, R., Gaitskell, J. N., and Gear, A. E. (1965). *Astrophys. J.* **141**, 1289.
Baldwin, I. E. (1961). *M. N. R. A. S.* **122**, 513.
Bernett, E. C., Wood, H. L., Jaffe, L. D., and Martens, H. E. (1963). *AIAA J.* **1**, 1402.
Bondar, L. N., Zelinskaya, M. R., Porfiryev, V. A., and Strezhneva, K. M. (1962). *Radiofizika* **5**, 802.
Bracewell, R. N. (1962). *In* "Handbuch der Physik" (S. Flugge, ed.), Vol. 52, pp. 42–129. Springer, Berlin.
Buettner, K. J. K. (1963). *Planet. Space Sci.* **11**, 135.
Davies, R. D., and Gardner, F. F. (1966). *Aust. J. Phys.* **19**, 823,
Davies, R. D. and Jennison, R. C. (1960). *Observatory* **80**, 74.
Dicke, R. H. and Beringer, R. (1946). *Astrophys. J.* **103**, 275.

Dmitrenko, D. A., Kamenskaya, S. A., and Rakhlin, V. L. (1964). *Radiofizika* **7**, 555.

Fedoseev, L. N. (1963). *Radiofizika* **6**, 4.

Gary, B., Stacey, J., and Drake, F. D. (1965). *Ap. J.* Suppl. Ser. **12**, 239.

Geoffrion, A. R., Korner, M., and Sinton, W. M. (1960). *Lowell Observ. Bull.* **5**, 1.

Gibson, J. E. (1958). *Proc. IRE* **46**, 280.

Gibson, J. E. (1961). *Astrophys. J.* **133**, 1072.

Golnev, V. J. and Soboleva, N. S. (1964). *Proc. Astron. Obs. Pulkova* **13**, 83.

Hagfors, T. and Hull, D. (1966). Unpublished data.

Hagfors, T. and Moriello, J. (1965). *J. Res. N.B.S.* **D69**, 1614.

Heiles, C. E. and Drake, F. D. (1963). *Icarus* **2**, 281.

Jacobs, E., King, H. E., and Stacy, J. M. (1964). Aerospace Corp., Los Angeles, California.

Jaeger, J. G. (1953). *Aust. J. Phys.* **6**, 10.

Kamenskaya, S. A., Semyenov, V. I., Troitsky, V. S., and Plechkov, V. M. (1962). *Radiofizika* **5**, 882.

Kamenskaya, S. A., Kislyakov, A. G., Krotikov, V. D., Naumov, A. I., Nikonov, V. N., Porfiryev, V. A., Plechkov, V. M., Strezhneva, K. M., Troitsky, V. S., Fedoseev, L. I., Lubiako, L. V., and Sorokina, E. P. (1965). *Radiofizika* **8**, 219.

Kislyakov, A. G. (1961). *Radiofizika* **4**, 433.

Kislyakov, A. G. and Plechkov, V. M. (1964). *Radiofizika* **7**, 1.

Kislyakov, A. G. and Salomonovich, A. E. (1963). *Radiofizika* **6**, 431.

Koshchenko, V. N., Losovsky, B. Ya., and Salomonovich, A. E. (1961). *Radiofizika* **4**, 596.

Krotikov, V. D. and Shchuko, O. B. (1963). *Sov. Astron. AJ (English Transl.)* **7**, 228.

Krotikov, V. D. and Shchuko, O. B. (1965). *Sov. Astron. AJ (English Transl.)* **9**, 113.

Krotikov, V. D., Troitsky, V. S., and Tseytlin, N. M. (1964). *Astr. Zh.* **41**, 5.

Linsky, J. L. (1966). *Icarus* **5**, 606.

Losovsky, B. Ya. (1967). *Astron. Zh.* **44**, 416.

Low, F. J. (1965). *Astrophys. J.* **142**, 806.

Low, F. J. and Davidson, A. W. (1965). *Astron. J.* **142**, 1278.

Matveev, Yu. G. (1967). *Astron. Zh.* **44**, 419.

Mayer, C. H. (1961). *In* "The Solar System" (G. P. Kuiper and B. M. Middlehurst, eds.), Vol. 3, pp. 442–472. Univ. of Chicago Press, Chicago, Illinois.

Medd, W. J. and Broten, N. W. (1961). *Planet. Space Sci.* **5**, 307.

Mezger, P. G. and Strassl, H. (1959). *Planet. Space Sci.* **1**, 213.

Moran, J. M. (1965). M. S. Thesis, Massachusetts Institute of Technology.

Moroz, V. I. (1965). *Astron. Zh.* **42**, 1287.

Murray, B. C. and Wildey, L. (1964). *Astrophys. J.* **139**, 734.

Naumov, A. P. (1963). *Radiofizika* **6**, 848.

Pawsey, J. L. and Bracewell, R. N. (1955). "Radio Astronomy." Oxford Univ. Press (Clarendon), London and New York.

Pettit, E. (1935). *Ap. J.* **81**, 17.

Pettit, E. and Nicholson, S. B. (1930). *Astrophys. J.* **71**, 102.

Piddington, J. H. and Minnett, H. C. (1949). *Aust. J. Sci. Res.* **A2**, 63.

Plechkov, V. M. (1965). Result quoted by Troitsky (1965).

Plechkov, V. M. and Porfiryev, V. A. (1965). Result quoted by Troitsky (1965).

Razin, V. A. and Fedorev, V. T. (1963). *Radiofizika* **6**, 5.

Rea, D. G., Hetherington, N., and Mifflin, R. (1968). *J. Geophys. Res.* **22**, 7009.

Saari, J. M. (1964). *Icarus* **3**, 161.

Salomonovich, A. E. (1958). *Astron. Zh.* **35**, 129.

Salomonovich, A. E. and Koshchenko, V. N. (1961). *Radiofizika* **4**, 591.

Salomonovich, A. E. and Losovsky, B. Ya. (1962). *Astron. Zh.* **39**, 1047.

Sinton, W. M. (1955). *J. Opt. Soc. Am.* **45**, 975.

Sinton, W. M. (1962). *In* "Physics and Astronomy of the Moon" (Z. Kopal, ed.), 1st ed., 407. Academic Press, New York.

Soboleva, N. S. (1963). *Sov. Astron. AJ (English Transl.)* **6**, 873.

Steinberg, J. L. and Lequeux, J. (1963). "Radio Astronomy." McGraw-Hill, New York.

Tolbert, C. W., and Coates, G. T. (1963). Texas Univ. Rept., pp. 7–24. Austin.

Tolbert, C. W., Krause, L. C. and Dickenson, R. M. (1962). *Proc. Nat. Aerospace Conf.*, p. 733.

Troitsky, V. S. (1954). *Astron. Zh.* **31**, 511.

Troitsky, V. S. (1961). *Astron. Zh.* **38**, 1001.

Troitsky, V. S. (1965). *J. Res. N.B.S.* **D69**, 1585.

Troitsky, V. S., Krotikov, V. D., and Tseytlin, N. M. (1967). *Astron. Zh.* **44**, 413.

Troitsky, V. S., Burov, A. B., and Aleshina, T. N. (1968). *Icarus* **8**, 423.

Waak, J. A. (1961). *Astron. J.* **66**, 7.

Watson, K. (1964). Thesis, Calif. Inst. Technol., Pasadena.

Wesselink, A. J. (1948), *Bull. Astron. Inst. Neth.* **19**, 351.

Zelinskaya, M. R., Troitsky, V. S., and Fedoseev, L. N. (1959). *Astron. Zh.* **36**, 643.

Radar Studies of the Moon[*]

JOHN V. EVANS AND TOR HAGFORS

Lincoln Laboratory
Massachusetts Institute of Technology
Lexington, Massachusetts

I

INTRODUCTION

Radio-echo studies of the Moon may be said to have begun in 1946 when radar engineering teams in Hungary and America demonstrated that lunar radio reflections could be detected on Earth (Bay, 1946; DeWitt and Stodola, 1949). Serious scientific study of the Moon by this means was, however, slow in developing and the largest part of the work to be reviewed here has been carried out since 1960. In large measure, a shortage of suitable radar equipment has been responsible for this slow growth and for the small number of groups

* The work reported in this document was performed at Lincoln Laboratory, a center for research operated by Massachusetts Institute of Technology, with support from the U.S. National Aeronautics and Space Administration under Contract No. NSR 22-009-106.

actively engaged in this type of work at any one time. Indeed, only a few organizations have been occupied in studies of the Moon by radar over an extended period. These include the U.S. Naval Research Laboratory, the University of Manchester's experimental station at Jodrell Bank, the Massachusetts Institute of Technology's Lincoln Laboratory, and the Arecibo Ionospheric Observatory operated by Cornell University.

Beginning in 1949, workers in Australia at the Commonwealth Scientific and Industrial Research Organization (Kerr et al., 1949; Kerr and Shain, 1951) studied the strength of the reflected signals, and were able to show that two fading mechanisms were in operation, one of which appeared to have an (terrestrial) ionospheric origin and the other they attributed to constructive and destructive interference between rays reflected from different parts of the lunar surface. From the character of this second fading pattern, they concluded (erroneously as it turned out) that the Moon is very rough on the scale of the exploring radar wavelength.

The first reliable observations concerning the scattering behavior of the Moon were obtained in the early 1950's, at the U.S. Naval Research Lab by using a high power, short pulse radar (Trexler, 1958). It was discovered that the Moon behaves remarkably like a specular reflector, indicating that locally the surface is unexpectedly smooth. These results were published only after similar findings had been obtained independently at Jodrell Bank (Evans, 1957) from a reexamination of the rapid fading of the echoes. Thereafter, radar studies of the Moon assumed considerable interest, since it was recognized that they are capable of providing information about the surface on the scale of a lunar landing vehicle, i.e., well below the limit of resolution of Earth-based telescopic observations.

Most of the conclusions reached from radar studies in the early 1960's have been confirmed in subsequent close-up photographic studies using the Ranger, (Trask 1970, and references cited therein) Surveyor (Shoemaker et al., 1969; Shoemaker and Morris, 1970; Morris and Shoemaker, 1970a, b) and Orbiter (Allenby, 1970) spacecraft. As a consequence of this unmanned (and later manned) exploration of the Moon, it is likely that the importance of ground-based radar studies will decline. Historically, they served their purpose in providing criteria for the design of lunar landing vehicles, when little else was available, and in the design of the radar altimeters aboard those vehicles.

The history of the early radar studies has been given in some detail by Evans (1962) and this will not be repeated here. Instead we shall review the major contributions of radar to lunar studies. These fall into three main areas, viz., (a) the distance and motion of the Moon, (b) the statistical nature of the lunar surface, and (c) studies differentiating portions of the surface. As far as is possible, we shall avoid any discussion of radar instrumentation or methodology. Other reviews (e.g., Evans, 1960, 1965, 1966; Evans and Pettengill, 1963a;

Hagfors, 1967) have discussed the types of measurement available to a radar set and presented results for each type. Wherever possible we shall refer to original papers when discussing experimental results and the reader may find in these details of the apparatus employed. Similarly, in a review of this type, it is not possible to rederive all the theoretical expressions that have been developed to describe the scattering properties of rough surfaces and the reader must seek these in the papers cited.

II

RADAR STUDIES OF THE MOON'S MOTIONS

A. *Range Measurements*

The mean center to center distance $\bar{D}_0$ between the Earth and the Moon is approximately 384,400 km. In each month, the actual value varies cyclically from the mean up to $\pm 30,600$ km. The radar distance to the Moon can readily be determined with high precision, but because the Moon's distance has been determined by optical methods over a long period of time, it is difficult to make a significant improvement by using radar. Only one group of workers (at the Naval Research Laboratory) has been actively engaged in these measurements over an extended period. This group has employed a 10 cm wavelength radar with a pulse length of 2 μsec (Yaplee *et al.*, 1958, 1959). The accuracy of measurement is limited by the fading of the echo, making it difficult to recognize with absolute certainty the reflection corresponding to the nearest point on the surface of the Moon.

Figure 1 shows the relevant geometry for these measurements. The quantities involved principally are the Earth's equatorial radius r and the equatorial horizontal parallax π. The measured radar distance BM in Fig. 1 is given by

$$BM = -a + (b^2 + D_0{}^2 - 2bD_0 \cos z)^{1/2} \tag{1}$$

in which

$$\cos z = \sin \phi' \sin \delta + \cos \phi' \cos \delta \cos HA,$$

where

$$a = \text{lunar radius to the subradar point,}$$
$$b = \text{observer's radius vector,}$$
$$D_0 = \text{predicted center-to-center distance } (r/\sin \pi),$$
$$r = \text{Earth's equatorial radius,}$$
$$\pi = \text{Moon's equatorial horizontal parallax,}$$
$$\phi' = \text{geocentric latitude of the station,}$$
$$\delta = \text{Moon's declination.}$$

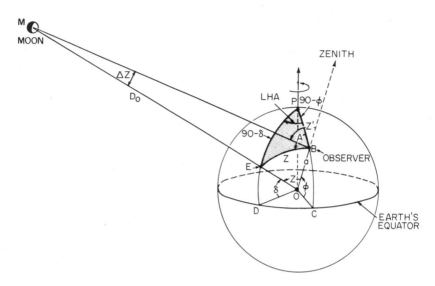

FIG. 1.   Geometry involved in radar range measurements of the distance to the Moon from a point on the Earth's surface.

The value of b can be obtained from the formula

$$b = \frac{4(1-f)}{[1 - (2f - f^2) \cos^2 \phi']^{1/2}} + H \qquad (2)$$

in which the oblateness constant f is approximately $1/298.3$ and H is the height of the antenna above sea level.

Yaplee *et al.* (1958) found that the value for the equatorial radius adopted by the U.S. Army Map Service was inconsistent with their observations, but that given by *The American Ephemeris and Nautical Almanac* ($r = 6,378,388$ m) was in better agreement. Later, Yaplee *et al.* (1959) showed that residual errors (observed–computed range) of about 3 km were encountered and that these were not the same from day to day. These discrepancies were found to be considerably reduced if the relative height of the terrain on the Moon nearest the radar was allowed for (Yaplee *et al.*, 1964); that is, the value for the effective lunar radius was adjusted according to the height of the subradar point as given by an equivalent optical map obtained by the Naval Research Laboratory from the Army Map Service (Fig. 2). However, the largest part of the uncertainty lay is the predicted center to center distance, and this was not removed until 1967 following the inclusion of a number of corrections to Brown's theory (Eckert *et al.*, 1966) and a precise redetermination of the mean center to center distance obtained by the Jet Propulsion Laboratory from range measurements

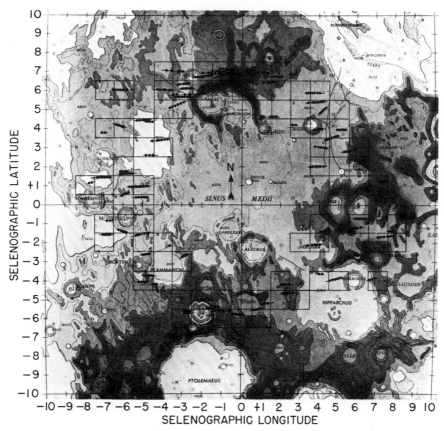

FIG. 2. Topographic map of the central portion of the lunar surface showing (as black dots) the location of subradar points in the measurements performed by Yaplee *et al.* (1964). Because of the Moon's libration (Section II), the subterrestrial point wanders over the lunar surface and in the course of a month approximately describes an ellipse (official U.S. Navy Photograph).

to the Lunar Orbiters I and II (Mulholland and Sjogren, 1967). Shapiro *et al.* (1968) concluded that the Moon's mean radius is between 1 and 2 km lower than the accepted value (1738 km) when the mean distance to the Moon is set equal to that [384,399.3 km] contained in the JPL Lunar Ephemeris No. 4 (Melbourne *et al.*, 1968). Subsequently, refinements have been made to this Ephemeris and an independent one has been developed at MIT (Slade, 1970) which is of comparable accuracy; both of these predict the range and velocity of the Moon as accurately as they can currently be measured by radar when allowance is made for the effects of the Earths atmosphere and the somewhat uncertain topographic height variations of the Moon's surface.

B. *Radial Velocity*

Because the Moon moves in an elliptical orbit, it will, in general, have a component of velocity toward the center of the Earth. This has a maximum value of the order of v_1 where

$$v_1 = \frac{2\pi \times 30{,}600}{28} \frac{\text{km}}{\text{days}} = \pm 80 \text{ m/sec}^{-1}. \tag{3}$$

In addition, an observer on the Earth will have a velocity component of the order of v_2 due to the rotation of the Earth where v_2 is given by

$$v_2 = \pm \frac{2\pi \text{ Earth's radius} \times \cos \phi' \times \cos \delta \times \sin HA}{\text{Earth's rotation relative to the Moon}} \tag{4}$$

where

$$\phi' = \text{the observer's latitude,}$$
$$\delta = \text{the Moon's declination,}$$
$$HA = \text{the local hour angle.}$$

The maximum value of v_2 (occuring on the equator at moonrise or moonset) is about ± 500 m sec^{-1}. Precise formulas for computing the Doppler shift of the echoes arising as a result of these velocity components have been developed by Fricker *et al.* (1958), among others.

Careful measurements of the Doppler shift of the echoes were first made by Blevis (1957) and by Fricker *et al*, (1958) but did not reveal any systematic differences between observed and predicted values. Thomas (1949) considered the possibility that the radius of the Earth might be determined from such observations. He concluded that the changes in the position of the effective scattering center of the Moon would be sufficient to prevent precise measurement of the Doppler shift. Pettengill (1960) has shown that this can be overcome by employing a coherent pulse radar (Section IV) and resolving the echo into different frequency components. The overall width of this power spectrum will be set by the apparent rotation rate (see Section II. C), but by bisecting the spectrum, Pettengill (1961) was able to measure mean Doppler shifts to an accuracy of ± 0.1 to ± 0.2 Hz with a radar operating at 440 MHz. Recently by employing a radar operating at 8000 MHz and a frequency resolution of ~ 0.5 Hz, the M.I.T. Lincoln Laboratory group has achieved radial velocity sensitivity of $\sim \pm 0.1$ cm sec^{-1}. At this level of precision discrepancies are observed between the measurements and the best available radar ephemerides, which, as noted above, may be attributed to effects of the Earth's atmosphere and the slope of the surface at the subradar point, together possibly with uncertainties in the figure of the Moon (Slade, 1970).

C. *Rotational Velocity*

The center of the Moon's disk as seen from the Earth is a point which wanders over the surface, generally following an elliptical path (Fig. 2). This motion has three principal causes which are illustrated in Fig. 3 (a–c). In (a), libration in

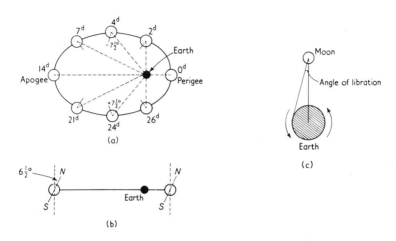

FIG. 3.   The three principal causes of the Moon's libration. (a) Libration in longitude (plan). (b) Libration in latitude (elevation). (c) Diurnal libration.

longitude is seen to result from the fact that because the Moon rotates on its own axis with constant angular velocity and moves in an elliptical orbit, it does not present exactly the same face to the Earth at all times. Thus, if a stick were placed on the surface at perigee pointing toward the Earth, it would do so again only at apogee. In between there would be an angle of up to $7\frac{1}{2}°$ between the line of sight to the Earth and the stick. The rate of libration l_L introduced by this component is at its maximum near apogee or perigee, when it has a value of about 4×10^{-7} radian sec^{-1}. It is direct in sense at apogee and retrograde at perigee. In (b), libration in *latitude* is shown to result from the fact that the Moon's axis of rotation is not perpendicular to the plane of its orbit, but inclined to it by $6\frac{1}{2}°$. This component has a period of one sidereal month and its rate l_B reaches a maximum value of 3×10^{-7} radian sec^{-1} when the Moon crosses the nodes of its orbit. In (c), the largest component, *diurnal libration*, is illustrated. An observer on the Earth sees a parallactic shift between the center of the Moon's disk and the true center of the Moon as the Earth rotates. This component is direct in sense when the Moon is in transit. The rate of diurnal libration, l_D, reaches a maximum value of $12 \cos \phi' \times 10^{-7}$ radian sec^{-1} at meridian transit when the Moon's declination $\delta \simeq 0°$ for an observer at a latitude ϕ'.

At any instant, the Moon may be considered to be spinning about an axis with angular velocity defined by the vector addition of the three main components l_D, l_L, and l_B. The values of l_L and l_B are approximately constant over a given day and can be obtained by interpolation from the figures listed in *The American Ephemeris and Nautical Almanac* for the Earth's selenographic longitude and latitude. The total rate of libration l_T may then be computed by resolving l_D, l_L, and l_B into components along and perpendicular to the line of sight. The component along the line of sight introduces no range changes and can be ignored. Thus at meridian transit

$$l_T{}^2 = (l_L + l_D \cos \delta_0)^2 + (l_B + l_D \sin \delta_0 \cos \Theta)^2 \tag{5}$$

where δ_0 is the greatest declination reached by the Moon during the month and Θ is an angle approximately equal to the difference between the longitude of the Moon and the mean longitude of the ascending node. More precise formulas have been developed by Fricker *et al.* (1958).

Signals which are returned from regions other than those along the apparent axis of librational rotation will exhibit an additional Doppler shift due to the apparent rotation of the Moon. The maximum Doppler shift of the signals will

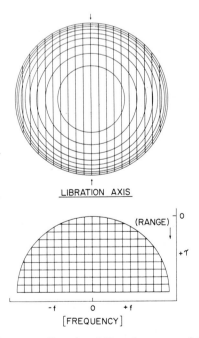

FIG. 4.   Contours of constant Doppler shift and constant delay at equal increments as seen from the Earth (top) and in plane view (bottom).

occur for reflections from the limb regions most distant from the libration axis and may be expressed as

$$f_0 = \pm 2l_T a/\lambda \tag{6}$$

where λ is the radio wavelength and a is the lunar radius. Other elements of the lunar surface will introduce a Doppler shift into the signals they reflect which is directly proportional to the distance of the element from the libration axis. Thus the Moon's disk may be imagined as ruled by a series of lines all parallel to the libration axis (Fig. 4), and each of which represents a locus of constant Doppler shift (Browne *et al.*, 1956).

It follows that if the Moon reflects as a uniformly bright scatterer, the *rf* power spectrum $\bar{P}(f)$ of the echoes will be semicircular in shape, because the amount of power at each frequency will be proportional to the length of the corresponding strip (Fig. 4). For any other scattering law, the echo power spectrum will be modified in a way that depends directly upon the distribution of scattering centers over the visible disk.

Very precise measurements of f_0 have been achieved by comparing measurements of the radial velocity of the Moon as a whole (above) with the radial velocity observed for known features (Section IV). Thus far no systematic discrepancies between the observed and computed values of the apparent rotation rate have been uncovered.

III

RADAR STUDIES OF THE STATISTICAL PROPERTIES OF THE LUNAR SURFACE

A. Reflection Coefficient

1. Observations

In radar observations of the Moon, the observable quantity that is related to the reflection coefficient is the intensity P_S of the reflected signals. The echo power, P_S, is related to the reflection properties of the target through the *radar equation* (Evans, 1962). The radar equation is normally stated in a manner that is proper for the observation of "point targets," i.e., objects which when viewed from the radar subtend an angular diameter much smaller than that of the antenna beam. In this case, the received echo power P_S may be stated as

$$P_S = \frac{P_T G_T A_R \sigma}{(4\pi R^2)^2} \text{ watts} \tag{7}$$

where

P_T = transmitted power (watts),

G_T = transmitting antenna gain,

A_R = receiving antenna aperture (m²),

σ = the cross section of the target (m²),

R = the target's range (m).

Of all the celestial objects detectable by means of radar, the Moon and perhaps the Sun are the only ones in which the target angular diameter is likely to be as large or larger than the antenna beam. The cross section σ observed during observations of the Moon depends, therefore, upon the distribution of the incident power over the surface. In practice, the Moon reflects preferentially from regions near the center of the disk and little loss in total reflected power will be observed until the antenna beamwidth (between half-power points) is made smaller than the angular extent of the Moon ($\frac{1}{2}°$). Thus, for most forms of simple antenna, this effect will not be important until the antenna diameter becomes larger than 100λ where λ is the radio wavelength.

Radar echoes from the Moon can readily be resolved also in range delay or in frequency. For example, a pulse of 11.6 msec is required to fully illuminate the surface, and if shorter pulses are employed, the effective or instantaneous cross section σ will never reach its maximum possible value (the CW cross section). Figure 5 shows how the peak instantaneous cross section falls as a function of pulse length for observations of the Moon at 68-cm wavelength. Because the largest part of the power is reflected from the nearest regions of the lunar surface, an appreciable reduction in cross section is not observed until pulses shorter than 1 msec duration are used. Figure 5 is applicable only for wavelengths of about 50 cm, or longer, as the scattering properties of the surface change markedly toward shorter wavelengths. Equally, where a CW radar is employed, the apparent rotation of the Moon may cause the reflected signals to be appreciably Doppler-broadened. If the receiver employs a narrowband filter, which does not accept all the frequency components of the reflected signal, then again the observed cross section will be lower than the full value.

In this review, we shall use the term σ to denote the total cross section of the Moon. This could be measured with a radar employing an antenna with a beamwidth $\geqslant\frac{1}{2}°$ by determining the peak echo power observed when pulses of $\geqslant$11.6 msec are transmitted. When these conditions are not met, Eq. (7) must be modified in the manner discussed by Evans (1965). In practice, echoes from the Moon are found to fade as a consequence of constructive and destructive interference between signals arriving from different parts of the lunar surface. Thus, an average value for the peak echo power (or mean square of the echo amplitude) must be obtained from many pulses to determine σ reliably. Alter-

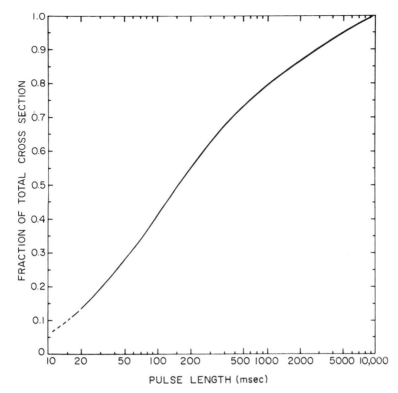

FIG. 5. The peak cross section of the Moon expressed as a percentage of the total cross section σ plotted as a function of pulse length. The radar depth of the Moon is 11.6 msec, and if pulses shorter than this are employed, the echo power will fall owing to the reduction in the instantaneous area illuminated by the pulse. This curve has been obtained from measurements at 68 cm.

natively, with a CW radar, many independent determinations of the echo power are required. In some of the earliest radar observations of the Moon (e.g., DeWitt and Stodola, 1949), this was not recognized and only the maximum value of the echo intensity was reported.

Many observers have reported values of σ and some of these are presented in Table I and plotted in Fig. 6. The values have been presented as fractions of the physical cross section of the Moon ($\pi a^2 = 9.49 \times 10^{12}\,\text{m}^2$), and span a range of over 10 octaves (from 8.6 mm to 22 m). The increase in cross section with increasing wavelength suggested in Fig. 6 depends largely on the three long-wave measurements reported by Davis and Rohlfs (1964). These measurements may have been subject to systematic errors introduced by ionospheric effects. If these three points are ignored, the remainder show no clear wavelength dependence. In part this is caused by the large error bars associated with each

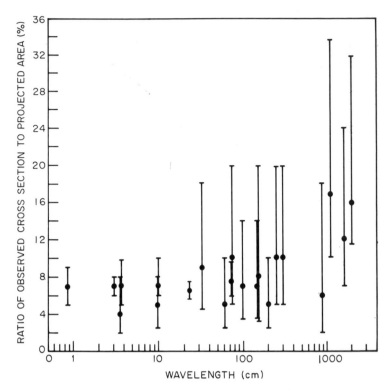

FIG. 6. Cross section of the Moon vs. wavelength according to measurements by the authors listed in Table I.

measurement which may conceal a marked dependence. The errors given in Table I are the reported values where these have been given, or ± 3 dB where no uncertainty was published. Absolute power measurements appear to be more difficult in radar astronomy observations than in radio astronomy for a number of reasons. Chief among these is the fact that common calibration targets of known wavelength dependence have not been available to radar astronomers in the way that certain sources can be taken as calibration points in optical and radio astronomy. Recently, however, carefully machined metal spheres have been placed in Earth orbit (e.g., the Lincoln Calibration Sphere) which serve as target standards. Unfortunately, these tend to be fainter targets than the Moon itself, and thus far only Evans and Hagfors (1966) have been able to obtain the cross section of the Moon (at $\lambda = 23$ cm) using a radar calibrated in this manner. This accounts for the extremely low probable error assigned to their measurement (Table I). Were the other measurements repeated with comparable accuracy, any wavelength dependence in the lunar cross section would be

TABLE I

VALUES FOR THE RADAR CROSS SECTION OF THE MOON AS A FUNCTION
OF WAVELENGTH REPORTED BY VARIOUS WORKERS

Author	Year	Wavelength (cm)	$\sigma/\pi a^2$	Estimated error (dB)
Lynn et al.	1963	0.86	0.07	±1.0
Kobrin	1963[a]	3.0	0.07	±1.0
Morrow et al.	1963[a]	3.6	0.07	±1.5
Evans and Pettengill	1963c	3.6	0.04	±3.0
Kobrin	1963[a]	10.0	0.07	±1.0
Hughes	1963[a]	10.0	0.05	±3.0
Evans and Hagfors	1966	23.0	0.065	±0.5
Aarons	1959[b]	33.5	0.09	±3.0
Blevis and Chapman	1960	61.0	0.05	±3.0
Fricker et al.	1960	73.0	0.074	±1.0
Leadabrand	1959[b]	75.0	0.10	±3.0
Trexler	1958	100.0	0.07	±4.0
Aarons	1959[b]	149.0	0.07	±3.0
Trexler	1958	150.0	0.08	±4.0
Webb	1959[b]	199.0	0.05	⊥3.0
Evans	1957	250.0	0.10	±3.0
Evans et al.	1959	300.0	0.10	±3.0
Evans and Ingalls	1962	784.0	0.06	±5.0
Davis and Rohlfs	1964	1130.0	0.19	+3.0 −2.0
Davis and Rohlfs	1964	1560.0	0.13	+3.0 −2.0
Davis and Rohlfs	1964	1920.0	0.16	+3.0 −2.0

[a] Revised value [privately communicated to Evans and Pettengill (1963c)].
[b] Reported by Senior and Siegel (1959, 1960).

obtainable with greater confidence. The mean of the values listed in Table I is close to 0.07, and using this mean value, the term $\sigma/(4\pi R^2)^2$ (Eq. 7) has a value of about 1.95×10^{-25} m^{-2} at the average distance of the Moon. This term represents the "path loss" encountered by a signal transmitted by an antenna of unit gain and received by one of unit aperture. Expressed in this fashion, the path loss is independent of wavelength and may be taken as 247 dB/m^2.

2. Theory

If the Moon were a perfect sphere with a surface composed of a metal or slightly lossy dielectric, the cross section would be

$$\sigma = \rho_0 \pi a^2 \tag{8}$$

where ρ_0 is the power reflection coefficient at normal incidence. It has been customary in the past (e.g., Evans and Pettengill, 1963b) to assume that the reflection coefficient ρ_0 is identical to the Fresnel reflection coefficient at normal incidence. This, of course, assumes that the surface can be represented as a sharp boundary between vacuum and a homogeneous material with certain electrical properties. On this assumption

$$\rho_0 = \left| \frac{\sqrt{\epsilon} - 1}{\sqrt{\epsilon} + 1} \right|^2 \tag{9}$$

where ϵ is the (complex) relative dielectric constant of the lunar surface material.

Other evidence suggests that the properties of the lunar soil change with depth (Scott et al., 1967). One may, therefore, suppose that there is a gradual change with depth in the electrical properties of the interface. This variation may take several different forms, two of which have been examined by Hagfors (1967).

If we have a uniform upper layer with a dielectric constant ϵ_1, and a depth b, supported by a semi-infinite homogeneous medium which has a dielectric constant ϵ_2, the power reflection coefficient at normal incidence is

$$\rho_0 = \frac{\epsilon_1(\sqrt{\epsilon_2} - 1)^2 - (\epsilon_1 - 1)(\epsilon_2 - \epsilon_1)\sin^2(\sqrt{\epsilon_1}kb)}{\epsilon_1(\sqrt{\epsilon_2} + 1)^2 - (\epsilon_1 - 1)(\epsilon_2 - \epsilon_1)\sin^2(\sqrt{\epsilon_1}kb)} \tag{10}$$

where $k = 2\pi/\lambda$ (λ is the wavelength in vacuo).

Reflections may be expected from a large number of area elements that are oriented to be normal to the ray path. If the depth b at the position of these elements is assumed to be distributed at random in accordance with a probability density $p(b)$, the mean normal reflectivity is found as

$$\bar{\rho}_0 = \int_0^\infty p(b)\,\rho_0(b)\,db \tag{11}$$

with $\rho_0(b)$ given by Eq. (10). The actual mean reflection coefficient will depend strongly on the form of the probability density function $p(b)$ and on the mean depth $\bar{b}$. When the mean depth is either much smaller than or much greater than $\lambda/4\sqrt{\epsilon_1}$, however, the form of $p(b)$ is no longer important and we obtain in the former case,

$$\bar{\rho}_0 = \left| \frac{\sqrt{\epsilon_2} - 1}{\sqrt{\epsilon_2} + 1} \right|^2 \quad \text{(thin layer),} \tag{12}$$

and in the latter case,

$$\bar{\rho}_0 = 1 - \frac{4\sqrt{\epsilon_1\epsilon_2}}{(\sqrt{\epsilon_2} + 1)(\sqrt{\epsilon_2} + \sqrt{\epsilon_1})} \quad \text{(thick layer)} \tag{13}$$

Hence, a thin layer is invisible and a thick layer serves to reduce the reflection coefficient from what it would have been in the absence of a top layer. For a given ϵ_2, minimum reflection occurs when $\epsilon_1 = \sqrt{\epsilon_2}$. Figure 7 shows graphically

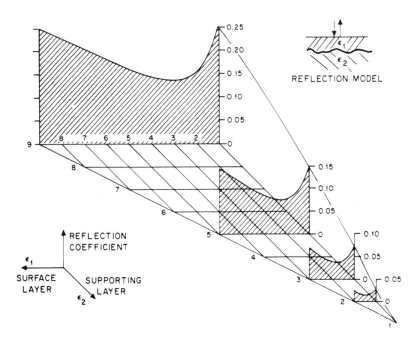

FIG. 7. Power reflection coefficient for a two-layer surface. The top layer has a dielectric constant ϵ_1 and the bottom layer ϵ_2 and the thickness is random.

the relationship between ϵ_1, ϵ_2, and $\bar{\rho}_0$. As can be seen, there are many combinations of ϵ_1 and ϵ_2 which can give rise to a required reflection coefficient. On the basis of this model, the increase in cross section seen at the longer wavelengths in Fig. 6, if real, may be understood if the depth of the surface layer equals $\lambda/4\sqrt{\epsilon_1}$ for a wavelength between 1 and 10 m. For a crude measure of the depth, one may substitute $\lambda = 5$ m and $4\sqrt{\epsilon_1} = 5$ to obtain a depth of 1 m.

On the other hand, a model involving a *gradual* transition in electrical properties through the upper layer may be more realistic. One that can be handled mathematically has a dielectric constant which varies linearly with depth from a value of ϵ_1 at the top to ϵ_2 at the transition to the underlying layer. The model is shown in Fig. 8. A linear variation in dielectric constant with depth leads to a Stokes differential equation having the two solutions $Ai(x)$ and $Bi(x)$. The reflection coefficient at normal incidence is given by

$$\rho_0 = |\, \text{Det}_-/\text{Det}_+ \,|^2 \qquad (14)$$

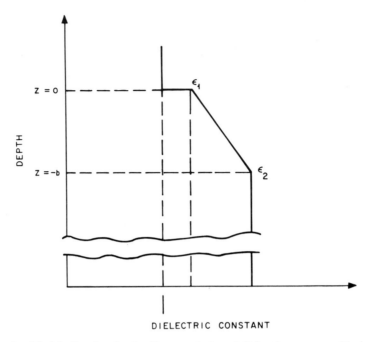

FIG. 8. Model of surface having linear variation of dielectric constant with depth.

where

$$\mathrm{Det}_{\pm} = \left\{ Bi\left(-\frac{\alpha b\epsilon_1}{\varDelta\epsilon}\right) \pm \frac{i\alpha}{k} Bi'\left(-\frac{\alpha b\epsilon_1}{\varDelta\epsilon}\right) \right\}$$

$$\times \left\{ Ai\left(-\frac{\alpha b\epsilon_2}{\varDelta\epsilon}\right) - \frac{i\alpha}{k\sqrt{\epsilon_2}} Ai'\left(-\frac{\alpha b\epsilon_1}{\varDelta\epsilon}\right) \right\}$$

$$- \left\{ Ai\left(-\frac{\alpha b\epsilon_1}{\varDelta\epsilon}\right) \pm \frac{i\alpha}{k} Ai'\left(-\frac{\alpha b\epsilon_1}{\varDelta\epsilon}\right) \right\}$$

$$\times \left\{ Bi\left(-\frac{\alpha b\epsilon_2}{\varDelta\epsilon}\right) - \frac{i\alpha}{k\sqrt{\epsilon_2}} Bi'\left(-\frac{\alpha b\epsilon_1}{\varDelta\epsilon}\right) \right\}$$

with

$$\alpha = (k^2 \varDelta\epsilon/b)^{1/3},$$
$$\varDelta\epsilon = \epsilon_2 - \epsilon_1,$$
$$b = \text{depth of layer},$$
$$k = 2\pi/\lambda,$$
$$\lambda = \text{wavelength \textit{in vacuo}}.$$

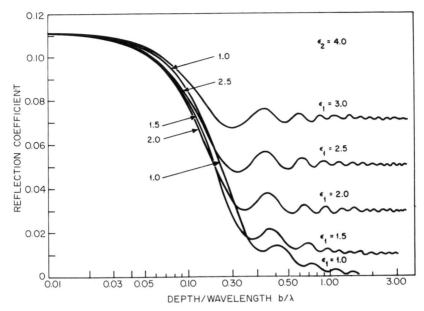

FIG. 9.   Reflection coefficient from surface having linear variation of dielectric constant with depth vs. layer thickness (Fig. 8).

Some computed results for this case are shown in Fig. 9. As can be seen, whenever λ is small compared with the transition region, the reflectivity is determined by the abrupt change in dielectric constant at the top, i.e., by $\epsilon_1 - 1$. We see that the effect of the subsurface will only be observable when $\lambda > 0.26b$.

Matveev (1967) has considered a case where the refractive index varies with depth in an exponential manner. He concluded that the Fresnel reflection coefficients may be used provided an effective dielectric constant ϵ_{eff} is inserted. This ϵ_{eff} will depend on wavelength and on the limiting values of the dielectric constant.

Other more complicated models can obviously be constructed. One might for instance have an abrupt transition in dielectric constant at $z = -b$ (see Fig. 8), and the depth to this transition may be random as in the homogeneous double-layer model. One may also have local variations in the electrical properties of the lunar surface material.

Thus far it has been assumed that the Moon's surface is completely smooth, i.e., we have not considered how the cross section depends upon surface irregularities. If the surface deviates from the spherical shape in a random manner *while remaining smooth locally*, the cross section is modified to

$$\sigma_S = (1 + s^2)\,\rho_0\pi a^2 \tag{15}$$

where s is a measure of the rms slope of the surface undulations. As explained in detail below, it has been found that in the case of the Moon, s is small enough so that it can be neglected in comparison with unity.

It has become evident—particularly from extensive polarization measurements (Section III,C)—that the return from the Moon cannot be completely described as a reflection from a locally smooth, gently undulating surface. Instead, the reflection must be thought of as originating in two different mechanisms: one which is associated with quasi-specular reflection from the gently undulating surface, and a second which arises as a result of a somewhat more vaguely defined small-scale diffuse scattering mechanism. In general, therefore, it is necessary to write

$$\sigma = g\rho_0\pi a^2 \tag{16}$$

where g is the gain of the Moon (over an isotropic reflector) in the direction back toward the radar. It follows that any measure of the cross section is a measure of the *product* $g\rho_0$.

An alternate approach to obtaining the scattering cross section is to consider the scattering properties of the target as specified by a function $\sigma_0(i, \phi, \theta)$ which defines the reflected intensity *per unit surface area* per steradian. This function is obtained by exploring the power reflected from an elemental area of surface illuminated at an angle of incidence i, and observed at an angle of reflection ϕ; the planes containing these two rays and the surface normal are at an angle θ as shown in Fig. 10. Provided there is no coherence between returns from different surface elements, the gain G_m of the Moon in the direction of backscattering is given by (Evans and Pettengill, 1963a)

$$G_m = \frac{4\pi \int_0^{\pi/2} \sigma_0(i) \sin i \, di}{\int_0^{\pi/2} \int_0^{\pi/2} \int_0^{2\pi} \sigma_0(i, \phi, \theta) \sin i \sin \phi \, d\theta \, d\phi \, di} \tag{17}$$

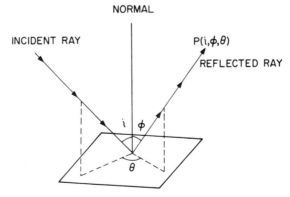

FIG. 10.   The geometry required for studying the complete scattering characteristics of an irregular surface in order to obtain a value for the gain of the whole Moon over an isotropic scatterer.

where $\sigma_0(i)$ is the particular case in which $i = \phi$, $\theta = 0$. From Earth-based observations alone, only $\sigma_0(i)$ can be determined and hence G_m as defined by Eq. (17) cannot be obtained. If, however, G_m were known either from theory or observation, the cross section for the whole sphere could be written as

$$\sigma = G_m \bar{\rho} \pi a^2. \tag{18}$$

Here $\bar{\rho}$ is the albedo averaged over the hemisphere, and this is usually different from the reflection coefficient at normal incidence ρ_0. This distinction has not always been recognized and the literature contains several instances where ρ_0 has been equated with $\bar{\rho}$ without proper explanation. If this is done, the directivity factor g is automatically made the same as G_m. Rea et al. (1964) have pointed out that these approximations are certainly not valid for the case of a smooth dielectric sphere.

3. Interpretation

At any single wavelength, it is found that a portion of the radar return may be associated with the smoother undulating portions of the surface, and the remainder with the small-scale diffuse scattering mechanism. This separation can be accomplished on the basis of studies of the power returned as a function of the angle of incidence (Section III,B). The amount of diffuse power has been estimated to be approximately as shown in Table II for the various wavelengths

TABLE II

COMPUTATION OF DIELECTRIC CONSTANT AT THREE WAVELENGTHS

λ (cm)	P Quasi-specular (%)	P Diffuse (%)	$1 - x$ (%)	x (%)	ρ_0	ϵ
68	80	20	8	92	0.063	2.8
23	75	25	9	91	0.061	2.7
3.6	70	30	14	86	0.057	2.6
0.86	15	85	68	32	0.033	2.1

examined. It must be pointed out that the estimates given in Table II are crude. Other extrapolation methods may give values considerably less than those quoted. For example, Pettengill and Thompson (1968) have arrived at 12% for their 70-cm results.

The presence of a nonspecular backscattering surface component will remove a fraction x of the surface available for quasispecular scattering. The total backscattering cross section may therefore be expressed as

$$\sigma = (1 - x)\sigma_s + x\sigma_d \qquad (19)$$

where the smooth cross section σ_s is given in Eq. (15).

Evans and Hagfors (1964) have discussed the solution of this equation for σ_s and pointed out that with only the ratio $(1 - x)\,\sigma_s/x\sigma_d$ known and two unknowns (x and σ_s/σ_d) it is not possible to proceed without some assumption. On may employ Eq. (18) to write the cross section of the diffuse component σ_d as

$$\sigma_d = G_m\bar{\rho}\pi a^2 \qquad (20)$$

from which

$$\sigma = [(1 - x)(1 + s^2)\rho_0 + xG_m\bar{\rho}]\,\pi a^2. \qquad (21)$$

The gain G_m of the diffuse component may then be estimated by assuming that this component scatters approximately according to the Lambert law, i.e.,

$$\sigma_0(i, \phi, \theta) \propto \cos^2\phi, \qquad (22)$$

from which $G_m = 8/3$ (Grieg *et al.*, 1948). Since also $1 + s^2 \to 1$, Eq. (21) becomes

$$\sigma = [(1 - x)\rho_0 + \tfrac{8}{3}x\bar{\rho}]\,\pi a^2. \qquad (23)$$

For the want of any better assumption, Evans and Hagfors (1964) next assumed that $\rho_0 = \bar{\rho}$, so that values of x could be obtained and these are given in the third column of Table II.

If it is further assumed that at all the wavelengths listed, $\sigma = 0.07\pi a^2$, the values for the reflection coefficient ρ_0 listed in Table II are obtained. Finally, interpreting these as reflections from a single interface leads to the values for the relative dielectric constant ϵ [via Eq. (9)] given in the last column. It is clear that if the dielectric constant increases with depth, these values represent an *upper limit* to the dielectric constant of the topmost material.

The assumptions involved in obtaining the values of ρ_0 listed in Table II are not serious where x is small. For example, were the diffuse component caused by boulders overlying a sandy-like material, it would seem reasonable that $\bar{\rho} > \rho_0$, in which case x would be reduced by the ratio $\bar{\rho}/\rho_0$ and ρ_0 would be raised somewhat. Yet, as long as the fraction x of the surface ascribable to diffuse scattering remains small, the scope for changing the values is limited. However, at $\lambda = 8.6$ mm where x is large, it is doubtful if the treatment outlined above has much validity, and thus the wavelength dependence in ρ_0 indicated in Table II should be viewed with caution. It may be real and as such be due to differing penetrations of the signals at different wavelengths, but this conclusion rests largely on other evidence (below).

Values of ϵ for some typical rock mineral samples are given in Table III. These have been taken from many values listed in a report by Brunschwig *et al.* (1960). The selection in Table III is somewhat arbitrary, but does indicate the wide

TABLE III

Some Typical Terrestrial Rocks and Their Values of Dielectric Constant[a]

Mineral	Type	Source	Dielectric constant ϵ
Andesite	Vesicular basalt	Chaffee County, Colo.	6.51
Olivine basalt, cellular		Washington	5.50
Basalt	Olivine basalt	Jefferson County, Colo.	8.89
Olivine basalt	Basalt	Lintz, Rhenish Prussia	17.4
Diabase		Mt. Tom, Mass.	10.8
Rhyolitic pumice	Pumice	Millard County, Utah	2.29
Rhyolite		Castle Rock, Colo.	4.00
Basaltic scoria	Scoria	nr. Klamath Falls, Oreg.	6.08
Trachytic tuff		nr. Cripple Creek, Colo.	5.32
Quartz sandstone	Sandstone	Columbia County, Pa.	4.84

[a] From Brunschwig *et al.* (1960).

scatter of values encountered. The basaltic specimens examined by Brunschwig *et al.* (1960) showed the greatest range of values (from 5.5 to 26.7) and an average value for the seven samples listed is 14. For the minerals listed as forms of andesite (5 samples), the mean was 8.8, while for the rhyolitic samples, it was, lower (4.1). Silicate materials, e.g., fused quartz, have dielectric constants in the range 4 to 7 and dry terestrial sand has a dielectric constant of about half this.

The chemical analyses of lunar samples carried out by the Surveyor vehicles (Jaffe, 1969) and Earth-based measurcments of the optical properties of the surface material (Hapke, 1968, 1970) suggest that most of the surface has a composition similar to terrestrial iron-rich basalts for which the dielectric constant has an average value of about 10. As the observed dielectric constant is considerably lower than this, it is evident that the surface is porous and broken in some way. Indeed this conclusion must be reached regardless of the chemical composition since no common solid rocks have a dielectric constant as low as is observed. Although the Apollo landings have dramatically confirmed this result, reached some ten years earlier from the radar observations, it is still of some interest to see how far it is possible to proceed on the basis of the radar evidence alone.

If it is assumed that the material forming the lunar surface is homogeneous to the depth penetrated by radar signals, then some constraints can be placed on the relative density D of the actual material in bulk to that of an extremely small sample. Campbell and Ulrichs (1969) have shown that the Rayleigh mixing formula

$$D \left[\frac{\epsilon - 1}{\epsilon + 1} \right] = \left[\frac{\epsilon_{\text{obs}} - 1}{\epsilon_{\text{obs}} + 1} \right] \tag{24}$$

accurately describes the effective dielectric constant ϵ_{obs} of powders of material having an initial dielectric constant ϵ. A theoretical basis for this formula was developed by Twersky (1962).

Adopting a value $\epsilon = 10$ and $\epsilon_{\text{obs}} = 2.7$, we conclude that

$$D \sim 0.55 \quad \text{(homogeneous medium).} \tag{25}$$

In the absence of any estimate of ϵ, we would have obtained limits for D by setting $\epsilon = 4.0$ (quartz), from which

$$D \sim 0.75 \quad \text{(homogeneous medium—upper limit),} \tag{26}$$

and $\epsilon = 20$, from which

$$D \sim 0.4 \quad \text{(homogeneous medium—lower limit).} \tag{27}$$

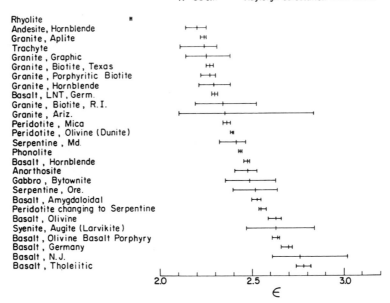

Fig. 11. Dielectric constants of powdered rock at 50% porosity (Campbell and Ulrichs, 1969).

These values are, of course, based upon the assumption of a homogeneous medium and it is unlikely that material with a relative density $D = 0.4$ could support itself for any considerable depth without being compacted under its own weight. Thus the model discussed in the previous section in which the dielectric constant increases linearly with depth over some interval b is probably more appropriate. It follows that the above estimates of D must represent *lower* limits and a "best estimate" based upon radar data alone would place $D \sim 0.5$, i.e., comparable to terrestrial sand at the depth of reflection. Figure 11 shows ϵ_{obs} for a number of rock powders at $D = 50\%$. It can be seen that the radar results for the Moon are consistent with a basaltic composition.

If the bulk density is as low as 50%, it follows that the waves must penetrate some distance into the surface. Campbell and Ulrichs (1969) find that decimeter waves penetrate ~ 10 wavelengths into rock powders of 1 gm/cm³ density before their intensity is reduced by a factor $\exp(-1)$. Thus, the decimeter radar results will be sensitive to the properties of the surface to a depth of the order of a meter.

B. Mean Surface Slope

1. Observations

In much the same way that the photometric function for the Moon (Fessenkov, 1961) has been used to infer the nature of the microstructure on the lunar surface, the radar reflection properties of the Moon may be employed to examine the terrain at scales roughly in the range $\lambda/10$ to 10λ where λ is the exploring wavelength. In the case of optical studies, the source (the Sun) and the observer (on Earth) are not colocated so that the complete function $\sigma(i, \phi, \theta)$ (Fig. 10) may be obtained. As already noted, it is only possible from Earth to make observations in which the incident and reflected rays lie approximately in the same direction. This difficulty can be overcome using orbiting satellites, and studies of this type have been attempted (*e.g.*, Tyler and Simpson, 1970) using Explorer 35, but a satellite really designed for this type of measurement has not yet been flown. In order to interpret the ground-based observations, it has been necessary to assume that the surface properties of the Moon are reasonably homogeneous and random. By that it is meant that on the scale to which the wavelength is sensitive, the relief does not vary in any systematic way over the Moon. This approximation serves to permit a useful interpretation of the results, but is obviously undesirable. There may well be, for example, marked differences between mare and highland ground, as indeed has been suggested by the satellite measurements (Tyler, 1968, Tyler and Simpson, 1970). In this event, the properties derived from radar studies would comprise some weighted mean of the two.

With modern radar equipment, echoes from the Moon may readily be resolved either in delay or in frequency. The libration of the Moon causes the lunar disk to appear to be rotating with respect to a terrestrial observer with a radial velocity that usually lies in the range 10^{-6} to 10^{-7} radian sec^{-1} (Section II,C). This gives rise to Doppler broadening of the signals so that the echo power at a given frequency offset is proportional to the reflectivity of a particular strip of the Moon's disk that is parallel to the apparent axis of rotation as was shown in Fig. 4 (Browne *et al.*, 1956). Thus the determination of the echo power spectrum $\bar{P}(f)$ yields the brightness distribution over the Moon's disk. Such measurements have been made by a number of workers (Evans, 1957; Evans and Ingalls, 1962; Daniels, 1963a, b). The best measurements of this kind have been performed using phase coherent radar systems, such as the Millstone Hill radar which was employed by Pettengill (1961) to obtain the results shown in Fig. 12. The limb-to-limb Doppler broadening introduced by the Moon is typically 2 Hz for a radar frequency of 100 MHz, and hence it is difficult to achieve good resolution by this technique. In Fig. 12, for example, the frequency resolution of the radar is one-fortieth of the total echo spectrum.

Much better resolution can be obtained by resolving the echoes with respect

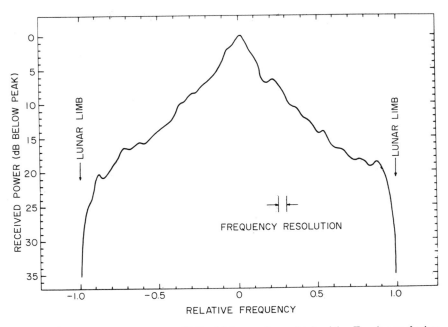

FIG. 12. The power spectrum $\bar{P}(f)$ of Moon echoes obtained by Fourier analyzing a chain of coherent pulses by means of a fast digital computer. These observations were made at 68-cm wavelength by Pettengill (1961) and show the strip brightness distribution across the lunar disk.

to delay. For example, pulse lengths of 12 μsec and shorter have been used to study the scattering behavior of the Moon (whose full radar depth is 11.6 msec). A short pulse illuminates at any instant an annulus on the surface as shown in Fig. 13. The amount of *actual* area illuminated by the pulse is given by $\pi a c \tau$

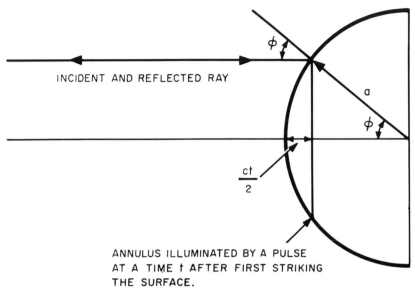

INCIDENT AND REFLECTED RAY

ANNULUS ILLUMINATED BY A PULSE
AT A TIME t AFTER FIRST STRIKING
THE SURFACE.

FIG. 13. The relation between the range delay t and the angle of incidence and reflection ϕ of the radio waves. The radius of the Moon is a; the velocity of light, c.

(where c is the velocity of light and τ is the pulse length) and is independent of delay t. However, the *projected* area will vary with the cosine of the angle of incidence ϕ where

$$\phi = \cos^{-1}\left[1 - (ct/2a)\right]. \tag{28}$$

Thus if the Moon behaved as a uniformly bright reflector (as it does optically), the average echo power versus delay function $\bar{P}(t)$ would have the form

$$\bar{P}(t) \propto \left[1 - (ct/2a)\right]. \tag{29}$$

Herein lies a second advantage of short pulse measurements; namely, there is a simple relation between the angle of incidence ϕ and the delay t [specified by Eq. (28)] as shown in Fig. 13. It follows that by determining the echo power as a function of delay $\bar{P}(t)$, one can determine directly the angular power spectrum $\bar{P}(\phi)$ of the echoes. We choose a function $\bar{P}(\phi)$ of the same form as the function $\sigma_0(i, \phi, \theta)$ used in Eq. (17), i.e., normalized to unit *actual* surface area. It follows that for a uniformly bright surface we have from Eq. (29)

$$\bar{P}(\phi) \propto \cos\phi \qquad \text{(Lommel–Seeliger)}, \qquad (30)$$

and for Lambert's law,

$$\bar{P}(\phi) \propto \cos^2\phi \qquad \text{(Lambert)}. \qquad (31)$$

In what follows, we shall largely base our discussion on short pulse observations of the Moon—these having been undertaken at the wavelengths listed in Table IV.

TABLE IV

ANGULAR POWER-SPECTRUM STUDIES OF THE MOON

Wavelength (cm)	Observer	Pulse length	Sample interval	Range of ϕ studied (deg)	Comments
0.86	Lynn *et al.* (1963)	2 sec	Once per pulse	8–60	Angular resolution achieved using 0.07° pencil beam
3.6	Evans and Pettengill (1963b)	30 μsec	20 μsec	5–80	
3.8	Evans (1968a)	10 μsec	10 μsec	2.5–90	
10	Hughes (1961)	5 μsec	20 μsec	2.5–14	Sensitivity limitations prevented higher values of ϕ from being examined
23	Evans and Hagfors (1966)	10 μsec	10 μsec	2.5–90	
68	Evans and Pettengill (1963b)	12 μsec	10 μsec	2.5–90	
600	Klemperer (1965)	100 μsec	?	7.5–90	
1130	Davis and Rohlfs (1964)	250 μsec	250 μsec	12–48	Echoes read from film

On any single sweep of the radar timebase, the echoes will show considerable structure as a function of delay owing to the constructive and destructive interference between reflections from different parts of the surface at the same delay. Thus in order to make meaningful measurements of the average scattering properties, it is necessary to sum many hundreds or thousands of sweeps. In the best measurements, this has been accomplished by sampling the signals with a digital voltmeter and summing the corresponding numbers in a fast digital computer.

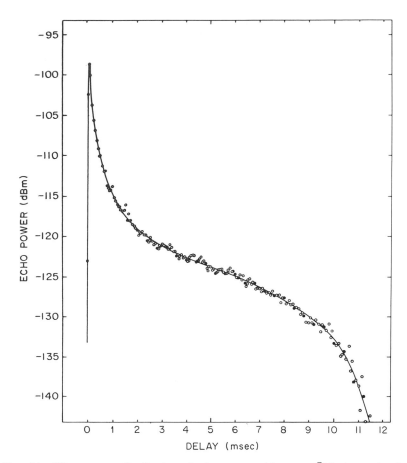

FIG. 14.   The average distribution of echo power with range $\bar{P}(t)$ observed using a transmitter pulse length of 65 μsec (Pettengill and Henry, 1962a).

The results of the first measurements made in this way (at 68-cm wavelength) are shown in Fig. 14 (Pettengill, 1960; Pettengill and Henry, 1962a) and these serve to illustrate the main features found subsequently at all other wavelengths. The echo power is seen to decay with delay initially very rapidly and then, beyond about 2 msec, more slowly. When Pettengill and Henry plotted the log of the echo power as a function of log cos ϕ [to see if laws of the type Eq. (30) or Eq. (31) would describe the behavior], they obtained the results shown in Fig. 15. In the range $50° < \phi < 80°$, the law $P(\phi) \propto \cos^{3/2} \phi$ was found to hold. It was on the basis of this (and additional evidence to be reviewed) that one could conclude that two scattering mechanisms are operating. For $\phi < 50°$, it is believed that the smoother elements of the surface are responsible for the

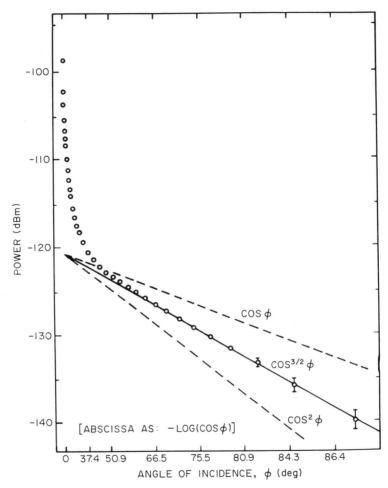

FIG. 15.   The angular power spectrum $\bar{P}(\phi)$ obtained from the data shown in Fig. 14 (Pettengill and Henry, 1962a).

reflections, and the distribution of echo power in this region is thus a measure of the distribution of slopes that these elements possess. At large angles ϕ, few elements of the surface are found tilted to be normal to the ray path, and in consequence we believe that it is the small scale structure on the surface (i.e., having vertical and horizontal scales comparable with the wavelength and therefore scattering almost isotropically) which dominates the reflections. This structure might take the form of protrusions (boulders) or depressions (craters or pits). The separation of $\bar{P}(t)$ into the two components proposed is illustrated in Fig. 16.

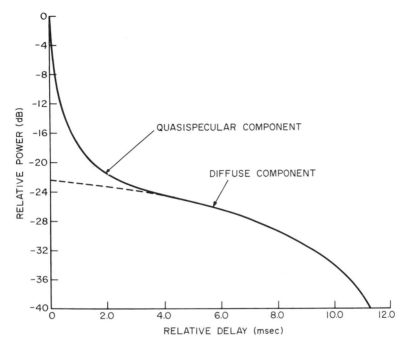

FIG. 16.   The average echo power vs. delay $\bar{P}(t)$ observed when 12 μsec pulses are reflected by the Moon at 68-cm wavelength (Evans and Pettengill, 1963b). This figure is intended to show the proposed separation of the echo power into two components representing different scattering mechanisms.

Figure 17 summarizes the curves later obtained (Table IV) at other wavelenths. At each of these wavelengths, it is possible to distinguish two regions where the dependence on ϕ seems markedly different, as illustrated in Fig. 18. We shall consider here the quasi-specular component leaving the diffuse component for later discussion where other observations relevant to its interpretation are reviewed.

The most precise observations of $\bar{P}(t)$ have been made at 3.8 cm (Evans, 1968a), 23 cm (Evans and Hagfors, 1966), and 68 cm (Evans and Pettengill, 1963b). These are summarized in Table V which presents the cross section per unit surface area (in dB above 1 m²) as a function of delay t at these three wavelengths. These data are plotted for small angles of incidence in Fig. 19, where the wavelength dependence of the scattering is clearly apparent.

Near normal incidence the dependence of cross section per unit area on wavelength varies as

$$\sigma \sim \lambda^{0.46} \qquad 3.8 \text{ cm} < \lambda < 23 \text{ cm}$$
$$\qquad\qquad\qquad\qquad\qquad\qquad\qquad (\phi \text{ near zero}). \qquad (32)$$
$$\sigma \sim \lambda^{0.32} \qquad 23 \text{ cm} < \lambda < 68 \text{ cm}$$

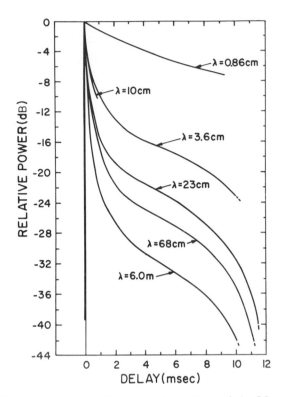

FIG. 17. This plot summarizes the radar observations of the Moon reported thus far in which a pulse length of 100 μsec or shorter was used (Table IV). All the curves have been normalized at zero delay. Although the resolution of the echo near $t = 0$ differed considerably in the experiments (and hence the relative positions of the curves is likely to be somewhat in error), the wavelength dependence in the scattering is clearly evident.

At grazing angles of incidence, the wavelength dependence goes in the opposite direction as will be discussed. The wavelength dependence of the scattering cross section σ_0 of unit surface area at a given angle of incidence ($i = \phi$) has been referred to as the "color" of the surface and has been discussed by Katz (1966). According to Katz, the different behavior for $\phi \to 0°$ and $\phi \to 90°$ is additional evidence for treating the returns for these cases as arising from entirely different mechanisms.

2. Theory

We have seen from the results presented in the previous section that the way in which the Moon scatters radio waves is distinctly unlike the manner in which it reflects light. For wavelengths of 1 m or longer, the Moon appears to be a very

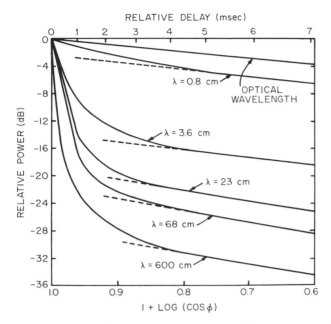

FIG. 18. The results plotted in Fig. 17 are here replotted as a function of log(cos ϕ). We find that the region beyond 4-msec delay (termed diffuse in the text) conforms to a simple dependence upon either cos ϕ (λ = 0.8 cm) or cos$^{3/2}$ ϕ (λ = 23, 68, and 600 cm).

limb-dark reflector whereas optically it is almost uniformly bright (Markov, 1948). One may conclude, therefore, that on a scale of 1 m the Moon is much smoother than on a scale of a few microns. The extent to which it is possible to deduce the statistical properties of the lunar surface from these results can only be reviewed briefly here.

A number of authors (Brown, 1960; Muhleman, 1964; Rea *et al.*, 1964) have developed theories involving only the principles of geometric optics. In this approach, only surfaces *normal* to the line of sight scatter back favorably, and their reflection coefficient is simply the Fresnel reflection coefficient for normal incidence ρ_0. Thus the angular power spectrum $\bar{P}(\phi)$ is directly related to the distribution of surface slopes in

$$\bar{P}(\phi)\, d\phi \propto p(\phi)/\cos\phi \cdot d\phi \qquad (33)$$

where $p(\phi)$ is the probability of finding an elemental area inclined to the mean surface with a slope in the range ϕ to $\phi + d\phi$.

Other workers have chosen to treat the problem employing diffraction theory. Starting assumptions are usually that the surface may be regarded as being plane, *perfectly conducting, smooth,* and undulating, and causing no shadowing. The

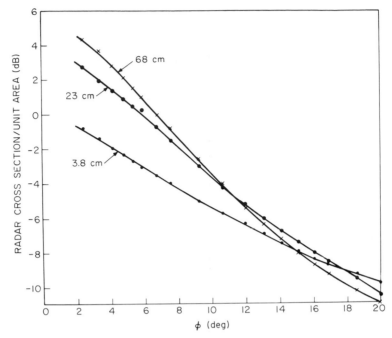

FIG. 19. Cross section per unit surface area versus angle of incidence for small angles ϕ; λ = 68, 23, and 3.8 cm.

requirement that the surface be perfectly conducting is invoked to avoid the difficulty of the reflection coefficient depending upon the angles of incidence and reflection. The word "smooth" is here intended to mean that the surface contains no structural components having horizontal and vertical dimensions comparable with the wavelength, since the boundary conditions are established locally by means of Fresnel's reflection formulas. It is next commonly assumed (e.g., Hargreaves, 1959; Daniels, 1961; Hagfors, 1961; Winter, 1962) that the departure of the true surface from the mean follows a Gaussian probability distribution; that is, the chance of finding a given point to be at a height h above the mean surface is proportional to $\exp[-1/2(h/h_0)^2]$ where h_0 is the rms height variation. Other forms of height distribution (Bramley, 1962) have been used, but the theory should not be very sensitive to this function provided $h_0 \gg \lambda$, (Hargreaves, 1959; Daniels, 1961, 1962). This follows because the phase variation in the reflected wave front will be many times 2π when $h_0 \gg \lambda$ and it becomes impossible to determine h_0 from the observations. Having described the vertical behavior of the surface, it remains only to describe its horizontal structure. This is done by means of an autocorrelation function $\rho(\Delta r)$ where

$$\rho(\Delta r) = h(r)\,h(r + \Delta r)/(h_0{}^2) \tag{34}$$

in which $h(r)$ is the height of the surface at a point r and $h(r + \Delta r)$ at a distance Δr away. The case where $\rho(\Delta r)$ is another Gaussian function,

$$\rho(\Delta r) \propto \exp[-\tfrac{1}{2}(\Delta r/d_0)^2], \tag{35}$$

in which d_0 is the horizontal scale size, has been treated by Hargreaves (1959) and Hagfors (1961). If the surface be considered plane and extending in one direction only, then each point on the surface introduces a phase change in the reflected wave of $(4\pi h/\lambda)$ radian. These phase fluctuations are said to be "shallow" if the rms phase fluctuation $\Omega = (4\pi h/\lambda)$ is less than 1 radian, and in this case the autocorrelation function describing the variation of radio phase with distance Δr over the surface will be the same as $\rho(\Delta r)$. If, on the other hand, $h_0 \gg \lambda$, the rms phase fluctuation, Ω becomes greater than 1 radian and the correlation in the reflected phase front will fall from d_0 to d_0/Ω (since 0, 2π, 4π,... radians are indistinguishable). At a large distance from the surface, the initial phase variations become modified due to the overlapping of many rays, and amplitude fluctuations appear. An rms phase fluctuation $\Omega/\sqrt{2}$ is then observed (Bowhill, 1957).

The angular power spectrum $\bar{P}(\phi)$ is given by the Fourier transform of the autocorrelation function describing the reflected phase front immediately after reflection and can be written as

$$\bar{P}(\phi) \propto \exp[-1/2(\phi/\phi_0)^2], \tag{36}$$

where $\phi_0 = \lambda/2\pi d_0$ for $h_0 \ll \lambda$, $\phi_0 = h_0/d_0$ when $\lambda \ll h_0$. It follows that where the wavelength is much larger than the vertical extent of the height fluctuations, the value of ϕ_0 yields directly the horizontal scale of the structure. When $\Omega \geqslant 1$ radian, ϕ_0 yields only the ratio h_0/d_0. Since h_0/d_0 is the rms slope, this indicates that information is available about the surface slopes but not the actual scale of the structure (horizontal or vertical). If observations could be made as the wavelength was increased, eventually (when $\lambda > h_0$) it would be possible to determine d_0. However, in the case of the Moon, this would require low-frequency radio waves which could not propagate through the Earth's ionosphere. Daniels (1961, 1962) has discussed at length this limitation in radar observations and shown that one cannot obtain information on the rms height fluctuation when this is many times the wavelength in size. It is also clear that small-scale height fluctuations $\leqslant \lambda/8$ will introduce only small phase changes in the reflected phase front and will be unimportant. Thus radar observations may be regarded as being sensitive to structure in the range, say, $\lambda/10$ to 10 or 100 wavelengths. Only by repeating the observations over a wide range of wavelengths can the true nature of the surface be determined.

Other forms than Gaussian for the autocorrelation function $\rho(\Delta r)$ have been explored.

These include the exponential

$$\rho(\Delta r) \propto \exp(-\Delta r/d') \tag{37}$$

(Daniels, 1961; Hayre and Moore, 1961; Hughes, 1962a, b), which closely approximates many terrestrial surfaces. Fung and Moore (1964) and Beckmann (1965) have employed a sum of such terms.

A number of papers have been written to extend the diffraction theory results allowing for the curvature of the Moon and the non-perfectly conducting nature of the surface. It is not possible here to summarize all these papers, and the reader is referred to the review by Barrick and Peake (1967). The treatment that follows is taken from the work of Hagfors (1961, 1964, 1966) who has shown that a reasonably good understanding of the angular dependence of the scattering can be obtained by assuming that the boundary condition on the reflecting surface is well approximated by the tangential plane assumption. This assumption involves replacing the surface at any point with a tangential plane and assigning to that point the boundary field which would exist on that tangential plane. The approximation is thought to be justifiable for the quasi-specular return, but may not be adequate for the diffuse return nor for the depolarized return. Beckmann (1968) has recently published a paper in which this view is contested. Using Kirchhoff theory, Beckmann derives a measure of the depolarization of the return from a range ring on the Moon. Since the theory presupposes a relatively smooth surface in order for convenient boundary conditions to be established and since the lunar surface possesses much small scale structure which depolarizes, one should regard the results of the theory with considerable suspicion.

When the surface is described as quasi-smooth with a reflectivity at normal incidence of ρ_0, and when the surface is assumed to deviate from its mean shape at a point $\mathbf{r}$ by an amount $h(\mathbf{r})$ which is a stochastic function of $\mathbf{r}$, we obtain

$$P_S = P_T \frac{G_R G_T}{8\pi R^4 \cos^2 \phi} \rho_0 \cdot \int_0^\infty d(\Delta r) \, \Delta r \, J_0(2k\Delta r \sin \phi) \tag{38}$$
$$\exp\{-4k^2 h_0^2 \cos \phi [1 - \rho(\Delta r)]\}$$

where

$$P_S = \text{received power,}$$
$$P_T = \text{transmitter power,}$$
$$G_R G_T = \text{product of receiver and transmitter antenna gains,}$$
$$R = \text{distance to the Moon,}$$
$$J_0(\ ) = \text{the zero-order Bessel function,}$$
$$k = 2\pi/\lambda,$$
$$\rho(\Delta r) = \langle h(r) \, h(r + \Delta r) \rangle_{\text{av}}/h_0^2,$$
$$h_0^2 = \langle h(r)^2 \rangle_{\text{av}}.$$

For the Moon it appears to be safe, based on optical observations, to assume that

$$2kh_0 \cos \phi \gg 1 \qquad (39)$$

for all angles and wavelengths of practical interest. In essence, this means that the surface undulations are much larger than the wavelength, thereby giving rise to "deep" phase fluctuation in the reflected phase front. As long as $\rho(\Delta r)$ can be expanded about $\Delta r = 0$ in a power series

$$\rho(\Delta r) = 1 - (\Delta r^2/2)[-\rho''(0)] + \cdots \qquad (40)$$

where

$$\rho''(0) = \partial^2 \rho / \partial \Delta r^2, \qquad \Delta r = 0,$$

and as long as this is a good approximation to the correlation function within the range over which the exponential term in Eq. (38) is appreciably different from zero, we obtain for the power scattered per unit area at an angle ϕ

$$P_S = \frac{1}{2 \cos^4 \phi h_0{}^2(-\rho'')} \cdot \exp[-\tan^2\phi/2h_0{}^2(-\rho'')] \frac{P_T G_T G_R \lambda^2}{64\pi^3 R^4} . \qquad (41)$$

This can be shown to be equivalent to the following expression:

$$P_S = \frac{1}{2 \cos \phi} p(\phi) \frac{P_T G_T G_R \lambda^2}{64\pi^3 R^4} \qquad (42)$$

where $p(\phi)$ is the probability density for the angle ϕ between the normal to a surface element and the mean normal to the surface (Hagfors, 1966; Fung and Moore, 1966; Fung, 1967). Under these conditions, which are equivalent to those of geometrical optics, it is a relatively straightforward affair to identify the mean slope and rms slope of the lunar surface from the mean echo power vs. delay data. Assuming, for example, that the autocorrelation function is Gaussian,

$$\rho(\Delta r) = \exp(-\Delta r^2/2d_0{}^2) \qquad (35)$$

and hence that

$$-\rho'' = 1/d_0{}^2, \qquad (43)$$

the rms slope along any direction on the surface is

$$(t_x)_{\mathrm{rms}} = (t_y)_{\mathrm{rms}} = h_0/d_0 . \qquad (44)$$

The "mean slope" or the mean value of the tangent of the angle between the normal to an arbitrary surface element and the normal to the mean surface becomes

$$\langle \tan \phi \rangle_{\mathrm{av}} = \left(\frac{\pi}{2}\right)^{1/2} \frac{h_0}{d_0} = 1.25 \frac{h_0}{d_0} . \qquad (45)$$

The rms of the tangent of this angle becomes

$$\{\langle\tan^2\phi\rangle_{\mathrm{av}}\}^{1/2} = \sqrt{2}\,\frac{h_0}{d_0} = 1.41\,\frac{h_0}{d_0}\,. \tag{46}$$

Let us next turn to cases where there is no such simple relationship between true and apparent surface slopes. This occurs whenever the conditions spelled out above are not met. Physically, the breakdown of these conditions means that the surface has a considerable amount of fine structure of a lateral scale not necessarily as small as the wavelength but of a vertical scale smaller than the wavelength of the exploring wave. This will bring about structural detail in the correlation function near the origin which will not appreciably influence the value of the integral [Eq. (38)] determining the backscattered power. This fine structure could, on the other hand, well be completely dominant in determining the true rms slope of the surface. In this situation, the slope distribution derived from a direct geometric-optics analysis can only apply to "apparent" slopes which differ from the true slopes as a result of the smoothing effect imposed by the finite electrical length of the incident waves. This smoothing scale, unfortunately, is a function of the angle of incidence of the radio waves on the surface. To see this, it is only necessary to refer again to the integral [Eq. (38)] giving the backscattering power per unit area as a function of angle of incidence ϕ. When ϕ is close to zero, the integral is determined primarily by a range of $\varDelta r$ extending from 0 to $\varDelta r_e$, where $\varDelta r_e$ is the solution of the equation

$$4k^2h_0{}^2\cos^2\phi[1 - \rho(\varDelta r_e)] = 1. \tag{47}$$

For somewhat larger angles of incidence ϕ, it will be the Bessel function that limits the range of $\varDelta r$ over which significant contributions to the integral are obtained. The range of $\varDelta r$ as determined by the "width" of the Bessel function extends from 0 to $\varDelta r_B$, and it is given, at least in order of magnitude, by

$$\varDelta r_B \approx 1/(k\sin\phi). \tag{48}$$

To relate the range of scales on the surface to the range of separations $\varDelta r$, we can expand $\rho(\varDelta r)$ into a power spectrum

$$\rho(\varDelta r) = (1/2\pi)\int_0^\infty \kappa\,d\kappa\,F(\kappa)\,J_0(\varDelta r,\kappa). \tag{49}$$

In writing Eq. (49), we have assumed that the surface deviations can be regarded as a sum of many waves of different frequency κ. The function $F(\kappa)$ is then a measure of the amplitude of each wave. For most random rough surfaces, the amplitude spectrum $F(\kappa)$ usually decreases monotonically with increasing frequency. As examples, consider the Gaussian autocorrelation function [Eq. (35)] for which we obtain

$$F(\kappa) = 2\pi d_0{}^2 \exp(-\kappa^2 d_0{}^2/2) \tag{50a}$$

and for the exponential autocorrelation function often used (Hughes, 1962a, b) [Eq. (37)] we obtain

$$F(\kappa) = 2\pi \, d_1^{-1}(d_1^{-2} + \kappa^2)^{-3/2}. \tag{50b}$$

Since the fine structure in the correlation function in the range of Δr either from 0 to Δr_e [Eq. (47)] or from 0 to Δr_B [Eq. (48)] cannot appreciably affect the integral in Eq. (38), we conclude that an approximation to the received power is obtained by "filtering" out the high frequency components in $F(\kappa)$ by writing

$$\rho_e(\Delta r) = (1/2\pi) \, B \int_0^{\kappa_M} \kappa \, F(\kappa) \, J_0(\Delta r, \kappa) \, d\kappa, \tag{51}$$

where B is a normalizing constant and

$$\kappa_M = \begin{matrix} 1/\Delta r_e & \Delta r_B > \Delta r_e \\ 1/\Delta r_B & \Delta r_B < \Delta r_e \end{matrix}. \tag{52}$$

This approximation can only be regarded as somewhat crude. A filtering function should have been applied to Eq. (49), and this filtering function would not necessarily have the rectangular shape implied by Eq. (51). For this reason, the procedure discussed serves only to illustrate the gradual decrease in the scale of the effective irregularities with increasing angle of incidence. Also, because of the truncation, $h_0{}^2$ must be adjusted somewhat to be an effective $h_e{}^2$ through

$$h_e{}^2 = h_0{}^2 \int_0^{\kappa_M} \kappa \, d\kappa \, F(\kappa) \Big/ \int_0^{\infty} \kappa \, d\kappa \, F(\kappa) = h_0{}^2/B. \tag{53}$$

For a Gaussian autocorrelation function [Eq. (35)],

$$\kappa_M = \begin{matrix} (\sqrt{2}kh_0 \cos\phi)/d_0 & \text{when} \quad \tan\phi < \sqrt{2}(h_0/d_0) \\ k \sin\phi & \text{when} \quad \tan\phi > \sqrt{2}(h_0/d_0) \end{matrix}. \tag{54}$$

As long as κ_M is large in comparison with $1/d_0$, there will be little effect of the truncation, since it makes little difference whether the integration in Eq. (51) is carried to infinity or to κ_M under these circumstances. This condition on κ_M is closely related to the assumption of a "deep" phase screen (i.e., $kh_0 > 1$) having gentle slopes (i.e., $h_0/d_0 < 1$). For the Gaussian autocorrelation function, it is, therefore, seen that the same range of scales will contribute to the scattered power at all angles of incidence.

In certain situations [e.g., the exponential autocorrelation function, Eq. (37)], the power spectrum $F(\kappa)$ decays sufficiently slowly with scale frequency κ that the effective autocorrelation function $\rho_e(\Delta r)$ appreciably changes form with the maximum frequency κ_M. Since κ_M is increasing with the angle of incidence ϕ, *the smallest scale of those components responsible for backscattering is decreasing with increasing angles of incidence.* This in turn brings about an increase in the effective slope with increasing angles of incidence. When this effect is appreciable, the intepretation of the backscattered power as a function of angle of incidence in terms of a geometric optics model (Rea *et al.*, 1964) is of doubtful value.

The effective slope of the filtered version of the surface may be defined in analogy with the Gaussian autocorrelation case as follows

$$\frac{h_e^2}{d_{e0}^2} = -h_e^2 \rho_e''(0) = \frac{h_0^2}{4\pi} \int_0^{\kappa_M} \kappa^3 F(\kappa)\, d\kappa. \tag{55}$$

Let us finally outline briefly a procedure which, if realizable, would lead to a somewhat more meaningful estimate of the characteristics of a rough surface. First, we observe that for very small angles ϕ, the scales involved do not change appreciably with ϕ [see Eq. (50)]. In the region of small ϕ, a reasonably well-defined apparent slope on the scale $2\pi\kappa^{-1}$ may therefore be defined by comparison with the Gaussian autocorrelation case. If the observation is repeated at a number of wavelengths in the small ϕ region, we will be able to define a slope function

$$f(\kappa_M) = \frac{h_e^2}{d_{e0}^2} = \frac{h_0^2}{4\pi} \kappa^3 F(\kappa)\, d\kappa. \tag{56}$$

The spectrum $F(\kappa)$ can then be determined from

$$F(\kappa) = \text{const } \kappa^{-3}\, df/d\kappa. \tag{57}$$

3. *Interpretation*

The lunar data (e.g., Fig. 19) are such that they will not fit a simple model involving the Gaussian autocorrelation function [Eq. (35)]. Instead a reasonably good fit is obtained by using the correlation function of Eq. (37) (Evans and Pettengill, 1963b). In this case, one obtains for the backscattering cross section per unit surface area

$$\sigma_0 = \frac{C \cdot \rho_0}{2\cos^6\phi} \left(1 + C\,\frac{\tan^2\phi}{\cos^2\phi}\right)^{-3/2} \tag{58}$$

where $C = (d_1\lambda/4\pi h_0^2)^2$.

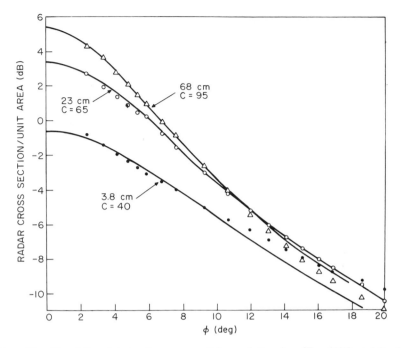

FIG. 20. Fit of the data to the exponential correlation law [Eq. (58)] at 3.8-, 23-, and 68-cm wavelength.

Figure 20 represents an attempt to fit the shape of the curves of expression (58) to the data of Table V by adjusting the reflection coefficient ρ_0 and the constant C for a best fit.

The data fit expression (58) quite well over a certain range of angles near normal incidence. Best fit is obtained with the parameters shown in Table VI (c.f., Table II for the values of ρ_0 derived). Whereas the fit at 68 and 23 cm is excellent, the fit at 3.8 cm is rather poor and the parameters quoted may therefore be somewhat in error. At 3.8 cm, it appears that the simple exponential correlation function is not a very good description of the surface scattering. Obviously, a better fit could be obtained by using a "composite" correlation function of the form (Beckmann, 1965)

$$\rho(\varDelta r) = \sum_{i=1}^{N} \alpha_i \exp(-|\varDelta r|/d_i) \qquad (59)$$

as long as a sufficient number of parameters α_i and d_i are allowed.

Since the correlation function which appears to fit the data best for the two longer wavelengths is such that a rms slope cannot easily be assigned to the

TABLE V

Radar Cross Section per Unit Surface Area[a]

Delay (μsec)	φ (deg)	Wavelength (cm)			Delay (μsec)	φ (deg)	Wavelength (cm)		
		3.8	23	68			3.8	23	68
10.00	2.38	—0.83	2.77	4.29	2250.00	36.29	—14.18	—16.78	—18.41
20.00	3.37	—1.43	1.92	3.69	2500.00	38.33	—14.58	—17.08	—18.81
30.00	4.12	—1.98	1.37	2.79	2750.00	40.28	—14.88	—17.43	—19.21
40.00	4.76	—2.38	0.87	2.09	3000.00	42.15	—15.23	—17.73	—19.51
50.00	5.32	—2.73	0.42	1.49	3250.00	43.96	—15.53	—18.08	—19.81
60.00	5.83	—3.03	0.02	0.99	3500.00	45.71	—15.83	—18.38	—20.01
70.00	6.30	—3.33	—0.38	0.49	3750.00	47.41	—16.13	—18.63	—20.21
80.00	6.73	—3.58	—0.78	—0.01	4000.00	49.07	—16.38	—18.93	—20.41
90.00	7.14	—3.83	—1.18	—0.51	4250.00	50.68	—16.63	—19.18	—20.61
100.00	7.53	—4.03	—1.53	—0.91	4500.00	52.26	—16.93	—19.58	—20.81
125.00	8.42	—4.58	—2.28	—1.91	4750.00	53.81	—17.13	—19.73	—21.06
150.00	9.22	—5.03	—3.03	—2.71	5000.00	55.32	—17.43	—19.98	—21.31
175.00	9.96	—5.38	—3.68	—3.41	5250.00	56.81	—17.73	—20.28	—21.61
200.00	10.65	—5.73	—4.23	—4.11	5500.00	58.27	—17.98	—20.53	—21.86
225.00	11.30	—6.03	—4.73	—4.71	5750.00	59.71	—18.33	—20.83	—22.16
250.00	11.92	—6.33	—5.23	—5.41	6000.00	61.13	—18.63	—21.08	—22.51
275.00	12.50	—6.68	—5.68	—5.86	6250.00	62.53	—18.93	—21.43	—22.86
300.00	13.06	—6.93	—6.08	—6.41	6500.00	63.92	—19.33	—21.78	—23.21
325.00	13.59	—7.23	—6.43	—6.81	6750.00	65.29	—19.73	—22.18	—23.56
350.00	14.11	—7.48	—6.78	—7.31	7000.00	66.64	—20.13	—22.58	—23.96
375.00	14.61	—7.73	—7.13	—7.61	7250.00	67.98	—20.53	—23.03	—24.36
400.00	15.09	—7.98	—7.43	—8.06	7500.00	69.30	—21.03	—23.43	—24.71
425.00	15.56	—8.23	—7.73	—8.41	7750.00	70.62	—21.43	—23.88	—25.16
450.00	16.01	—8.43	—8.03	—8.71	8000.00	71.92	—21.88	—24.33	—25.66
475.00	16.45	—8.63	—8.43	—9.01	8250.00	73.21	—22.38	—24.83	—26.16
500.00	16.88	—8.78	—8.58	—9.31	8500.00	74.50	—22.88	—25.38	—26.66
600.00	18.51	—9.33	—9.53	—10.31	8750.00	75.78	—23.38	—25.93	—27.21
700.00	20.01	—9.83	—10.43	—11.11	9000.00	77.05	—23.98	—26.53	—27.81
800.00	21.40	—10.23	—11.23	—11.91	9250.00	78.31	—24.53	—27.18	—28.46
900.00	22.72	—10.63	—11.83	—12.61	9500.00	79.57	—25.33	—27.83	—29.06
1000.00	23.97	—10.93	—12.58	—13.31	9750.00	80.82	—26.03	—28.58	—29.76
1100.00	25.15	—11.08	—13.18	—13.96	10000.00	82.07	—26.93	—29.43	—30.61
1200.00	26.29	—11.58	—13.63	—14.61	10250.00	83.32	—27.93	—30.48	—31.61
1300.00	27.39	—11.83	—14.03	—15.21	10500.00	84.56	—29.33	—31.58	—32.61
1400.00	28.44	—12.08	—14.38	—15.71	10750.00	85.80	—30.83	—33.08	—34.06
1500.00	29.46	—12.33	—14.68	—16.11	11000.00	87.04	—32.03	—34.93	—35.81
1750.00	31.88	—13.33	—15.88	—17.51	11250.00	88.27	—32.83	—37.58	—38.41
2000.00	34.15	—13.73	—16.33	—18.01					

[a] Values are $\log_{10} \sigma$ where σ is the cross section (m²) per unit area illuminated. Resolution in delay is approximately 10 μsec.

TABLE VI

Wavelength (cm)	C	ρ_0
68	95	0.0697
23	65	0.0648
3.8	40	0.0500

surface [$\rho''(0)$ being infinite], we apply the theory just outlined to derive a measure of slopes and an amplitude spectrum for the surface undulations.

For the 68 and 23 cm data where the exponential correlation function appears to provide an accurate description, the solution to Eq. (47)—assuming deep phase modulation—is

$$\frac{1}{\Delta r_e} = \frac{4k^2 h_0^2 \cos^2 \phi}{d_1} = \frac{2k \cos^2 \phi}{\sqrt{C}} = \frac{4\pi \cos^2 \phi}{\lambda \sqrt{C}}. \tag{60}$$

Hence, even for normal incidence, the scales of importance extend down to the size of the wavelength of the impressed signal. The slope function defined in Eq. (56) now becomes

$$f(\kappa_M) = \frac{h_0^2 \cdot \kappa_M}{2d_1} [\sqrt{1 + (d_1\kappa_M)^{-2}} - (d_1\kappa_M)^{-1}]^2. \tag{61}$$

When $d_1\kappa_M$ is large, as appears to be the case here, we obtain approximately

$$f(\kappa_M) = \frac{h_e^2}{d_{e0}^2} \approx \frac{\cos^2 \phi}{2C(\kappa_M)}. \tag{62}$$

This appears to imply that the angles corresponding to the rms slopes at the scale of 68 and 23 cm, respectively, are 6° and 7°. These values are noticeably lower than the 10° to 12° one would tend to derive on the basis of geometrical optics (e.g., Rea *et al.*, 1964). The discrepancy only reflects the difficulty of assigning a slope parameter to the lunar surface structure.

Next we proceed to fit a power law to the spectral function $F(\kappa)$. This can be done with somewhat greater confidence than the assignment of a slope to the surface since the limit of integration κ_M enters in a less critical manner. The fitting procedure results in a law of the form

$$F(\kappa) = \text{const } \kappa^{-3.7} \tag{63}$$

for the amplitude spectrum of the surface undulations in the lateral size region 20 to 70 cm. We are, unfortunately, not aware of any data with which this amplitude spectrum can be compared.

C. Small Scale Structure

1. *Observations*

The experiments described above were ones in which the polarization of the receiving antenna was controlled to accept the signal that would be reflected by a plane mirror. Thus, if a left-hand circularly polarized wave was transmitted, the receiver was connected for right-hand circular polarization. As noted above in the tail region, the angular power spectrum of the echoes $(50° < \phi < 80°)$ is found to conform to a law of the type $\bar{P}(\phi) \propto \cos^{3/2} \phi$. This dependence is illustrated in Fig. 21 for the data presented in Table V. There is

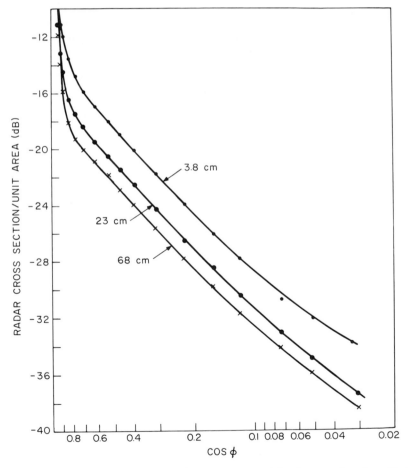

FIG. 21.   Cross section per unit surface area vs. $\cos \phi$ for large angles; $\lambda = 68, 23$, and 3.8 cm.

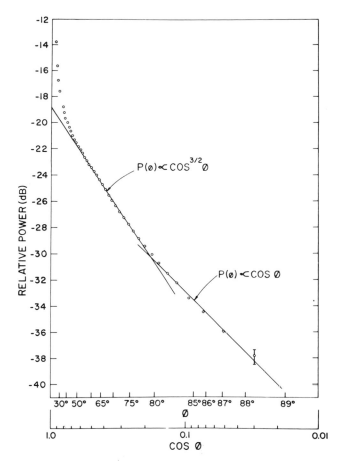

Fig. 22. The results obtained for the expected component of the echoes at 23 cm have been plotted as a function of $\log(\cos \phi)$. We find that in the region $\phi > 80°$, the echo power is proportional to $\cos \phi$ and for $50° < \phi < 80°$ to $\cos^{3/2} \phi$. Similar behavior was observed at 68-cm wavelength (Evans and Pettengill, 1963b), but the reason for this dependence is not clearly understood.

some evidence that at grazing angles ($\phi > 80°$), the dependence changes to the form $\bar{P}(\phi) \propto \cos \phi$ as shown in Fig. 22 for the 23-cm results.

The cross section per unit surface area near grazing incidence ($\phi \to 90$)° is found to increase with decreasing wavelength. The dependence observed is

$$\begin{aligned} \sigma \sim \lambda^{-0.32} & \quad 3.8 \quad \text{cm} < \lambda < 23 \quad \text{cm} \\ \sigma \sim \lambda^{-0.26} & \quad 23 \quad \text{cm} < \lambda < 68 \quad \text{cm} \end{aligned} \quad \phi \to 90°. \qquad (64)$$

This dependence should be compared with the opposite effect found for the surface when viewed near normal incidence [Eq. (32)].

Additional information concerning the diffuse scattering component can be obtained from experiments in which the cross-polarized component of the echoes is examined. Earliest of this type of measurement were studies by Pettengill and Henry (1962a), and later Evans and Pettengill (1963b) and Evans and Hagfors (1966) of the power contained in the sense of circular polarization orthogonal to that which would be reflected by a plane mirror. The echo power vs. delay $D(t)$ for this component was loosely termed the "depolarized" component and was found to differ markedly from that observed for the expected component $P(t)$ as illustrated in Fig. 23 (c.f., Fig. 14).

The angular dependence found here is $D(t) \propto \cos \phi$, implying that for this component the surface is uniformly bright as would be expected if the surface elements that are capable of converting the signals into the orthogonal mode

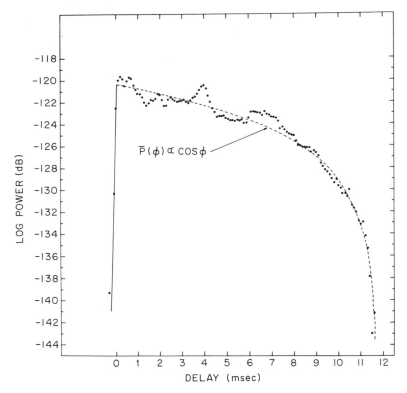

FIG. 23.   The average echo intensity versus range delay $\bar{D}(t)$ for the depolarized component of the signals (Evans and Pettengill, 1963b). The dotted curve indicates the expected behavior for a uniformly bright moon. (The departures from the smooth curve are thought to represent the existence of variations in the reflectivity over the Moon's disk. These might be expected to be more easily seen in the depolarized than in the polarized component).

are small and thus behave as isotropic scatterers. The random departures from the $\cos\phi$ dependence seen in Fig. 23 are real and result from the existence of anomalously bright scattering regions on the surface (Section IV). (The peak at about 4 msec delay, for example, is associated with the crater Tycho.) The ratio of the polarized $P(t)$ to depolarized $D(t)$ components is plotted for the 23- and 68-cm results in Fig. 24. There seems to be somewhat greater depolarization

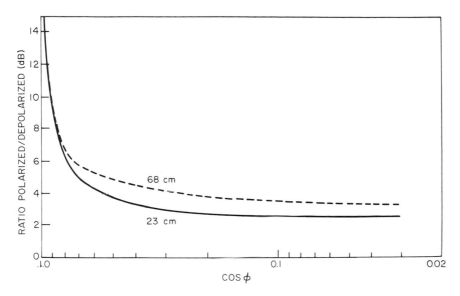

FIG. 24. The log of the ratio of the intensity of the "depolarized" and "polarized" circular returns at 23 and 68 cm vs. $\cos\phi$.

at 23 cm but the errors associated with the measurements are comparable with the difference shown in this figure.

The most extensive study of the polarization properties of the lunar surface has been undertaken by Hagfors *et al.* (1965) and Hagfors (1967), at 23-cm wavelength. In these, the depolarization of linear signals was examined, necessitating special precautions to avoid difficulties with Faraday rotation. Linearly polarized signals were transmitted and orthogonal linear signals received. The orientation of the received linear polarizations with respect to the plane containing the transmitted signal was changed between runs so that the output power in each channel would vary sinusoidally about a mean level. The least-mean-square sine wave was fitted to the data, and the mean and the depth of modulation were determined to give the total power and the power ratio between orthogonal components. Figure 25 shows the polarized and the depolarized linear components plotted as a function of $\cos\phi$ and their ratio is shown in

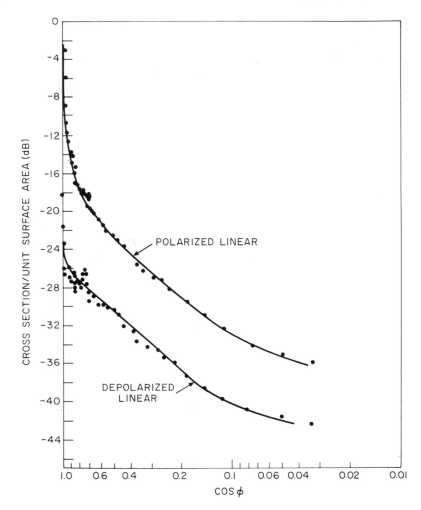

Fɪɢ. 25.   Cross section per unit surface area at 23 cm for the "polarized" and the "depolarized" linear returns vs. cos ϕ.

Fig. 26. A comparison of these results with those obtained with circularly polarized waves is made below in the discussion. Unfortunately, comparable results at other wavelengths are not available.

For linear and circularly polarized waves, the total received power, i.e., the sum of the power in the "polarized" and "depolarized" components, should be identical functions of the angle of incidence. As a check, the 23-cm total power for the two cases is plotted against cos ϕ in Fig. 27 and, as may be seen, the agreement

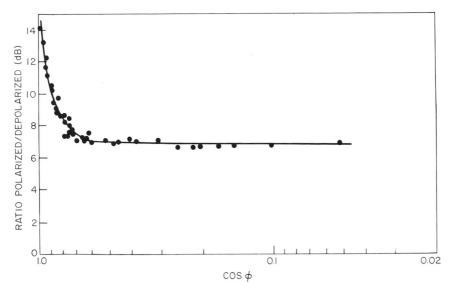

FIG. 26. The log of the ratio of the intensity of the polarized and depolarized linear returns at 23 cm for linear polarization vs. cos ϕ.

is reasonable. For $\phi \geqslant 40°$, the total power varies as $\cos\phi^{3/2}$ to a very good approximation in either case, except near grazing incidence where the angular dependence tends toward $\cos\phi$ as noted previously.

The depolarization measurements described so far combine the power returned from a complete ring of constant delay and hence include all possible directions of the local plane of incidence. It could be that there is little or no depolarization of a *linearly* polarized wave with electric field either *in* or *perpendicular* to the local plane of incidence and that the depolarization observed arises as a result of different backscattering coefficients for these two principal planes. To test this hypothesis, Hagfors (1967) arranged that the transmitted E-field was aligned with the libration axis of the Moon as shown in Fig. 28. By frequency analyzing the data to select the Doppler frequency component corresponding to the center of the lunar disk, together with gating the echoes as a function of delay, areas such as the ones shown shaded in Fig. 28 may be examined. In these two areas the E-field lies *in* the local plane of incidence. The polarized/depolarized power ratio observed under these circumstances is shown as a function of cos ϕ in in Fig. 29 for some 23-cm observations. As can be seen, there is little difference between these results and the ones obtained as an average over the complete range ring.

The *ratio* of the two principal backscattering coefficients for *linearly* polarized waves has been measured at both 23 and 3.8 cm. In both sets of observations,

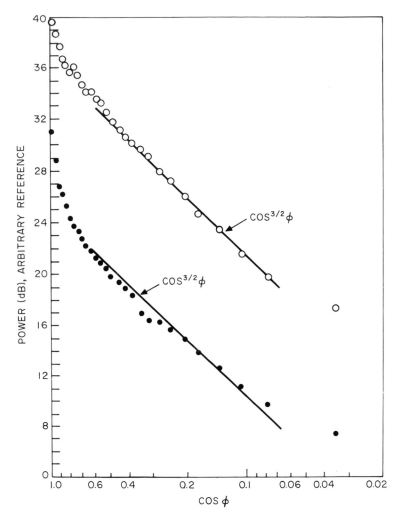

Fig. 27. Comparison of the angular variation of total backscattered power (i.e., the sum of the two orthogonal components) at 23 cm when illumination is circularly and linearly polarized. ○—linear; ●—circular.

the transmitted wave was circularly polarized and two orthogonal linearly polarized components of the echo were studied separately. At 23 cm, the necessary resolution on the Moon was achieved by an application of the delay–Doppler technique as described above [see Hagfors *et al.* (1965) for more details]. For 3.8 cm, sufficient resolution was afforded by the small angular diameter of the radar beam to provide the necessary discrimination between different areas.

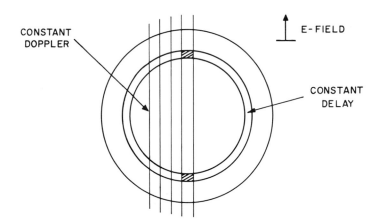

FIG. 28.   Method of selecting areas using delay–Doppler resolution where the E-field is aligned with the plane of incidence.

The results of these measurements are shown in Figs. 30 and 31. It can be seen that the ratio of power in the component polarized normal to the plane of incidence to that polarized in the plane of incidence approaches 0.5 for the 23-cm data and 0.75 for the 3.8-cm data near grazing angles of incidence ($\phi \to 90°$).

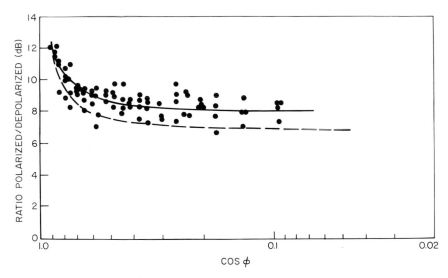

FIG. 29.   The ratio of the backscattered power at 23 cm in two orthogonal, linearly polarized components for linearly polarized illumination with the polarization parallel to the plane of incidence. The dashed curve shows the depolarization when the polarization of the illumination is averaged over all angles for the same data.

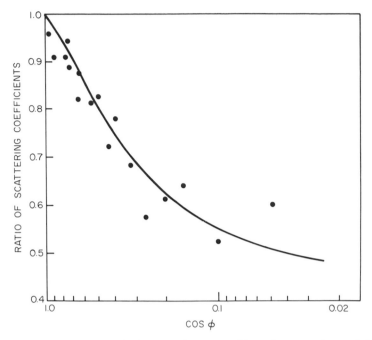

F_{IG}. 30.   The ratio of the backscattered power at 23 cm in two orthogonal, linearly polarized components from a small region of the lunar surface for circularly polarized illumination.

These results imply preferential reflection of waves whose electric field lies *in* the plane of incidence.

2. *Theory*

For oblique angles of incidence (i.e., angles ϕ in excess of 40°), it does not appear that the smooth undulating surface model (discussed earlier) provides an adequate description of the scattering mechanism. In particular, the polarization data are difficult to explain without invoking the scattering from discrete, individual, wavelength-sized objects strewn over the surface.

The depolarization of circularly polarized waves incident on the surface may be thought of as arising in at least one of two different ways. There may be a systematic difference in the backscattering coefficients for waves polarized *in* or *perpendicular* to the local plane of incidence (the two principal linear polarizations). Alternatively there may be a conversion of energy in either of the two principal linear components into orthogonal modes, in the sense that illumination in one principal linear polarization gives rise to scattered power in the orthogonal linear polarization also. The results above show that both phenomena appear to be present. In order to evaluate the relative importance of these mecha-

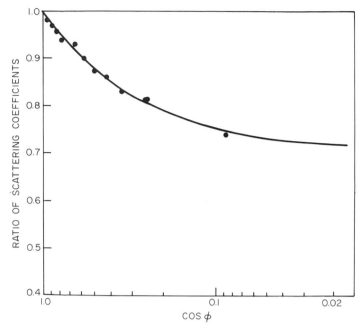

FIG. 31. The ratio of backscattered power at 3.8 cm in two orthogonal, linearly polarized components from a small region of the lunar surface for circularly polarized illumination.

nisms in causing depolarization of circularly polarized waves, we may argue as follows.

Let the backscattering matrix of a surface element be

$$S = \begin{pmatrix} r_{11} & r_{12} \\ r_{21} & r_{22} \end{pmatrix}, \tag{65}$$

so that the linearly polarized fields in and across the local plane of incidence, E_1 and E_2, respectively, are related to the incident fields E_1' and E_2' through

$$\begin{Bmatrix} E_1 \\ E_2 \end{Bmatrix} = \begin{Bmatrix} r_{11} & r_{12} \\ r_{21} & r_{22} \end{Bmatrix} \begin{Bmatrix} E_1' \\ E_2' \end{Bmatrix}. \tag{66}$$

The corresponding connection between circularly polarized waves is

$$\begin{Bmatrix} E_r \\ E_l \end{Bmatrix} = \frac{1}{2} \begin{Bmatrix} r_{11} + r_{22} - i(r_{12} - r_{21}), & r_{11} - r_{22} - i(r_{12} + r_{21}) \\ r_{11} - r_{22} + i(r_{12} - r_{21}), & r_{11} + r_{22} + i(r_{12} - r_{21}) \end{Bmatrix} \begin{Bmatrix} E_r' \\ E_l' \end{Bmatrix}. \tag{67}$$

The ratio of depolarized to polarized received circular components when illumination is circular hence becomes

$$\frac{\text{Depol}}{\text{Pol}} = \frac{\langle|\, r_{11} - r_{22} - i(r_{12} + r_{21})|^2\rangle_{\text{av}}}{\langle|\, r_{11} + r_{22} + i(r_{12} - r_{21})|^2\rangle_{\text{av}}} . \tag{68}$$

In the particular case when there is no conversion of energy from one linear mode to one orthogonal, i.e., $r_{12} = r_{21} = 0$, we obtain

$$\frac{\text{Depol}}{\text{Pol}} = \frac{\langle|\, r_{11} - r_{22}\,|^2\rangle_{\text{av}}}{\langle|\, r_{11} + r_{22}\,|^2\rangle_{\text{av}}} . \tag{69}$$

A systematic phase difference of the two principal linear power backscattering coefficients $\rho_{\parallel}$ and $\rho_{\perp}$ would lead to a preferential circular polarization of the scattered wave for linearly polarized illumination. This possibility must be excluded at the outset as being physically implausible. Thus we must take the phases of the two reflection coefficients to be the same, and the circular depolarized-to-polarized power ratio may be expressed as

$$\frac{\text{Depol}}{\text{Pol}} = \left(\frac{\sqrt{\rho_{\parallel}} - \sqrt{\rho_{\perp}}}{\sqrt{\rho_{\parallel}} + \sqrt{\rho_{\perp}}}\right)^2 \tag{70}$$

Figure 32 shows a plot of the expected ratio of depolarized and polarized power for circular polarization as a function of the ratio $\rho_{\parallel}/\rho_{\perp}$. Note that in this case there would be no depolarization of the two principal linearly polarized components.

Figure 32 shows that the ratio of the reflection coefficients $\rho_{\parallel}/\rho_{\perp}$ actually observed ($\sim$0.5–0.7—see Figs. 30 and 31) is inadequate to account for the depolarization of circularly polarized waves (Fig. 24). It is, therefore, concluded that the fact that r_{12} and r_{21} are nonzero, as indicated by the experimental results shown in Fig. 29, is probably the most important factor in producing depolarization. In other words, although the surface preferentially reflects linear waves whose plane of polarization is perpendicular to the plane of the surface (i.e., $\rho_{\perp} > \rho_{\parallel}$), this effect is not adequate to account for the large depolarization of circular (and linear) waves in the region $\phi > 40°$. Instead it is the conversion of energy from the incident to orthogonal modes that is the most important effect.

For a dielectric medium, one would expect preferential scattering of waves whose electric field is *perpendicular* to the plane containing the ray and surface normal. Thus the observed dependence (preferential reflection when the wave lies in the plane of incidence) can be accounted for if it is assumed that the diffuse echo component is reflected from within the lunar surface and thus has an intensity depending upon the *transmission* coefficient into the surface. For the

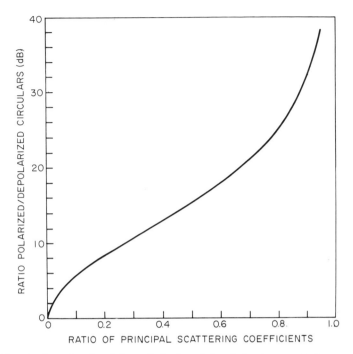

FIG. 32. A plot of ratio of the polarized and depolarized power to be expected for circularly polarized illumination as a function of the ratio of the backscattering coefficients of the two principal linear polarizations.

case where the electric field lies in the local plane of incidence, the transmission coefficient of a dielectric medium is

$$T_{\parallel} = \frac{4\epsilon \cos \phi (\epsilon - \sin^2 \phi)^{1/2}}{[\epsilon \cos \phi + (\epsilon - \sin^2 \phi)^{1/2}]^2} \tag{71}$$

and for perpendicular

$$T_{\perp} = \frac{4 \cos \phi (\epsilon - \sin^2 \phi)^{1/2}}{[\cos \phi + (\epsilon - \sin^2 \phi)^{1/2}]^2} \tag{72}$$

The ratio $T_{\perp}/T_{\parallel}$ varies with angle of incidence ϕ as shown in Fig. 33 for two values of the dielectric constant ϵ. At $\phi = 0$, $T_{\perp}/T_{\parallel} = 1.0$ and at $\phi = 90°$, $T_{\perp}/T_{\parallel} = 1/\epsilon$.

3. Interpretation

In order to continue the discussion, it is convenient at this point to introduce a specific model which may be adjusted to reproduce the observed data. Let us suppose that in the region $\phi \geqslant 40°$ the backscattering arises in part from

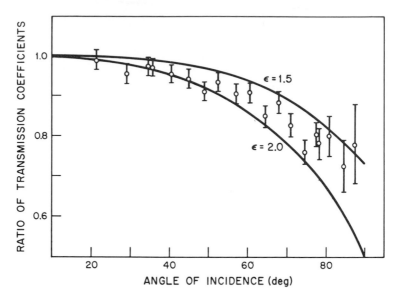

FIG. 33.   Ratio of square root of echo powers observed for two linear components
plotted in Fig. 30. If all the power is scattered from within the surface at large angles
of incidence, and further, if there is no spill-over from one linear component into
the other, then this is the ratio of transmission coefficients into surface. Also shown
is the expected variation for two values of the dielectric constant of the top layer
through which the waves have been assumed to penetrate [Eqs. (71) and (72)]. The
data were obtained 18 June 1965.

reflectors which do not depolarize at all. This type of mechanism clearly is
dominant near the subradar point, as indicated previously. With increasing angles
of incidence, we postulate that the scattering occurs increasingly from a discrete
structure which acts as single scatters. These discrete scatterers may, as a first
approximation, be thought of as linear dipoles of more or less random orientation.
The assumption of single scattering rather than multiple scattering to account
for the polarization effects is well justified by the very low reflectivity of the lunar
surface material.

A collection of linear dipoles with random orientation will depolarize a circularly polarized wave completely, i.e., the energy scattered in right and left
polarization will be of equal strength. By observing the ratio of polarized to
depolarized power as above, it is, therefore, possible to estimate the relative
amount of power P_1 scattered by the reflection mechanism and the power P_2
by the dipole scatter mechanism. The ratio of polarized to depolarized power
for *circular* illumination becomes

$$\frac{\text{Pol}}{\text{Depol}} = \frac{P_1 + \frac{1}{2}P_2}{\frac{1}{2}P_2} = 2\frac{P_1}{P_2} + 1. \tag{73}$$

A collection of randomly oriented dipoles illuminated with a *linearly* polarized wave will return 25% of the scattered power in the orthogonal mode. In this case, therefore, the ratio of polarized to depolarized power for linearly polarized waves becomes

$$\frac{\text{Pol}}{\text{Depol}} = \frac{P_1 + \frac{3}{4}P_2}{\frac{1}{4}P_2} = 4\frac{P_1}{P_2} + 3. \tag{74}$$

The results shown in Fig. 24 require that about 65% of the power returned at oblique angles of incidence be ascribable to the dipole scattering mechanism while those of Fig. 26 require 70% of the power to be attributed to the dipoles. This agreement between the conclusions drawn from linearly and circularly polarized data lends credence to the validity of the model postulated.

When asked what structure on the lunar surface can act as a dipole, we are forced to consider those portions (e.g., rocks) in which the local radii of curvature are markedly different so that induced currents can flow preferentially in a given direction. A case in point would be the edge of a block. Only if the edge lies in the plane of the incident wave will it reradiate all of the induced energy in the same plane.

In the model outlined above, there is no preferred reflection behavior for linearly polarized signals. This could be introduced, for example, by arranging that the dipoles are not oriented entirely at random but that a portion are oriented perpendicular to the local mean surface.

The explanation proposed initially (Hagfors *et al.*, 1965) was that the effect depends upon transmission through a top surface layer to rough material lying beneath. Support for this interpretation was derived from the fact that few rocks lying upon the surface could be seen in the Ranger photographs (Heacock *et al.*, 1965). Thus in order to account for the diffuse component of the echoes, it appeared necessary to suppose that the discrete scatterers were buried under a layer of light material. Thus Hagfors *et al.* (1965) supposed that at oblique angles of incidence, where the diffuse component predominates, the reflections are entirely from a rubble-like layer underlying an overburden of lighter porous material. As such, the ratio of the powers for the two linear components per unit area of surface would be expected to vary as

$$\frac{P_\perp}{P_\parallel} = \left(\frac{T_\perp}{T_\parallel}\right)^2 \tag{75}$$

where $T_\perp$ and $T_\parallel$ are the transmission coefficients in the two cases [Eqs. (71) and (72)]. Equation (75) assumes that the reflection coefficient of the underlying material is itself independent of the plane of the electric field, which is not unreasonable for small discrete scatterers. Thus by plotting $(P_\perp/P_\parallel)^{1/2}$ observed at various angles of incidence, Hagfors *et al.* (1965) were able to test the model as shown in Fig. 33. The observed points suggested that the dielectric constant of the top material is of the order of 1.7 to 1.8. This value is in good agreement with estimates based upon the polarization of the thermal emission of the Moon (*e.g.*, Davies and Gardner, 1966), thereby lending greater credibility to the model. Indeed this two-layer surface model served to explain the long standing discrepancy between radar and radiometric values for the dielectric constant.

Beckmann (1968), who is successful in predicting the depolarization of returns from a single range ring from theory, fails to point out that the backscattering coefficients of his theory predict $P_{\parallel}$ to be smaller than $P_{\perp}$, whereas the observations show the reverse to be true. The apparent success of Beckmann's theory thus must be coincidental.

An explanation invoking preferentially aligned surface structure (*e.g.*, orienting most of the dipoles to be perpendicular to the local surface) was first suggested by Gold (1965) but dismissed at the time as being altogether too artificial. However, its acceptability increased when it was recognized that on a smooth surface such as the Moon's, objects would be seen at grazing incidence with an image beneath (Fig. 34). This model gained further support from the

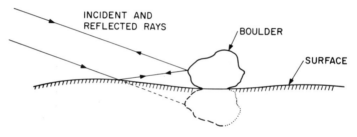

FIG. 34. The imaging property of a smooth surface seen at grazing incidence. This effect will cause a structure lying on the surface (such as boulders) to appear longer in the vertical extent than the horizontal and this may account for the preferential reflection of waves polarized perpendicular to the local surface.

Surveyor I and III pictures which showed a far greater abundance of surface rocks than had been supposed from an anlysis of the Ranger pictures. If the Surveyor pictures are typical of the lunar surface as a whole, they can be employed to test between the two possible interpretations we have considered. The most recent analysis of Surveyor I pictures indicates that the cumulative rock distribution is of the form (Shoemaker *et al.*, 1967)

$$N_1 = 5 \cdot 10^5 \cdot y^{-2.11} \tag{76}$$

whereas the Surveyor III pictures appear to fit a law of the form

$$N_3 = 3 \cdot 10^6 \cdot y^{-2.56} \tag{77}$$

where N is the cumulative number of grains per 100 m^2 and y is the diameter of grains in millimeters. The number density of rocks or grains per square meter with diameter between y and $y + dy$ is

$$n_1(y)\, dy = 1.055 \cdot 10^4 \cdot y^{-3.11}\, dy, \tag{78}$$

$$n_3(y)\, dy = 7.68 \cdot 10^4 \cdot y^{-3.56}\, dy, \tag{79}$$

while the geometric cross section of each grain expressed in square meters is

$$\sigma_g = (\pi/4)\, y^2 \cdot 10^{-6}. \tag{80}$$

Let us next assume that each grain has a radar cross section which is a certain constant fraction R of its geometrical cross section, when the diameter exceeds the wavelength, and zero otherwise. The cross section per unit area is, therefore, found to be

$$\sigma_1 = (\pi R/4) \cdot 0.096\{y_{min}^{-0.11} - y_{max}^{-0.11}\}, \tag{81}$$

$$\sigma_3 = (\pi R/4) \cdot 0.137\{y_{min}^{-0.56} - y_{max}^{0.56}\}. \tag{82}$$

For y_{min} we may choose the wavelength of observation and for y_{max} we substitute a number large enough to make the last terms in Eqs. (80) and (81) vanish. The results are (λ in millimeters)

$$\sigma_1 = R \cdot 0.075 \cdot \lambda^{-0.11}, \tag{83}$$

$$\sigma_3 = R \cdot 0.108 \cdot \lambda^{-0.56}. \tag{84}$$

This is to be compared with the wavelength dependence observed in Eq. (64), which is intermediate between the results of Eqs. (83) and (84). Assuming the reflectivity R to be the same as that of the Moon as a whole (i.e., 0.065 from Table VI), we obtain the results shown in Table VII, which also shows the cross section per unit surface area actually observed at an angle of incidence of 40°. It can be seen from this table that the number of rocks on the surface as derived

TABLE VII

COMPARISON OF OBSERVED AND COMPUTED CROSS SECTIONS
PER UNIT SURFACE AREA

λ (cm)	Computed σ_1 (dB)	Computed σ_2 (dB)	Observed at $\phi = 40°$
68	—26.22	—37.42	—19.21
23	—25.70	—34.78	—17.43
3.8	—24.83	—30.39	—14.88

from both the Surveyor I and Surveyor III pictures is inadequate to account for the return at oblique angles of incidence. Hence, if the areas photographed are typical of the lunar surface, the earlier interpretation of the radar data in terms of buried single scatterers (Hagfors *et al.*, 1965) may still hold true. Thompson *et al.* (1970) contend that Eq. (80) grossly underestimates the scattering cross section and that rocks lying *on* the surface can account for most or all of the diffuse component of the echoes. More recent calculations made by the authors appear to bear out the conclusions of Thompson *et al.* (1970).

In sum, it appears that a combination of rock fragments lying on and within the first 1 meter of the surface would seem to account for the diffuse component of the echoes.

IV

RADAR MAPPING OF THE LUNAR SURFACE

A. Introduction

In the previous section dealing with the average scattering behavior of the Moon, the assumption was made that the lunar surface is sufficiently homogeneous that it can be treated as an ensemble of elements statistically the same but viewed at different angles of incidence. The fact that most of the measured functions when plotted are found to vary smoothly with ϕ supports this gross assumption. In this section, we review the various attempts to explore the possibility of regional differences over the surface. These may be distinguished from the statistical studies by the fact that they call for detailed maps of the radar brightness distribution whereas the latter depend only upon average properties, usually implying cylindrical symmetry about the center of the disk.

The first attempt to examine the disk for regional differences was made by Pettengill and Henry (1962b). Because the antenna beam that they employed was larger in angular extent than the Moon itself, Pettengill and Henry were forced to rely upon the range and frequency (i.e., Doppler) resolution of the radar set. Using these two parameters, they were able to subdivide the surface into small regions as shown in Fig. 28. This technique involves the use of a coherent radar, i.e., one in which the phases of the transmitted pulses are related, and this in turn requires that the pulses be obtained by coherent amplification of signals provided by a stable, continuously running oscillator. Similarly the receiver local oscillators must be derived from this same stable oscillator. The resolution in range is accomplished by transmitting short pulses which are sampled upon reception, and the frequency discrimination is accomplished by Fourier analyzing these samples in a digital computer (Pettengill, 1964; Evans, 1968b). The power observed in given delay and frequency intervals corresponds to the sum of the powers reflected from two regions on the surface as shown in Fig. 28. Despite this ambiguity, Pettengill and Henry (1962b) were able to demonstrate the existence of one major anomalously reflecting feature on the surface, namely the crater Tycho. This crater was found to be brighter than its environs by about a factor of six both for the expected (i.e., polarized) and depolarized circular components. This last observation precluded any explanation based solely upon favorably oriented surfaces (located, for example, on the crater walls) since these would not necessarily be expected to depolarize. Instead it was necessary to suppose that Tycho is rougher than its surroundings and possibly has a higher intrinsic reflection coefficient also.

This pioneering work was carried out at the M.I.T. Lincoln Laboratory using the $\lambda = 68$ cm Millstone Radar, and was of limited application because of the two-fold ambiguity between given range and Doppler coordinates and the surface of the Moon (Fig. 28). Later, workers at the Arecibo Ionospheric Observatory ($\lambda = 70$ cm) and at the M.I.T. Lincoln Laboratory Haystack Radar ($\lambda = 3.8$ cm) were able to obtain unambiguous high resolution reflectivity maps and this work is reviewed below. In this work, the reflections from one of the pair of regions are suppressed by employing a narrow beam antenna. Figure 35 indicates how this is accomplished at the M.I.T. Lincoln Laboratory using the beam of the Haystack radar.

Although most radar mapping activity has centered on the delay–Doppler technique, a second method has been developed that also yields unambiguous reflectivity maps. As noted in Section II, the apparent spin axis (and rate of spin) of the Moon presented to a terrestrial observer changes during the course of the day. On a number of days, the angle θ between the instantaneous spin axis as seen by a terrestrial observer and the prime selenographic meridian changes during the course of a day by 180° or more. This behavior means that the lines of constant frequency shift (Fig. 4) take up all possible positions across the face of the Moon during the course of the day. It is then possible to map the radar reflectivity of the Moon employing only the frequency resolution of the radar.

In this technique, one explores the scattered field pattern over the ground in a similar manner to the interferometer measurements discussed by Evans and Pettengill (1963a). However, only a single antenna is needed because the phase reference is provided by the coherent oscillator in the radar, and the equivalent of different orientations of the interferometer is provided happily by the motion of the Moon. In this respect, the technique resembles that of aperture synthesis employed in radio astronomy. Alternatively, the method may be regarded as range-Doppler mapping in which the range resolution has been dispensed with and the additional information is obtained by repeating the Doppler processing for all possible positions of the spin axis.

The range–Doppler technique mentioned above has been described extensively in the literature (Evans and Pettengill, 1963a; Pettengill, 1964; Evans, 1968b)

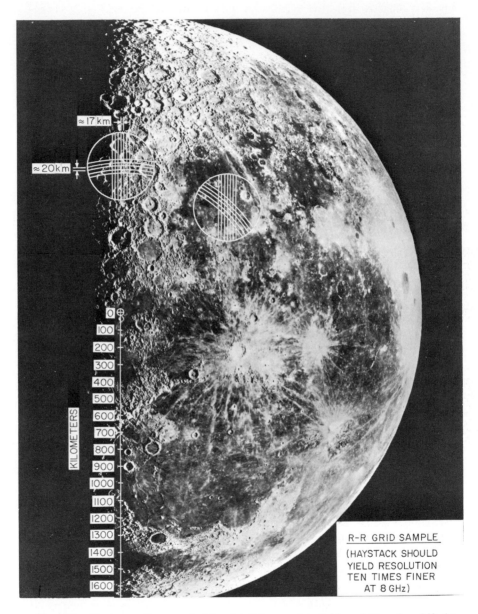

R-R GRID SAMPLE
(HAYSTACK SHOULD
YIELD RESOLUTION
TEN TIMES FINER
AT 8 GHz)

FIG. 35. Contours of constant delay (at 0.1 msec intervals) and Doppler shift (at 1 Hz intervals for a wavelength $\lambda = 3.8$ cm) drawn on the disk of the Moon. Also shown is the angular diameter subtended by the Haystack radar beam for two positions. In mapping the Moon with this radar, it has been possible to subdivide each of the small squares shown here into $\geqslant 100$ smaller squares.

but the Doppler-only method has been little discussed. The technique was first proposed by Thomson and Ponsonby (1968) but thus far has seen greatest use at Millstone where Hagfors has employed it to map the Moon at 23 cm. In practice, one measures the (complex) echo amplitude autocorrelation function $R(t)$ (or the echo power spectrum and then Fourier transforms to obtain the complex autocorrelation function) at various times throughout the day. Since the rate of libration l_T is known, the correlation as a function of time t can be transformed to one of distance over the ground. By repeating the measurements at different times, the two dimensional spatial correlation function $R(u, v)$ can be examined. The brightness distribution over the disk of the Moon $P(\xi, \eta)$ is then given by the double Fourier integral (Bracewell, 1962)

$$P(\xi, \eta) = \int\int R(u, v) \exp[-2\pi i(u\xi + v\eta)] \, du \, dv. \tag{85}$$

The distance in the uv plane over which the correlation is measured cannot be made large because θ and l_T are changing with time. Further, only a limited number of angles θ can be examined. Thus the two dimensional spatial correlation function $R(u, v)$ can be determined only along certain directions through the uv plane. In order to perform the Fourier transformation [Eq. (85)], it is then necessary to smooth the autocorrelation results in order to obtain continuous and smoothly varying values for $R(u, v)$.

Hagfors has considered in some detail the consequences of these effects and concludes that for $\lambda = 23$ cm, the best that can be achieved is a reflectivity map in which the disk of the Moon is resolved into elements $\sim 50 \times 50$ km in size, when 10% reflectivity accuracy is sought in the course of a single day's observations. Even this can only be achieved by arranging that energy from the bright central region of the disk does not enter the sidelobes of the synthesized antenna beam. For maps made for the depolarized component, this can best be achieved by good isolation between orthogonal modes in the antenna feed system. For the polarized component, one can gate out or suppress the central region if short pulses are transmitted. In the limit (at a very high radar frequency), the resolution obtainable by this technique would be restricted by the wandering of the selenographic latitude and longitude of the subradar point during the course of the day, since the method depends upon the assumption that the libration axis always passes through the same central point.

B. *Observations*

The Arecibo Ionospheric Observatory radar ($\lambda = 70$ cm) was the first to be applied to the study of the Moon in which the beamwidth (10 arc min) was less than the angular diameter of the disk. In the first observations, the beam was displaced from the Moon's center, and employing simple range integration, the scattering behavior was examined for various positions of the beam around the limb (Thompson and Dyce, 1966). These measurements clearly resolved differences in the scattering properties of mare and highland ground—the latter being significantly more reflective at 70 cm. Thus the highlands of the southwest quadrant of the Moon were found to backscatter $1\frac{1}{2}$ to 2 times as much power as the mare regions of the east and northeast quadrants. Similarly, the mountain regions surrounding individual circular mare were found to be $1\frac{1}{2}$ to 2 times more reflecting. This differentiation between mare and highland ground is also evident in the reflectivity maps produced by Hagfors at 23 cm, employing the Doppler-only technique described above. Figure 36 shows a low resolution map produced by Hagfors for the depolarized component at 23-cm wavelength, and Fig. 37 shows a map in which the resolution $\sim$80 × 80 km is close to the maximum obtainable. In Fig. 36, the contours extend over the edge of the disk due

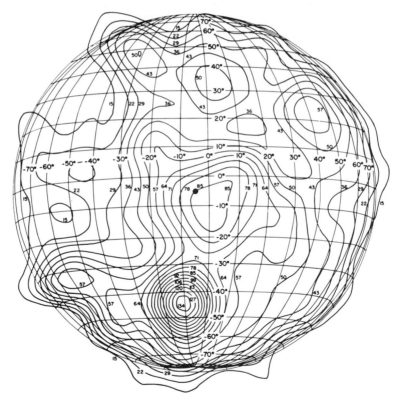

Fig. 36. A contour map of lunar reflectivity for the depolarized circular component at 23-cm wavelength. The contours are labeled in relative power. This map was obtained using the aperture synthesis technique described in the text.

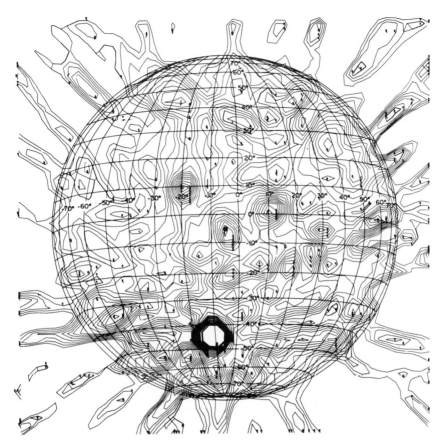

FIG. 37. A contour map of lunar reflectivity for the depolarized circular component at 23-cm wavelength. The contours are labeled in relative power. This map has higher resolution (~80 × 80 km) than that shown in Fig. 36 and consequently higher sidelobes also. This accounts for the power appearing in regions away from the lunar disk.

to the finite size of the synthesized beam. In Fig. 37, on the other hand, the energy shown beyond the disk is that appearing in the sidelobes of the synthesized beam. These are increased as the resolution is increased. In both figures, the subradar point is shown with a black spot. Imperfect isolation between the polarized and depolarized components in the feed system has allowed some of the polarized signal to leak through to give the bright central feature. The other features are believed to represent real variations of reflectivity.

The most striking feature on the maps, not unexpectedly, is the crater Tycho at 43°S, 11°W). In the high resolution map (Fig. 37), not all of the brightness contours associated with Tycho have been drawn since the peak intensity for the crater Tycho is ten times higher than what it is in the regions of Mare

Imbrium. The craters Copernicus (10N, 20W), Theophilus (12S, 27E), and Aristarchus (24N, 48W) are also recognizable as bright features.

It is obvious that some of the mare regions appear quite dark, notably Serenitatis, Tranquilitatis, Imbrium, and Procellarum. It is also interesting to note some peculiarities in the map. High reflectivity appears to be associated with Sinus Iridium or possibly with the Jura Mountains immediately to the north of it. The reflectivity in the southern hemisphere is higher than in the northern hemisphere. Furthermore, in the southern hemisphere, reflectivity is higher in the western part than in the eastern part. The depolarized reflectivity from the areas to the southwest of Mare Humorum are anomalously bright, although there are no spectacular craters in this region. On the whole, there seems good correlation between high depolarized reflectivity and bright mountainous regions and between low reflectivity and dark mare regions.

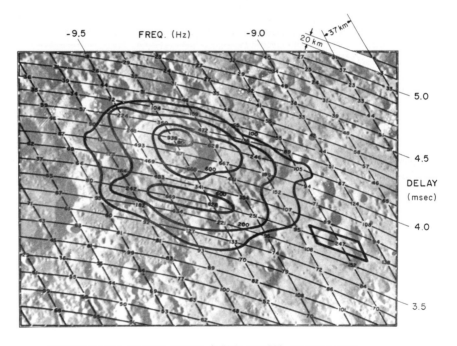

ARECIBO RADAR, CRATER TYCHO (DEPOLARIZED), 24 SEPT. 1964

FIG. 38.   Relative echo intensity (numbers) as a function of position in the vicinity of the crater Tycho (Thompson, 1966). Contours enclose regions with intensities greater than 100, 200, 600, and 800 units. The lines running nearly horizontally are contours of increasing delay (at 0.1 msec intervals) and those in the other direction are lines of constant Doppler shift (in 0.1 Hz steps). The resolution achieved is indicated by the cell at the top left. Note that the strongest reflections come from the walls of the crater most favorably oriented toward the radar.

Shorthill *et al.* (1960) and Shorthill and Saari (1965) have observed that virtually all the rayed craters have anomalous thermal properties in the infrared. During the waning cycle, they cool less rapidly than their surroundings and the reverse is true during the waxing phase. Even more spectacular differential cooling is observed during an eclipse (Saari and Shorthill, 1963, 1965). This behavior could be attributed to the absence in these craters of an appreciable dust layer which overlies most of the remainder of the surface or the presence of a large number of exposed rock fragments. The extent to which this effect is observed seems to bear a direct relation with the estimated age of the craters concerned. (Tycho is the most conspicuous example.)

Working on the hypothesis that these craters would also be anomalously reflecting, Thompson (1966) and Thompson and Dyce (1966) examined the major rayed craters for enhanced scattering using the delay–Doppler technique. Figures 38 and 39 provide examples of the results obtained. The resolution obtained (20 × 40 km approximately) is considerably better than that of Pettengill and Henry (1962b) and shows that in Tycho, for example, there are regions which are a factor of ten times as reflecting as an equivalent area of the environs. These observations confirmed that the rayed craters are indeed anomalously reflecting in essentially the same ordering that they are observed to be anomalous in the infrared. In addition, they showed that all new or fresh appearing craters (judged to be new chiefly by their high optical albedo) are brighter than their environs. An example of such a small crater is shown in Fig. 38.

It is noticeable for both Tycho and Langrenus (Figs. 38 and 39) that the strongest reflections are returned from the inside wall of the crater that is most favorably oriented so as to be nearly normal to the ray path. Another region of enhanced reflectivity is associated with the outside wall of the crater nearest the radar. Evidently the high mean slope of some of the surfaces found in these craters does play some part in making them brighter than their environs, but as noted above this is incapable of explaining the entire effect.

In more recent work at Arecibo (Pettengill and Thompson, 1968; Thompson *et al.*, 1970), maps have been obtained in which the resolution has been improved to yield a cell size of approximately 10 × 10 km. Meanwhile, at M.I.T. Lincoln Laboratory, the Haystack radar has been employed to map large regions of the lunar surface at 3.8 cm wavelength to a resolution of approximately 2 × 2 km (Lincoln Laboratory, 1968). Examples in which some of these radar maps are compared with photographs taken from the Rectified Lunar Atlas (1963) and the Orthographic Atlas of the Moon (1960) are presented in Figs. 40 and 41. Recent measurements in this program have achieved a resolution of 1 × 1 km and Fig. 42 shows the crater Tycho when mapped at the higher resolution. This figure may be compared with the lower resolution map and the optical photograph of the same region (Fig. 41).

The general variation across the map of the brightness contours in Figs. 38

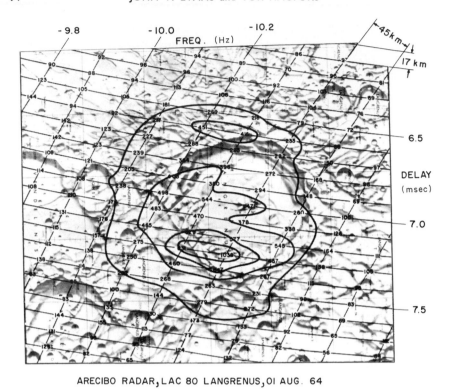

ARECIBO RADAR, LAC 80 LANGRENUS, 01 AUG. 64

FIG. 39. A plot similar to Fig. 42 for Langrenus. Here the contours overlie an LAC sheet and not a photograph. Again the parts of the walls of the crater that are tilted most nearly normal to the ray path are the best reflectors.

and 39 is due to the variation in the mean surface slope with increasing delay. Since the mean scattering law [i.e., $\bar{P}(\phi)$] is known, the 3.8-cm measurements have been corrected for this effect. Two other corrections which must be applied are the removal of the two-way weighting imposed by the antenna beam and the different amounts of projected surface area contained in each delay–Doppler cell. All three of these effects are allowed for as routine parts of the M.I.T. data processing program. The transformation of the echo power from a matrix of delay–Doppler cells into the corresponding selenographic coordinates is also performed by the computer. Because the libration axis and libration rate change continuously, this coordinate transformation may be allowed to proceed for only a limited amount of time before a new coordinate conversion must be computed. If this is not done, some resolution will be lost as there will be a smearing of the echo power into several selenographic cells. The images shown in Figs. 40–42 were generated upon a computer display by presenting the power reflected from each surface element as a given level of brightness. This is accomplished

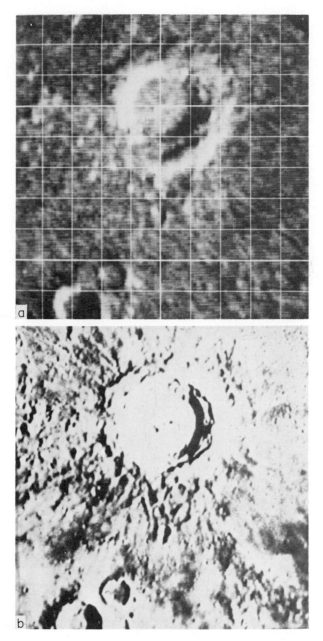

FIG. 40. (a) 3.8-cm radar map of region surrounding lunar crater Copernicus. Map is nominally centered at 8N, 20W and the grid lines are at 1° intervals (30 km). (b) Optical photograph of same region (Rectified Lunar Atlas, University of Arizona Press, 1963).

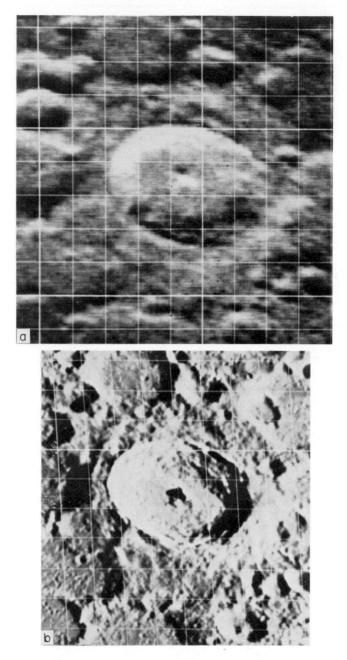

Fig. 41. (a) High-resolution 3.8-cm radar map of lunar crater Tycho, direction cosine projection. Grid lines are spaced approximately 17 km. (b) Optical photograph of same region (Orthographic Atlas of Moon, University of Arizona Press, 1960).

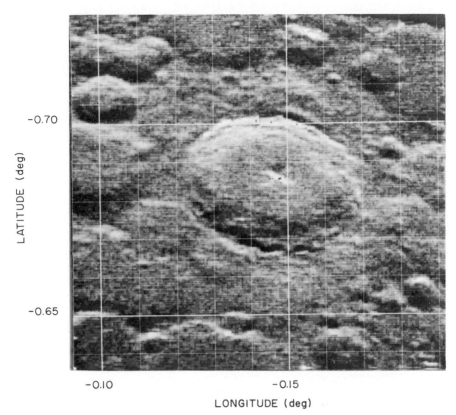

-0.70

LATITUDE (deg)

-0.65

-0.10   -0.15

LONGITUDE (deg)

Fig. 42. Radar map of crater Tycho using 5-μsec resolution. Here, a small grid square is 0.01 × 0.01 lunar radius units and resolution is 0.0005 × 0.0005 or approximately 1 km.

by assigning to each surface element a corresponding cell on the cathode ray tube face, and returning the spot to this position a number of times (up to 31) in proportion to the local reflectivity. Thus the film was employed to integrate these levels and the brightness in Figs. 40–42 is porportional to the logarithm of the echo power. The computer routines available for this work (which is being performed for the National Aeronautics and Space Administration) permit one to call for the reflectivity to be displayed as a function of seleno-graphic Cartesian coordinates (normalized to the lunar radius) or in Mercator (or Lambert Conic) projection. In the latter case, the maps taken of the limb region equal or exceed the best ground-based telescopic resolution. An example of such a map is shown in Fig. 43.

In order to interpret the figures shown here, it should clearly be understood that the light and dark portions of the *optical* pictures represent illuminated

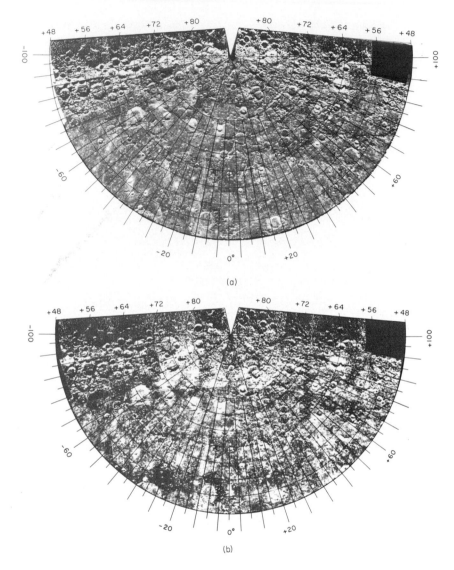

FIG. 43.   Radar maps of the northern polar region of the Moon in Lambert conic projection. The resolution is approximately 2×2 km. (a) Polarized component; (b) depolarized component (Lincoln Laboratory 1970).

regions and regions in shadow. The radar pictures, on the other hand, represent *variations in reflectivity*. All portions of the surface are being illuminated, since with the radar transmitter and receiver colocated on Earth we have the equivalent of the full moon situation in telescopic observations. A second point to be made is that the contrast available in the radar pictures far exceeds that in full moon photographs. In the latter case, the albedo difference between the brightest and darkest portions of the surface amounts only to a factor of two. In the radar case, enhancements in reflectivity above the mean as large as a factor of ten are observed. This is demonstrated in Fig. 44 where the brightness observed in

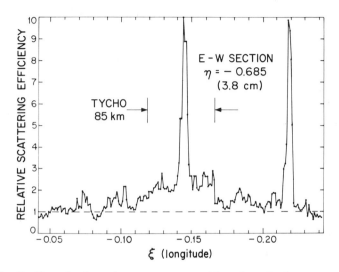

Fig. 44. Profiles of echo enhancement vs. position through the center of Tycho drawn from the data shown in Fig. 42 for the longitude $\eta = -0.685$ (east-west). Only data in the polarized mode were available. The intensity peak near the center of Tycho is believed to be from the steep sides of the central peak.

Fig. 42 along latitude $\eta = -0.685$ is plotted. The two reflection peaks are the reflections from central peak in Tycho and the inner wall of much smaller crater some distance away.

C. Interpretation

A wealth of data has been gathered by the mapping programs being carried out at Arecibo and Haystack which remains to be interpreted. Preliminary examinations of the 70-cm maps confirm the higher reflectivity of the highland ground over the mare. This is more noticeable in the depolarized than polarized component implying that highland ground is rougher (on the scale of the wavelength). However, differences in the dielectric constant between the highlands and mare either due to chemical differentiation or density differences cannot be ruled out. The rayed craters (and especially Tycho) are extremely bright with respect to their surroundings both in the polarized and depolarized components. Pettengill and Thompson (1968) conclude that this can largely be explained in terms of increased local slopes and local roughness though again it is possible that some of the increased reflectivity may be caused by increased density. Elsewhere small new craters are found to reflect strongly over the crater floor and inner walls, while in the rayed craters the increased brightness spills over the crater walls onto the surrounding terrain. Older large craters (e.g., Plato) are found to be bright all around their walls, implying that despite erosion these have remained rough (or dense). This latter observation is extremely interesting and implies that the erosion of the walls proceeds in such a manner that the debris never succeeds in burying them, i.e., there is a transport of fine debris away from the crests of the walls onto the crater floor and onto the ground surrounding the crater. In certain instances, increased reflectivity is observed for some areas that seems unrelated to any visible structure. This implies the existence of fields of small rocks in a few areas, the rocks either lying on or just beneath the unconsolidated surface.

The interpretation of the 3.8-cm results has barely begun, in part because the depolarized component at this wavelength has only recently been measured [Lincoln Laboratory (1970)]. The radar reflectivity pictures (Figs. 40-43) show a great resemblance to the corresponding optical photographs. This suggests that it is largely changes of the surface slope that govern the reflectivity of the polarized component of the echoes at this wavelength. In regions typical of the average lunar surface, local enhancements in reflectivity could be related to the tilt angle through the known lunar scattering law. Unfortunately, many of the craters as seen by radar probably have atypical surface roughness as well as atypical small-scale slope distributions. However, if the crater profiles are known from optical shadow measurements, it should be possible to build up a family of such "crater" scattering laws from fine-grained radar data. This, in turn, might prove to be correlated with relative surface roughness (determined independently from depolarization mode observations at one or more wavelengths) and could then be used to deduce slopes for those craters where optical shadows are not available. A worthwhile application for such a program would be in the limb region where the radar resolution can exceed that available to Earth-based photography.

The comparison of the 3.8-cm and 70-cm data may prove particularly fruitful in view of the different penetration depths of the two waves. Features observed at 70 cm but absent at 3.8 cm may be attributable to structure lying a little below the immediate surface. In the one instance (Pettengill and Thompson, 1968) where a detailed comparison of the 3.8 and 70-cm maps has been made, the enhancement of the crater Tycho at both wavelengths, on the basis of certain assumptions made, was found explainable simply in terms of increased roughness. Figure 45 indicates the amount of surface that must be presumed rough to account for the observed reflectivity increases. This is comparable at both wavelengths, but so close to 100% that it seems probable that some of the increased reflectivity is due to increased density of the material. Ultimately it is hoped that a curve similar to Fig. 45 can be obtained which has been based

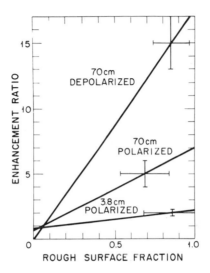

FIG. 45. Plot of the enhancement ratio to be expected for craters in the vicinity of Tycho as a function of their fractional surface roughness. Mean values of roughness for the lunar surface of 0.057 at 70-cm wavelength and 0.146 at 3.8 cm have been assumed.

upon the examination of a large number of cases, and which can simply be used to determine the local roughness of any anomalously bright region of the surface simply by reading from a table of the observed enhancement.

V

SUMMARY

A. *Mean Surface Slopes*

The mean surface slope determined over an interval of 1 m ($\sim$5λ at $\lambda = 23$ cm) by radar studies is somewhat less than about 10° and appears to increase as the wavelength is shortened and vice versa. This estimate should be taken to apply chiefly to mare ground since this tends to occupy the subradar region, which controls the scattering behavior $\bar{P}(\phi)$ for small ϕ. Similar estimates have been obtained from the Ranger pictures.

B. *Rough Structure*

A fraction of the surface is covered with structure having horizontal and vertical scales comparable with the wavelength. The percentage increases as the size of the wavelength is decreased and is approximately 10% for the 23-cm wavelength. This structure appears to give rise to most of the depolarization of the returned signals. By studying the nature of this depolarization, it is concluded that the scatterers have angular shapes and may be identified with the stones and blocks seen in Surveyor pictures.

C. *Surface Topography*

Enhanced reflectivity (with respect to mare ground) has been found to be associated with rayed craters, smaller "new" (high albedo) craters, and highland ground. The bulk of this increased reflectivity has been attributed to increased roughness in these areas; however, it is quite possible that the intrinsic reflectivity is higher also either because of a more compact or chemically different surface material (or both).

The radar maps upon which these conclusions are based have a resolution in the range 5×5 to 1×1 km and seem to show also the existence of small boulder fields not visible from Earth, through recognizable on pictures taken with the lunar orbiters. A great deal of additional interpretive work remains before the full significance of all the features that can be seen in the radar maps can claim to have been recognized.

The correspondence between the reflectivity and thermal anomalies implies that the rayed craters (and the smaller new craters) are regions in which rock is exposed to view at the depths to which infrared and radiowaves may penetrate. This in turn implies that over the rest of the surface porous material extends to a considerable depth (centimeters). The bright appearance of the walls of older craters (like Plato) requires that the erosion process operate in a way which preserves the roughness of the walls despite the fact that these are more liable to attack by debris from neighboring impacts than level ground. This in turn seems to require a mechanism to transport the finely eroded material to lower levels.

D. *Density and Dielectric Constant of the Surface Material*

The interpretation of the reflection coefficient in terms of a dielectric constant depends upon assumptions concerning the nature of the interface between free space and the surface. If it is assumed that the material is homogeneous to the depth penetrated by the wave, then a dielectric constant of ~ 2.8 is obtained, implying a density of the order of 50% solid if the primary material is iron rich basalt. On this model, it is probable that there is a transition to denser material at a depth of the order of a meter.

REFERENCES

Allenby, R. J. (1970). *Space Sci. Revs.* 11, 3.
Barrick, D. E. and Peake, W. H. (1967). *Battelle Mem. Inst. Rept.* **197A-10-3**.
Bay, Z. (1946). *Hung. Acta Phys.* 1, 1.
Beckmann, P. (1965). *J. Geophys. Res.* **70**, 2345.
Beckmann, P. (1968). *J. Geophys. Res.* **73**, 649.
Blevis, B. C. (1957). *Nature* **180**, 138.
Blevis, B. C. and Chapman, J. H. (1960). *J. Res. N. B. S.* **D64**, 331.
Bowhill, S. A. (1957). *J. Atmos. Terr. Phys.* **11**, 91.
Bracewell, R. N. (1962). *In* "Handbuch der Physik" (S. Flügge, ed.), Vol. 54, p. 42. Springer, Berlin.
Bramley, E. N. (1962). *Proc. Phys. Soc. London* **80**, 1128.
Brown, W. E. (1960). *J. Geophys. Res.* **65**, 3087.
Browne, I. C., Evans, J. V., Hargreaves, J. K., and Murray, W. A. S. (1956). *Proc. Phys. Soc. London* **B69**, 901.
Brunschwig, M., Fensler, W. E., Knott, E., Olte, A., Siegel, K. M., Ahrens, T. J., Dunn, J. R., Gerhard, F. B., Jr., Katz, S., and Rosenholtz, J. L. (1960). *Univ. Mich. Rept* **3544-1-F**, Ann Arbor, Michigan.
Campbell, M. J. and Ulrichs, J. (1969). *J. Geophys. Res.* **74**, 5867.
Daniels, F. B. (1961). *J. Geophys. Res.* **66**, 1781.
Daniels, F. B. (1962). *J. Geophys. Res.* **67**, 895.
Daniels, F. B. (1963a). *J. Geophys. Res.* **68**, 449.
Daniels, F. B. (1963b). *J. Geophys. Res.* **68**, 2864.
Davies, R. D. and Gardner, F. F. (1966). *Aust. J. Phys.* **19**, 283.
Davis, J. R. and Rohlfs, D. C. (1964). *J. Geophys. Res.* **69**, 3257.
DeWitt, J. H. and Stodola, E. K. (1949). *Proc. Inst. Radio Eng.* **37**, 229.
Eckert, W. J., Walker, M. J., and Eckert, D. (1966). *Astron. J.* **71**, 314.
Evans, J. V. (1957). *Proc. Phys. Soc. London* **B70**, 1105.
Evans, J. V. (1960). *Contemp. Phys.* **2**, 116.
Evans, J. V. (1962). *In* "Physics and Astronomy of the Moon" (Z. Kopal, ed.), 1st ed. Chapter 12, pp. 429–479. Academic Press, New York.
Evans, J. V. (1965). *J. Res. N. B. S.* **D69**, 1637.
Evans, J. V. (1966). *Ann. N. Y. Acad. Sci.* **140**, 196.
Evans, J. V. (1968a). Unpublished observations.
Evans, J. V. (1968b). *In* "Radar Astronomy" (J. V. Evans and T. Hagfors, eds.), Chap. 9, pp. 499–545. Academic Press, New York.
Evans, J. V., Evans, S., and Thomson, J. H. (1959). *In* "Paris Symposium on Radio Astronomy" (R. N. Bracewell, ed.), p. 8. Stanford Univ. Press, Stanford, California.
Evans, J. V. and Hagfors, T. (1964). *Icarus* 3, 151.
Evans, J. V. and Hagfors, T. (1966). *J. Geophys. Res.* **71**, 4871.
Evans, J. V. and Ingalls, R. P. (1962). *M.I.T. Lincoln Lab. Rept.* **288**, Lexington, Massachusetts.
Evans, J. V. and Pettengill, G. H. (1963a). *In* "The Moon Meteorites and Comets; The Solar System" (B. M. Middlehurst and G. P. Kuiper, eds.), Vol. IV, Chapter 5, pp. 129–161. Univ. of Chicago Press, Chicago, Illinois.

Evans, J. V. and Pettengill, G. H. (1963b). *J. Geophys. Res.* **68**, 423.

Evans, J. V. and Pettengill, G. H. (1963c). *J. Geophys. Res.* **68**, 5098.

Fessenkov, V. G. (1961). *In* "Physics and Astronomy of the Moon" (Z. Kopal, ed.), 1st Ed. Chap. 4, pp. 99–130. Academic Press, New York.

Fricker, S J., Ingalls, R. P., Mason, W. C., Stone, M. L., and Swift, D. W. (1958). *M.I.T. Lincoln Lab. Rept.*, **187**, Lexington, Massachusetts.

Fricker, S. J., Ingalls, R. P., Mason, W. C., Stone, M. L., and Swift, D. W. (1960). *J. Res. N. B. S.* **D64**, 455.

Fung, A. K. (1967). *Radio Sci.* **2**, 1525.

Fung, A. K. and Moore, R. K. (1964). *J. Geophys. Res.* **69**, 1075.

Fung, A. K. and Moore, R. K. (1966). *J. Geophys. Res.* **71**, 2939.

Gold, T. (1965). Private communication.

Grieg, D. D., Metzger, S., and Waer, R. (1948). *Proc. Inst. Radio Eng.* **36**, 652.

Hagfors, T. (1961). *J. Geophys. Res.* **66**, 777.

Hagfors, T. (1964). *J. Geophys. Res.* **69**, 3779.

Hagfors, T. (1966). *J. Geophys. Res.* **71**, 379.

Hagfors, T. (1967). *Radio Sci.* **2**, 445.

Hagfors, T. Brockelman, R. A., Danforth, H. H., Hanson, L. B., and Hyde, G. M. (1965). *Science* **159**, 1153.

Hapke, B. (1968). *Science* **159**, 76.

Hapke, B. (1970). *Radio Science* **5**, 293.

Hargreaves, J. K. (1959). *Proc. Phys. Soc. London* **B73**, 536.

Hayre, H. S. and Moore, R. K. (1961). *J. Res. N. B. S.* **D65**, 427.

Heacock, R. L., Kuiper, G. P., Shoemaker, E. M., Urey, H. C., and Whitaker, E. A. (1965). *Jet Propulsion Lab. Rept.* **JPL-TR-32-700**, Pasadena, California.

Hughes, V. A. (1961). *Proc. Phys. Soc. London* **78**, 988.

Hughes, V. A. (1962a). *Proc. Phys. Soc. London* **80**, 1117.

Hughes, V. A. (1962b). *J. Geophys. Res.* **67**, 892.

Jaffe, L. D. (1969). *Space Sci. Revs.* **9**, 491.

Katz, I. (1966). *J. Geophys. Res.* **71**, 361.

Kerr, F. J. and Shain, C. A. (1951). *Proc. Inst. Radio Eng.* **39**, 230.

Kerr, F. J., Shain, C. A., and Higgins, C. S. (1949). *Nature* **163**, 310.

Klemperer, W. K. (1965). *J. Geophys. Res.* **70**, 3798.

Lincoln Laboratory (1968). "Radar Studies of the Moon," Final Report Vol. 2, 18 April, (NASA Contract NSR 22-009-106), Lincoln Laboratory, Lexington, Massachusetts.

Lincoln Laboratory (1970). "Radar Studies of the Moon," Final Report, 28 February, (NASA Contract NAS-9-7830), Lincoln Laboratory, Lexington, Massachusetts.

Lynn, V. L., Sohigan, M. D., and Crocker, E. A. (1963). *M.I.T. Lincoln Lab. Rept.* **331**, Lexington, Massachusetts. See also *J. Geophys. Res.* **69**, 781 (1964).

Markov, A. V. (1948). *Astron. Zh.* **25**, 172.

Matveev, Yu. G. (1967). *Sov. Astron.-AJ (English Transl.)* **11**, 332.

Melbourne, W. G., Mulholland, J. D., Sjogren, W. L., and Sturms, F. M., Jr. (1968). Jet Propulsion Laboratory Technical Report 32-1306, Pasadena, California.

Morris, E. C. and Shoemaker, E. M. (1970a). *Icarus* **12**, 167.

Morris, E. C., and Shoemaker, E. M. (1970b). *Icarus* **12**, 173.

Muhleman, D. O. (1964). *Astron. J.* **69**, 34.

Mulholland, J. D. and Sjogren, W. L. (1967). *Science* **155**, 74.

Pettengill, G. H. (1960). *Proc. Inst. Radio Engr.* **48**, 933.

Pettengill, G. H. (1961). Unpublished M.I.T. Radar Astronomy Summer School Course Notes.

Pettengill, G. H. (1964). *J. Res. N. B. S.* **D68**, 1025.

Pettengill, G. H. and Henry, J. C. (1962a). "The Moon" (Z. Kopal and Z. K. Mikailov, eds.), p. 519. Academic Press, New York.

Pettengill, G. H. and Henry, J. C. (1962b). *J. Geophys. Res.* **67**, 4881.

Pettengill, G. H. and Thompson, T. W. (1968). *Icarus* **8**, 457.

Rea, D. G., Hetherington, N., and Mifflin, R. (1964). *J. Geophys. Res.* **69**, 5217.

Saari, J. M. and Shorthill, R. W. (1963). *Icarus* **2**, 115.

Saari, J. M. and Shorthill, R. W. (1965). *Nature* **205**, 964.

Scott, R. F., Roberson, F. I., and Clary, M. C. (1967). *NASA, Rept.* **SP-146**, 61 Washington D. C.

Senior, T. B. A. and Siegel, K. M. (1959). *In* "Paris Symposium on Radio Astronomy" (R. N. Bracewell, ed.), p. 29. Stanford Univ. Press, Stanford, California.

Senior, T. B. A. and Siegel, K. M. (1960). *J. Res. N. B. S.* **D64**, 217.

Shapiro, A., Uliana, E. A., Yaplee, B. S., and Knowles, S. H. (1968). *In* "Moon and Planets" II (A. Dollfus, ed.), pp. 34–36. North-Holland, Amsterdam.

Shoemaker, E. M., Batson, R. M., Holt, H. E., Morris, E. C., Rennilson, J. J., and Whitaker, E. A. (1967). *NASA, Rept.* **SP-146**, 9, Washington, D. C.

Shoemaker, E. M., Morris, E. C., Batson, R. M., Holt H. E., Larson, K. B., Montgomery, D. R., Rennilson, J. J., and Whitaker, E. A. (1969). NASA Report **SP-184**, 19, Washington, D. C.

Shoemaker, E. M. and Morris, E. C. (1970). *Radio Science* **5**, 129.

Shorthill, R. W. and Saari, J. M. (1965). *Ann. N. Y. Acad. Sci.* **123**, 776.

Shorthill, R. W., Borough, H. C., and Conley, J. M. (1960). *Publ. Astron. Soc. Pacific* **72**, 481.

Slade, M. (1970). Private communication.

Thomas, A. B. (1949). *Aust. J. Sci.* **11**, 187.

Thompson, T. W. (1966). *Center Radiophys. Space Res. Rept.* **64**, Cornell University, Ithaca, New York; see also *J. Res. N. B. S.* **D69**, 1667 (1965).

Thompson, T. W. and Dyce, R. B. (1966). *J. Geophys. Res.* **71**, 4843.

Thompson, T. W., Pollack, J. B., Campbell, M. J., and O'Leary, B. T. (1970). *Radio Science* **5**, 253.

Thomson, J. H. and Ponsonby, J. E. B. (1968). *Proc. Roy. Soc.* **A303**, 477.

Trask, N. J. (1970). *Radio Science* **5**, 123.

Trexler, J. H. (1958). *Proc. Inst. Radio Eng.* **46**, 286.

Twersky, V. (1962). *J. Math. Phys.* **3**, 724.

Tyler, G. L. (1968). *J. Geophys. Res.* **73**, 7609.

Tyler, G. L. and Simpson, R. A. (1970). *Radio Science* **5**, 263.

Winter, D. F. (1962). *J. Res. N. B. S.* **D66**, 215.

Yaplee, B. S., Bruton, R. H., Craig, K. J., and Roman, N. G. (1958). *Proc. Inst. Radio Eng.* **46**, 292.

Yaplee, B. S., Roman, N. G., Craig, K. J., and Scanlon, T. F. (1959). *In* "Paris Symposium on Radio Astronomy" (R. N. Bracewell, ed.), p. 19. Stanford Univ. Press, Stanford, California.

Yaplee, B. S., Knowles, S. H., Shapiro, A., Craig, K. J., and Brouwer, D. (1964). *Naval Res. Lab. Rept.* **6134**, Washington D. C.

Cratering and the Moon's Surface*

E. J. ÖPIK

Armagh Observatory, Armagh, Northern Ireland
and
Department of Physics and Astronomy
University of Maryland, College Park, Maryland

* Supported by National Aeronautics and Space Administration Fund NsG-58-60.

INTRODUCTION

This monographic chapter has grown out of a planned, much less expanded, review article on the Moon's surface. Instead, a complete mechanical and statistical analysis of the lunar surface has been drawn along new quantitative lines, without, however, attempting anything like a complete review of the existing literature. Also, during the two years that have passed since the original 1966 deadline, much new factual material has been provided by the American Ranger, Surveyor, and Orbiter, as well as by Russian spacecraft; some of these data have been incorporated into the framework of this analysis, although incompletely. The material is too voluminous and still increasing, awaiting exhaustive treatment at a later date. Yet, as things stand now, the selected data used here appear to be sufficient to characterize the mechanical and other properties of the lunar soil and surface, so that not much substantial change except in some details can be expected from a comprehensive discussion of the entire material. The success in predicting, statistically from first principles, the observed distribution of crater numbers over a wide range of sizes, from 3 cm to 5 km, lends support to the reliability of the theoretical basis of cratering and erosion, which forms the backbone of this chapter.

The phenomenal growth of lunar literature, while contributing to the knowl-

edge of our satellite, has not removed the occurrence of contradictory interpretations even in such basic questions as that of the origin of lunar craters.
In this respect an old tendency manifests itself: to make hypotheses about
astronomical objects that are based on only one aspect of the problem, while
overlooking or ignoring contradictory evidence. Hence an impression is created
that astronomers always disagree among themselves, an impression that persists
even when reading the best review articles on the Moon (Baldwin, 1964a).[1]

Undoubtedly, the difficulty of a direct proof and impossibility of experi-
mentation were conducive to such a state of affairs, surprisingly, even in the
case of the nearest of all celestial bodies. Also, lunar physical study has for too
long been neglected by professional astronomers and left in the hands of
amateurs—whose merits, however, are by no means to be underestimated.

At present, space research has brought the Moon, so-to-speak, within an
arm's length so that many theories can be proved or disproved as in a laboratory,
and the moon is increasingly becoming the object of professional study. Yet a
new source of misinterpretations is becoming troublesome. Previously, the
astronomer had time for the study of all the relevant literature and for a critical
assessment of the available evidence. Nowadays, with the enormous supply of
scientific publications, it becomes progressively more difficult to master the
entire literature, or even the details of one narrow branch of science. This has
led to the ever increasing habit of trusting authority, second- and even third-
hand. Statements are repeated that never would have been made upon critical
study of the evidence. Erring may be human, but too much reliance on unchecked
authority may lead to unwarranted perpetuation of error, as has happened with
the much publicized so-called gaseous eruption from the crater Alphonsus.
The spectrogram was not studied properly, or it would have become obvious
that no gas was emitted, but that luminescence of the solid peak of the crater
was responsible for the phenomenon. As a classical case of repeated misinter-
pretation, the Alphonsus "eruption" is specially dealt with in Section I.

When this, and similar unfounded or one-sided interpretations are discarded,
the picture of the lunar surface becomes much less controversial. As a powerful
instrument for interpretation, too little used until now, the quantitative theory
(and experiment) of solid-body impact (hypervelocity and low-velocity) helps
to resolve the most relevant problems of crater formation and erosion, dust
formation and transport, bearing on the strength of lunar soil and rock and the
mechanical structure of the upper few kilometers of the lunar crust. The
quantitative approximation is of the order of 10–20% in absolute linear measure,

---

[1] Symbolically, in this context, there exists a purely formal ambiguity in defining
selenographic directions. In this article the directions are reckoned "astronomically"
as for the terrestrial telescopic observer. When South is above, West is to the left, so
that Mare Crisium is in the western hemisphere. In the "astronautical" reckoning, the
directions are inverted as for an observer standing on the moon.

thus far better than an order-of-magnitude approach. With another little used instrument, the theory of planetary encounters as developed by the author, it is possible to remove much (if not all) of the ambiguity relating to the origin and internal structure of the Moon, which is also directly related to the present structure and properties of the lunar surface.

I

THE ALPHONSUS EVENT AND FLUORESCENCE ON THE LUNAR SURFACE

Instigated by some observations of Dinsmore Alter in California, the Russian astronomer Kozyrev kept the crater Alphonsus under observation in October and November of 1958. As stated in his report (Kozyrev, 1959a), he was intentionally in search of volcanic phenomena on the Moon. In the early morning of November 3, 1958, he noticed an unusual brightening on the peak of the crater and, while the brightening lasted, a spectrogram taken with the 50-inch Crimean reflector (linear scale: 10 sec of arc or 18.4 km to the mm; dispersion 23 Å/mm at $H\gamma$; exposure 30 min) showed strong banded emission over the peak. The emission was no longer visible on the next spectrogram, nor was it visible in previously taken spectra. The Moon was one day before last quarter; the altitude of the sun over Alphonsus was 18°, and about 31° over the illuminated slope of the peak. Reproductions of the spectrograms, with photographs of the crater itself, were published repeatedly (Kozyrev, 1959b, 1962) but no essential points were added to the first discussion (Kozyrev, 1959a) that appeared under the challenging title of "Volcanic Activity on the Moon." Essentially, Kozyrev— and others—identified the band structure of the observed emission with that of the cometary radical C_2, as fluorescent in sunlight.

Yet the details of the spectrum along the slit, or at right angles to the dispersion, show without the least trace of doubt that the luminescence was strictly confined to the illuminated portion of the peak, and that therefore no eruption of gas ever did take place. This was pointed out by Öpik (1962a, p. 252; 1962b, p. 218; 1963b) but somehow was overlooked. Distinguished authors, trusting Kozyrev's announcement and without taking a critical look at the published spectrograms, have been led to discussions of the "gaseous eruption" (e.g., Baldwin, 1963, pp. 415–419). Actually, Kozyrev did measure the distribution of monochromatic brightness of the spectrum along the slit and his measurements did show indeed—what was also obvious from a direct inspection of the spectrograms—that the increase in brightness did not affect the shadow of the peak. [See Kozyrev (1962, Fig. 2); obviously the linear scale there should be kilometers, not seconds of arc, and the orientation is inverted relative to the spectrogram.] However, he did not see the consequences of this fact; everyone else then accepted Kozyrev's interpretation on his authority.

Kozyrev's announcement was hailed as the first definite proof of gaseous phenomena on the Moon. After some doubts and questioning, chiefly concerned with the band structure of the spectrum, the astronomical community seems to have accepted this interpretation. Nobody seems to have worried about the second dimension of the spectrogram that reproduced the surface features and showed a puzzling detail. The emission was spatially restricted to the bright peak, about 4–5 km wide, without trespassing into the shadow, about the same width. The transition was abrupt at the border of the shadow and took place over a distance of about 1 km, which corresponds to the resolving power of the photograph. The neutral C_2 gas could not have been restricted by a magnetic field and, with a molecular velocity in excess of 0.5 km/sec, the gas would have spread over a radius of some 900 km during the exposure, covering both the peak and its shadow. Gases emitted from a point source (the peak) would have formed something similar to a comet's head (coma), with a strong central condensation and an intensity decreasing inversely as the first power of distance. The average intensity over the shadow would then have been equal to about one-half the average intensity over the peak. Nothing of this sort was shown in the spectrogram.

There nevertheless appears to be some similarity between the emission from Alphonsus' peak and the cometary or Swan bands of C_2. In this respect, Kozyrev (1959a, p. 87) points out a strange detail (translation from Russian): "The Swan bands should be completely sharp on the long-wave side, yet they turned out to be washed out over about 5 Å." Here seems to be the clue to the interpretation: bands originating in a solid lattice must be washed out, on account of perturbation by other nearby atoms. Kozyrev proposes another interpretation (to fit into his concept of a gas), namely, that the radiation was created *in statu nascendi* when C_2 was produced from its parent molecules. However, this would mean that each C_2 molecule radiated only once, not repeatedly (5–10 times per second) fluorescing in sunlight (what could have prevented it from doing so?), and the brightness could then never have been "10^4 times as intense as in comets" (Kozyrev's estimate).

Clearly, Kozyrev's phenomenon can be interpreted only as emission, probably fluorescent, from a solid surface and not from an expanding gas. Most that has been written about this event is, therefore, not valid; also, the identification of the emitting molecules can hardly be made with any degree of reliability, although there may have been blurred emission from C_2 somehow present in the solid lattice.

Experimentally, it has been shown that meteoritic enstatite ($MgSiO_3$, $FeSiO_3$, as distinct from the more usual olivines, Mg_2SiO_4, Fe_2SiO_4, $MgFeSiO_4$) emits fluorescent light under proton bombardment (40 keV), and also that certain regions on the Moon, around Aristarchus and Kepler in particular, may become fluorescent, apparently in response to bursts of corpuscular radiation from solar flares (Kopal, 1966a). There is a grave difficulty in describing the source of the observed lunar fluorescence in terms of the energy of the proton stream, which falls short by many orders of magnitude, as follows from the observed intensities of solar wind. Focusing effects of the Earth's magnetosphere have been suggested, which would hardly work. It seems that the only explanation is to ascribe the fluorescent radiation to direct sunlight (as for C_2 in comets), whereas the role of the corpuscular bursts would be to raise the molecules to a metastable state capable of fluorescence. The ground state of the C_2 molecule is a singlet, while the lowest level of the Swan bands is a triplet state, only about 0.09 eV above the ground state. The transition from triplet to singlet is forbidden and can be achieved efficiently only by collisions. A similar situation may obtain in the case of lunar luminescence; the emission from the metastable state would then derive from direct sunlight, which is amply sufficient as it is in comets, and not from the inadequate energy of the corpuscular stream acting only as a trigger.

II

CRATERING RELATIONSHIPS

A. Destructive Impact and Volcanism

There are no signs of continuing volcanism on the Moon. Extensive lava flows, as witnessed by the maria, flooded craters, and small "domes," must have happened early in the history of the Moon, during the first one million, even the first 20,000 years of its existence. On Earth, volcanism is related to mountain building and this in turn is the consequence of powerful erosion cycles leading to recurrent imbalance in the Earth's crust. On the Moon, erosion from inter-planetary dust is about 2000^{-1} times as efficient as in terrestrial deserts (Öpik, 1962a); if, on Earth, the major orogenic cycles followed at intervals of the order of 2×10^8 years, on the Moon the interval should be of the order of 10^{12} yr: volcanism could never occur.

The lunar surface markings, from craters down to the compacted dust layer, are undoubtedly produced or evolved under the bombardment of interplanetary bodies and particles, as well as that of the secondary ejecta from the surface itself. The quantitative study of cratering contains, therefore, the most important clue to the structure and history of the lunar surface.

Usually, the term "hypervelocity" is applied to cratering impacts. This refers to the case where the initial velocity of the projectile exceeds the velocity of sound in the target and/or the case where the frontal pressures at penetration exceed the strength of both the projectile and the target, so that the projectile itself is destroyed and flattened while entering the target.

Actually, cratering cannot be presumed to be a purely hypervelocity phenomenon when the whole of the crater volume is considered. Hypervelocity phenomena may occur only in the heart of the crater. Destruction and ejection of the target material take place over most of the crater volume when the shock-front velocity is less than the velocity of sound while the shock pressure still exceeds the strength of the material, or when the energy density of vibration is more than can be borne by the elastic forces in the target. From this standpoint, a uniform quantitative theory of destructive cratering, applicable also to low-velocity impact, has been worked out by Öpik (1936, 1958a, 1961a). The theory, based on the consideration of *average* pressure and momentum transfer over shock fronts, from first principles and without experimental adjustment of the parameters, gives an approximation to experiment within 10–20% in linear dimensions and can effectively substitute for the huge amount of experimental material accumulated and not yet properly systematized.

B. *Destructive Impact: Mechanical Theory*

Full or mutually destructive impact is the common "hypervelocity" case when both the target and the projectile are destroyed during the penetration. Formulas for direct application to lunar or similar cases are given below; they are partly new developments, as a sequel to the latest published paper (Öpik, 1961a).

In Fig. 1, a schematic half-section of a cratering event is represented. The

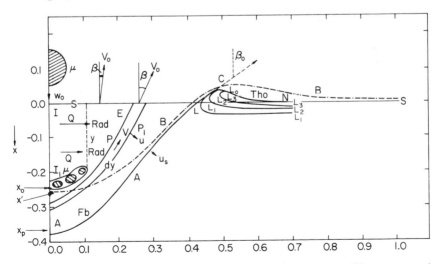

Fig. 1. Vertical semi-cross section, to scale, of an impact crater [the prototype is the nuclear explosion "Teapot" crater in Nevada (Shoemaker, 1963)]. The linear scale unit is B_0, the rim-to-rim diameter of the crater.

relative dimensions are partly kept to scale of the "Teapot" nuclear crater in the desert alluvium of Nevada (Shoemaker, 1963). A meteorite of "equant shape" (whose linear dimensions in different directions do not differ more than in a ratio of about 2 to 1), mass μ, density δ, and initial velocity w_0 normal to the target surface $ISLS$, penetrates into the target and, while itself flattened and deformed or broken up, stops at I_1 with its front surface reaching a depth x_0 below the surface. If the velocity were sufficiently high, the meteorite with a "central funnel" Q (20–25 times the mass of the meteorite) might be completely or partly vaporized and backfired. The forward passage of the meteorite, combined with the backfiring, creates a destructive shock wave that stops at A, at a depth x_p in the frontal direction and propagates laterally as a *radial momentum* (labeled Rad in Fig. 1) either with the shock velocity u or the sound velocity, whichever is greater. In the crater bowl the material is crushed, pulverized, or even melted (near Q) and after having been stopped at a bedrock surface AAL, as conditioned by a limiting "crushing" value of $u = u_s$, is partly ejected upward (velocity vector v inside, v_0 at the surface under an angle β to the normal). The bedrock surface AAL is itself displaced outward, producing a raised lip LL_0N, with the underlying strata L_1, L_2, and L_3 characteristically bent over into the lip. Part of the debris falls back into the crater, part is thrown out over the lip, forming the apparent crater and surrounding surface BCB with the rim at C and an apparent depth x' (as distinct from x_0 and x_p). The volume of the crater bowl $AALL_0$, below the bedrock rim level L_0, is close to

$$V = 0.363 x_p B_0{}^2, \tag{1}$$

where B_0 is the rim to rim diameter of the crater.

The "mass affected" is assumed equal to

$$M = \rho V, \tag{2}$$

where ρ is the original target density; it depends on the radial momentum,

$$M = k\mu w_0/u_s, \tag{3}$$

where

$$u_s{}^2 = s/\rho. \tag{4}$$

Here s (dyne/cm^2) is the lateral crushing strength of the target, and k a coefficient of radial momentum varying between 2 and 5, depending on the degree of vaporization and backfiring, and defined by the quadratic equation (Öpik, 1961a)

$$k = nw_0{}^2(1 - 0.04k^2)^{1/2} + 2, \tag{5}$$

where $n = 6.5 \times 10^{-13}$ for iron impact into stone, and 4.2×10^{-13} for stone impact into stone when w_0 is given in cm/sec.

Using numerical integration (Öpik, 1936), an interpolation formula for the relative depth of penetration can be set up as

$$p = x_{\mathrm{p}}/d = 1.785(\delta/\rho)^{1/2}(w_0^2/s_{\mathrm{p}})^{1/30}\cos\gamma, \tag{6}$$

and from Eqs. (1)–(4) the relative crater diameter results as

$$D = B_0/d = 1.20[(kw_0\delta)/p]^{1/2}/(\rho s)^{1/4}. \tag{7}$$

Here the nondimensional numerical factor 1.20 allows for the funnel-shaped crater profile and differs from the factor of unity formerly used (Öpik, 1961a). Equation (6) tentatively allows for oblique incidence, γ being the angle of incidence relative to the normal to SS, and s_{p} is the compressive strength or frontal resistance (dyne/cm^2) of the target material (usually an order of magnitude greater than s). The reduced spherical equivalent diameter of the projectile is

$$d = (6\mu/\pi\delta)^{1/3} = 1.241(\mu/\delta)^{1/3}, \tag{8}$$

and p and D are the depth and diameter of the crater, respectively, in units of d. The ratio of depth to diameter becomes

$$x_{\mathrm{p}}/B_0 = p/D = 1.99(\cos\gamma)^{1.5}(s\delta)^{1/4}/[(k\rho)^{1/2}\,w_0^{0.4}s_{\mathrm{p}}^{0.05}]. \tag{9}$$

The numerical coefficients in Eqs. (6) and (9) are dimensionally adapted to cgs units [in Eq. (9) the dimension of the coefficient is cm$^{0.15}$ gm$^{-0.05}$, and in Eq. (6) it is cm$^{0.1}$ gm$^{-1/30}$].

Typical parameters can be assumed: for silicate stone of a planetary upper crust, $\rho = 2.6$, $s = 9 \times 10^8$, $s_{\mathrm{p}} = 2 \times 10^9$; for nickel iron, $\delta = 7.8$; $s_{\mathrm{p}} = 2 \times 10^{10}$. Table I[2] contains some relative crater dimensions calculated with these constants.

The equations are supposed to be valid when the aerodynamic pressure $K_{\mathrm{a}}\rho w_0^2$, with the drag coefficient $K_{\mathrm{a}} \sim 0.5$, greatly exceeds s_{p}, the compressive strength of both the target and the projectile. For a sixfold safety margin, $w_0 > 3$ km/sec for iron impact into stone, and $w_0 > 1$ km/sec for a hard stone projectile impact into stone. In such a case, aside from backfiring, a radial momentum equal to μw_0 is generated both in the target and the projectile, adding up in k as a component equal to 2; backfiring due to explosive vaporization increases the value of k as the velocity increases [cf. Eq. (5)].

Only the aerodynamic component of frontal pressure generates radial momentum, whereas the "dead resistance" s_{p} does not participate. Hence at smaller velocities k further decreases, [Eq. (5) being no longer valid], in proportion to the ratio of aerodynamic to total resistance, down to a value of unity

---

[2] All tables will be found in the Appendix at the end of the chapter.

and even less. This value is reached at the lower velocity limit w_m for the applicability of the model when the projectile is no longer subject to lateral expansion, or when

$$\tfrac{1}{2}\rho w_m{}^2 = s_p \quad \text{(projectile)}. \tag{10}$$

For hard stone impact into stone, $w_m = 0.39$ km/sec; for iron impact into stone, $w_m = 1.24$ km/sec.

In large-scale phenomena friction generates an additional component of lateral resistance depending on the weight of the overlying mass and the coefficient of friction f_s :

$$s = s_c + f_s g \rho x_c . \tag{11}$$

Here s_c is the component of lateral strength due to cohesion, g is the acceleration of gravity, and x_c is the half-depth of radial momentum that, approximately, can be set equal to

$$x_c = 0.610 x_0 \tag{12}$$

with

$$x_0 = 0.800 x_p ; \tag{13}$$

these values represent more or less overall averages for destructive impact (cf. Fig. 1).

In some cases s_c itself may depend on depth; an effective depth corresponding to x_c is then to be adopted.

To compare the preceding formulas with experiment would require laborious study, because of the amount and complexity of the experimental material accumulated. It is also unnecessary at this stage because it turns out that the formulas describe the experiments with an accuracy that is not inferior to that of the parameters involved when they are known, such as the strength characteristics of the material; and, in many cases the parameters are unknown and can only be derived best from the very formulas as given above. This especially applies to the Moon.

Two examples may illustrate the approximation to experiment obtained by the application of Eqs. (3), (4), (6), and (7).

Table II summarizes experiments with aluminum spherical pellets that were accelerated *in vacuo* with a light-gas gun and fired into aluminum targets of different tensile strength (s_t) as determined in the laboratory (Rolsten *et al.*, 1966). For ductile metallic solids, $s_p = 5s_t$, and $s = 3s_t$ can be assumed, and much of the mass affected will stick to the crater, making its diameter smaller than predicted by Eq. (7). This expectation is borne out by the last line of the table, although the systematic difference is but slight. The observed penetrations include the height of the lip and, to make the data comparable, the calculated penetrations x_p were increased by an average factor of 1.16. With this factor, there is a perfect—and rather unexpected—agreement between theory and observation.

In another set of experiments (Comerford, 1966), the results were compared with Öpik's theory with the conclusion that "theory and experiment agree reasonably well for brittle materials, but there is only partial agreement when theory is compared with measurements on ductile materials." The latter point, referring to the behavior of ductile materials, has also been anticipated theoretically (Öpik, 1958a, p. 32). "The discrepancy is shown to be attributable to the ability of ductile materials to deform plastically without fracturing" (Comerford, 1966). Planetary crustal or surface materials are predominantly of the brittle type and the theory should apply well here.

Not all of the mass affected is demolished; part of it is plastically displaced into the rim or lip (cf. Fig. 1, LL_0N). The crushed volume of debris, as contained between the basic rock ($AALL_0NS$) and the apparent surface ($x'BBCB...$), equals 0.669 of the total volume affected for the typical crater contour. Hence the mass crushed can be assumed to be

$$M_c = 0.669k\mu w_0/u_s , \tag{14}$$

and the volume crushed

$$V_c = 0.244x_pB_0{}^2. \tag{15}$$

Part of this material falls back into, or stays in the crater ("fallback" F_b in Fig. 1) and part is ejected over the rim ("throwout" Tho in Fig. 1).

C. Ejection Velocity, Heating, and Crater Ellipticity

The modification of the target in cratering events is basically of two types (apart from the hypervelocity phenomena in and around the central funnel Q), with possible transitions: the destruction of the target over the volume of the crater bowl ($IELAA$ in Fig. 1); and the plastic compression and deformation of the bedrock surface ($AALL_0N$). Most of the debris of the bowl are fanning out into an expanding volume, being crushed as in one-sided compression. In "normal" fragmentation, for fragments of "finite" dimensions, only moderate heating takes place because the excessive shock required for frictional and compressional heating would pulverize the material. However, some of the target material, especially around the central funnel, may become locked in all-sided compression, which, at pressures of 10^5–10^6 atm, may be subject to pressure modifications of its crystal structure (coësite) and to more intense heating without, however, acquiring considerable ejection velocities. The amount of such material, subject to hyperpressures without ultimate fragmentation, is relatively small. In what follows we shall concern ourselves only with the massive debris and ejecta of the crater bowl which are the product of crushing, leading to "normal" fragmentation. With a few reservations (central funnel, vaporization) the formulas of this section apply also to semidestructive impact (cf. next subsection).

Let P (Fig. 1) represent a surface of constant shock velocity u, and the mass enclosed in it being yM, so that y is the "fractional mass affected." The shock velocity at P is then

$$u = kw_0\mu/yM = u_s/y,$$ (16)

valid outside the central funnel (Q) at which, approximately,

$$y_Q = 25\mu/M,$$ (17)

where

$$u_Q = 0.04kw_0.$$ (18)

The kinetic energy at P is released and converted mainly into heat and partly into the kinetic energy of ejection. If λ_x is the kinetic efficiency of the shock at depth x, the transverse velocity at the shock front is

$$v = \lambda_x u,$$ (19)

and the heat release in erg/gm of the crater material becomes (Öpik, 1958a)

$$q = \tfrac{1}{2}u^2(1 - \lambda_x^2).$$ (20)

In the central funnel, turbulent mixing is supposed to lead to a uniform heating and impulse ejection velocity $0.2\lambda_c w_0$, so that the heat release becomes (Öpik, 1958a, 1961a)

$$q_c = 0.02w_0^2(1 - 0.02k^2)(1 - \lambda_c^2).$$ (21)

If the heat released is sufficient to cause vaporization, the velocity of ejection from the central funnel increases to be greater than the value $0.2\lambda_c w_0$ and the recoil momentum increases; this has been taken into account in Eq. (5). The fraction vaporized in the central funnel is then (Öpik, 1961a)

$$f_g = 3.3 \times 10^{-13}[(1 - 0.02k^2)\,w_0^2 - 10^{12}] \leqslant 1.$$ (22)

($k^2 \to 20$ for high velocities and stone impact into stone.) When $w_0 = 24$ km/sec ($k^2 = 15$), $f_g = 1$. For higher velocities, shock vaporization at the expense of released heat q becomes possible outside the central funnel. Vaporization can take place only when $w_0 > 10.4$ km/sec according to this equation.

An element of mass dy between two shock surfaces P and P_1 (Fig. 1) streams out in a manner analogous to hydrostatic flow of a liquid from the bottom opening of a vessel, the velocity decreasing according to $h^{1/2}$, where h is the fluid level of the vessel. For a vessel of constant width the mass is $\sim dh$, the kinetic energy is $v^2 \sim h$, and the frequency of v^2 to $v^2 + dv^2$ is proportional to dh or to $d(v^2)$: the kinetic energy has a constant frequency law:

$$f(v^2)\, d(v^2) = d(v^2) \times \text{const.} \tag{23}$$

We assume the same law for the distribution of v inside dy, the ejection velocity decreasing linearly with depth x. Conventionally, we assume all dy elements to reach to the same depth x_0, so that

$$v^2 = v_0^2[1 - (x/x_0)], \tag{24}$$

also

$$v_0 = \lambda u, \tag{25}$$

and the relative (normalized) frequency of v^2 or the fraction of v^2 between v^2 and $v^2 + dv^2$ to be

$$dn = dv^2/v_0^2 = dx/x_0. \tag{26}$$

Although the hydrostatic analogy is remote, the accepted velocity distribution accounts, qualitatively at least, for the loss of kinetic energy in collisions and turbulent friction while a mass element makes its way outward, so that the loss will be the greater the deeper the point from which it had started. The assumptions are justified by the application to lunar and terrestrial crater profiles (cf. Sections II,F and V,A) and probably are not far from reality even quantitatively.

A similar rough assumption is to be made for the distribution of the exit angles β of the ejecta (Fig. 1). The condition of continuity and near incompressibility of the target material, over the relevant major fraction of the mass affected, requires that the ejection vectors must be all in "meridian" planes directed outward, and that the exit angles form a continuous sequence from $0°$ at the center to β_0 at the rim, at $y = 1$, where the direction is tangent to the crater lip at L. An interpolation formula

$$\sin \beta = y \sin \beta_0 \tag{27}$$

is here proposed, without further justification, to represent the fanning-out of the ejection angles at the original target surface.

Equation (7) defines an average crater diameter which, in the case of ellipticity, can be assumed to be the mean of the maximum and minimum diameters. In a homogeneous target, the crater ellipticity $\epsilon = (a - b)/a$, in the direction of motion of the projectile, should depend on the angle of incidence γ as follows (Öpik, 1961a):

$$\epsilon = 2\{\sec \gamma + [(p \tan \gamma)/3] - 1\}/3D \tag{28}$$

in former notations. The formula should be valid for angles less than $\gamma = 60°$, and roughly up to 75°.

D. Semidestructive Impact

This is the case of a hard projectile entering a softer target with a velocity below w_m [Eq. (10)]. Essentially, the projectile retains its shape and, to some extent, also its aspect relative to the direction of motion, while the target yields, being crushed and forced into hydrodynamic flow. The equation of motion (for $\gamma = 0°$) is

$$m \, dw/dt = -(K_a \rho w^2 + s_p) \tag{29}$$

in former notations, with the mass load per square centimeter cross section being defined as

$$m = \mu/\pi R^2,$$

where R is the equivalent radius of the cross-section (σ) contour at right angles to the direction of motion, $\sigma = \pi R^2$.

The drag coefficient depends on the shape of the projectile. For a flattish angular front surface, $K_a = 0.75$ can be assumed as an overall mean characteristic value (while a value of 0.5 better suits a hemispherical front, as well as the case of full destructive impact with a hydrostatically deforming projectile).

With

$$dw/dt = w \, dw/dx$$

Eq. (29) can be integrated for the specific case of $s_p = $ const to give

$$w^2 = (w_1{}^2 + N) \exp(-Px) - N, \tag{30}$$

where w_1 is the initial entry velocity and

$$N = s_p/(\rho K_a), \qquad P = 2K_a \rho/m.$$

For $w = 0$, the depth of penetration x_0 is determined by

$$Px_0 = \ln[1 + (w_1{}^2/N)]. \tag{31}$$

Equation (7) remains valid, as well as other equations of Sections II,B and II,C except Eqs. (5), (6), (9), (13), (17), (18), and (22). Instead of Eqs. (17) and (18) $y_Q = \mu/M$ and $u_Q = kw_0$ can be set.

At first contact of the projectile with the target, there is a shock forcing a hydrodynamic flow pattern on the target; only after the flow is established is Eq. (29) valid. For an incompressible model with a blunt front, the shock momentum transmitted to the target is close to $\rho w_1 \cdot \frac{1}{2} R \sigma$, and this must equal the loss of momentum by the projectile, $(w_0 - w_1) m\sigma$, where w_0 is the initial velocity before contact. Hence the shock ratio of velocities becomes

$$\psi = w_1/w_0 = [1 + (\tfrac{1}{2}\rho R/m)]^{-1}. \tag{32}$$

The coefficient of radial momentum is less than unity and is obtained by integrating the first (hydrodynamic) term of Eq. (29),

$$k_1 = \int K_a \rho w^2 \, dt / 2 K_a m w_1$$

(K_a not cancelled purposely),

$$k = [(1 - \psi)/2K_a] + \psi k_1 . \tag{33}$$

The momentum integral is rather inconvenient for quick use. Instead, the work integral yields the fraction of hydrodynamic work [$\int dx$ of the first term in Eq. (29), with Eq. (30) substituted] to total work as

$$\kappa_1 = 1 - (N/w_1^2) \ln[1 + (w_1^2/N)]. \tag{34}$$

For very varied conditions and parameters an empirical relation has been established by numerical integration, i.e.,

$$k_1 = \kappa_1 (0.65 + 0.35\kappa_1^6)/2K_a , \tag{35}$$

which represents the momentum transfer within a few percent.

Also, instead of Eq. (13), which refers to the mutually destructive impact, the effective depth of disturbed material in the target can be assumed equal to

$$x_p = x_0 + \tfrac{1}{2}R. \tag{36}$$

The above equations are for an idealized case of a flat evenly loaded front surface of the projectile parallel to the target surface and $\gamma = 0^0$. The actual shape of the projectile and orientation of the front surface would introduce complications, including rotational couples which are disregarded here. For oblique impact under moderate angles, a symbolical improvement would consist of replacing x by $x \sec \gamma$ in the preceding equations and taking the encounter cross section at right angles to the direction of motion.

E. Impact of Rigid Projectile into Granular Target

In the preceding, the frontal resistance from target cohesion s_p was assumed to be constant. On granular surfaces, i.e., dust, sand, and gravel, including partly consolidated material, the resistance is variable, increasing with the depth of penetration; this is obvious from the experience that a heavy load sinks more deeply into sand than a light one. Experiments as reported below have shown that for an upper thin layer of sand (0.5–15 cm) the cohesive frontal resistance can be represented quite well by a small constant term plus a main quadratic term of the depth x,

$$s_p = S_p(x^2 + a^2). \tag{37}$$

Natural sand or gravel, containing an unsifted variety of grain sizes, and natural stony projectiles with a flattish bottom were preferred to artificially normalized laboratory conditions. Experiments on a natural beach (Almuñécar, Spain, October 1966 and 1967), although showing considerable local differences, generally conformed to Eq. (37). There were no obvious differences between the static values of s_p (pure pressure without motion) and the dynamic values computed from cratering impact experiments, according to the modified cratering formulas as given below. It was surprising to find that similarly conducted experiments by Surveyor spacecraft yielded even quantitatively similar mechanical characteristics of the lunar top soil.

When Eq. (37) is substituted into Eq. (29), integration yields, for this particular case of variable resistance,

$$w^2/Q = [(w_1^2/Q) + 2 + a^2P^2] \exp(-\xi) - (2 - 2\xi + \xi^2 + a^2P^2), \tag{38}$$

with $\xi = Px$, P being the same as in Eq. (30), and with

$$Q = S_p m^2/(4K_a^3 \rho^3) \quad (\text{cm/sec})^2. \tag{39}$$

The ultimate depth of penetration $x_0 = \xi_0/P$ is obtained from Eq. (38) with $w = 0$, or from

$$(2 - 2\xi_0 + \xi_0^2 + a^2P^2) \exp \xi_0 = (w_1^2/Q) + 2 + a^2P^2. \tag{40}$$

Equations (32), (33), and (35) further determine w_1, the entry velocity, and k, the coefficient of radial momentum transfer, but instead of Eq. (34) the dynamic work ratio now becomes

$$\kappa_1 = 1 - Q(a^2P^2\xi_0 + \tfrac{1}{3}\xi_0^3)/w_1^2. \tag{41}$$

When

$$\xi_0 \to O(<0.20), \qquad \tfrac{1}{3}\xi_0^3 \to (w_1^2/Q) - a^2P^2\xi_0$$

and

$$\kappa_1 \to (\tfrac{1}{4}\xi_0^4 + \tfrac{1}{2}a^2P^2\xi_0^2) Q/w_1^2.$$

With the total radial momentum defined as

$$J_r = k\mu w_0 ,$$

and on the provisional assumption that crater volume is determined through dynamic action alone according to Eqs. (1)–(4), an "apparent" average lateral strength s_a is defined as

$$s > s_a = J_r^2/\rho V^2 = 7.59 J_r^2/\rho x_p^2 B_0^4 \quad \text{dyne/cm}^2. \tag{42}$$

This is an apparent value and a lower limit because, at the low velocities and energies involved, the static work of penetration (against cohesive strength) also participates appreciably in producing a crater, as has been shown by experiments in gravel.

A satisfactory representation of the experiments by a law of cohesive resistance in the form of Eq. (37) has been arrived at by trial and error. Surprisingly, no dependence of the static or dynamic resistance (bearing strength) on the linear dimension could be detected except the inevitable shock interaction [Eq. (32)]. Shape of the contact surface (projectile or slug) is, of course, of decisive importance in the dynamic interaction as it determines K_a and k; the use of flattish surfaces throughout has given a degree of homogeneity to the experiments that should also correspond to low-velocity impacts of throwout boulders on the Moon.

The constant component in Eq. (37) was too small and variable for exact determination, but its form, $a^2 S_p$ (thus proportional to the strength at greater depth and not an absolute constant), was preferable, with an overall value of the parameter $a^2 = 2 \pm 0.5$ cm^2. (The same constant worked well also in interpreting the Surveyor experiments on the Moon.) The constant term is, of course, of importance only at small loads and penetrations.

Static tests of cratering, interpreted according to a certain rational model, gave the clue to the ratio s/s_a, i.e., the true to apparent average lateral strength. If μ is the static mass load, g the acceleration of gravity, σ the frontal cross section (of stone, slug, or rod), and x_0 the equilibrium depth attained by gradual loading so that the velocity is kept near zero, the maximum resistance equals $s_p(\text{max}) = \mu g/\sigma$, and the resistance parameter of Eq. (37), with $a^2 = 2$ and cgs units becomes

$$S_p = \mu g/\sigma(x_0^2 + 2) \quad \text{dyne/cm}^4.$$

The resistance averaged ever x_0, the entire depth of penetration, is then

$$\bar{s}_p = S_p(\tfrac{1}{3}x_0^2 + 2).$$

A "pressure crater" is formed, of diameter B_0 and depth $x' \ll x_0$ (unlike the impact craters in sand or gravel where $x' = x_0$ and invariably reach the bottom contact surface). The volume displaced by slug and crater then equals [cf. Eq. (1)]

$$V_p = (0.363x'B_0^2 - \sigma x') + \sigma x_0 = V_1 + V_2.$$

The total work of "static" penetration evidently is

$$E_p = \sigma \bar{s}_p x_0 \quad \text{erg},$$

a fraction F of which is assumed to be transmitted at right angles on lateral work of cratering as measured by the product of lateral strength and volume displaced,

$$FE_p = sV_p .$$

In an incompressible medium, the volume displaced is ultimately lifted up, the average lifting height being $\frac{1}{2}(x' + h)$ for V_1 and $\frac{1}{2}(x_0 + h)$ for V_2, h denoting the rim elevation of the crater. This involves work done against gravity,

$$E_g = g\rho[\tfrac{1}{2}V_1(x' + h) + \tfrac{1}{2}V_2(x_0 + h)].$$

The work of uplift against gravity is transmitted at right angles from the lateral expansion in the same manner as that in which the lateral work originated from the downward work of penetration. It is sensible then to assume (and the numerical applications amply support the assumption) that

$$E_g = FsV_p = F^2E_p \quad \text{or} \quad F = (E_g/E_p)^{1/2},$$

whence a value for the average lateral strength in static penetration becomes

$$s = FE_p/V_p = (E_gE_p)^{1/2}/V_p . \tag{43}$$

For example, in the "static" Experiment 2 of Table III(a)[3], with a round rod of $\sigma = 3.63$ cm^2 and a final load of 7.28×10^4 gm, $x_0 = 15.0$ cm, $x' = 1.7$ cm, $h = 1.0$ cm, and $B_0 = 12.5$ cm. Hence $V_1 = 114.0$ cm^3, $V_2 = 53.7$ cm^3, $s_p(\text{max}) = 1.97 \times 10^7$ dyne/cm^2, $S_p = 8.65 \times 10^4$ dyne/cm^4, $\bar{s}_p = 6.66 \times 10^6$ dyne/cm^2; $E_p = 3.62 \times 10^8$ erg, $E_g = 9.73 \times 10^5$ erg, $F = 0.0518$, and $s = 1.12 \times 10^5$. With $f_s = 0.63$ (determined from the angle of repose) and $g = 980$/cm^2, the contribution from friction becomes [Eqs. (12) and (13)] $640x_0$ dyne/cm^2, and the lateral strength, according to Eq. (11), is then $s_c = 1.12 \times 10^5 - 9600 = 1.02 \times 10^5$ dyne/cm^2. Unlike s_p, this is an average or effective value, to be compared with the average bearing strength: $\bar{s}_p/s_c = 65.3$ (an unusually high ratio). Although variable, there did not seem to be a systematic dependence of this ratio on penetration, whence Eq. (37), properly modified, can also be adapted to represent the *average* lateral strength,

$$s_c = S_c(\tfrac{1}{3}x_0^2 + a^2), \tag{37a}$$

the value of $a^2 = 2$ cm^2 being used throughout.

---

[3] In the tables, abbreviated numerals are often used, substituting for powers of ten; e.g., $1.58^{-10} = 1.58 \times 10^{-10}$ and $6.0^8 = 6.0 \times 10^8$.

In interpreting the dynamic (impact) experiments, it was assumed that the independently calculated dynamic (V_d) and static (V_p) cratering volumes were additive. Let $V = V_d + V_p$ denote the total crater volume, proportional to $x_p B_0{}^2$ of Eq. (42). According to this equation, evidently

$$s/s_a = (V/V_d)^2 = V^2/(V - V_p)^2 = [1 - (FE_p/sV)]^{-2}.$$

A quadratic equation with respect to s is obtained which ultimately yields

$$s_a/s = A(1 + \tfrac{1}{2}A) - (A + \tfrac{1}{4}A^2)^{1/2}, \tag{44}$$

where

$$A = s_a V/FE_p .$$

An average value of $F = 0.118$, obtained from the first six static tests [Table III(a)], was used. Table III contains a summary of the experiments. No two experiments (static with gradual loading) were made on the same spot; only the ultimate load and penetration were recorded.

Because of the large dispersion, logarithmic mean values were quoted as being more significant. The "probable deviation ratio" of a single experiment, corresponding to 0.845 of the absolute deviation in the logarithm,

$$\sum |\varDelta|/[n(n - 1)]^{1/2},$$

would indicate that 50% of the deviations, if Gaussian, would be expected to be within this ratio of the logarithmic mean and its reciprocal. Experiment 25 was made at oblique incidence $\gamma = 24°6$ from the vertical and is interpreted with corresponding modification, $\xi = Px \sec \gamma$ using $S_p \cos^2 \gamma$ instead of S_p in Eq. (39), while S_c remains unaffected. Experiments 30–32 yield anomalously high frontal resistance while the s_c values are normal despite high s/s_a ratios. These were the only experiments where a longish slug was made to impact on its narrow end, and every time it would tilt over and be found lying overturned on its long side after impact; possibly, in this twisting movement, the area of resistance was increased, which could account for the abnormal values calculated on the assumption of the small area of encounter. The lateral resistance was not affected, depending on total momentum, actual volume of crater, and penetration, without direct intervention of aspect or area of contact.

Although measurements of x_0 and B_0 on sand craters cannot be very accurate, the dispersion in the inferred S_p and S_c values is much greater than that which could be due to straightforward errors of observation. A natural granular layer possesses an intrinsic nonhomogeneity to which the dispersion testifies, and which, in particular, may depend on accidental configurations of the larger grains or pebbles.

The increase of strength with depth depends on two factors: the tighter packing of the deep layers, and the strengthening effect of the weight of the overlying layers. In Experiment 2a, when these layers in Experiment 2 were removed, the surface laid bare in such a manner was unable to support a load of $s_p = 1.97 \times 10^7$ dyne/cm^2, which it had withstood under the weight of the former layer of 15 cm, with $S_p = 8.65 \times 10^4$ dyne/cm^2. With reapplication of the former load, a new penetration of $x_0 = 5.0$ cm was achieved, yielding a strength coefficient $S_p = 7.30 \times 10^5$ dyne/cm^2, which was 8.4 times larger than the former value—a characteristic of the intrinsic tightening of the granular matrix with depth. With the layer unremoved, the resistance at $x_0 = 15.0 + 5.0 = 20.0$ cm, according to Eq. (37), would have been $s_p = 8.65 \times 10^4 \times 402 = 3.48 \times 10^7$ dyne/cm^2, or 1.77 times the value when the layer was removed; this indicates the degree of additional reinforcement below the depth of 15 cm due to the weight of the overlying 15 cm.

All these details are brought out because of their close qualitative and quantitative analogy with similar experiments made on the lunar surface by Surveyor spacecrafts, so that the results of these terrestrial experiments can be applied with some confidence, by way of extrapolation, to the mechanical properties of the lunar soil.

Although having good working value, it would be wrong to extrapolate Eqs. (37) and (37a) to greater depths without limitation. The quadratic law of increasing strength can be valid only in a top layer; at greater depth it should merge into a constant value corresponding to compacted granular material. For the "Teapot" nuclear crater in desert alluvium (cf. Shoemaker, 1963), at $x_0 = 3150$ cm, $s_c = 4.0 \times 10^7$ dyne/cm^2 (cf. Section II,F); a value of $S_c = 4 \times 10^3$, which is about the mean in Table III(c) for gravel, would reach the observed value at $x_0 = 100$ cm, and this should not be surpassed in a granular matrix except at very much greater depths when plastic compaction into solid rock takes place. The same must be true of the frontal strength; with $s_p < 2 \times 10^8$ (as for sandstone) and $S_p = 5 \times 10^4$ as a mean value in Table III(c), the limiting depth for the quadratic term is only about $x_0 = 63$ cm. Thus, it can be assumed provisionally that Eqs. (37) and (37a) are probably valid to a depth of about 60–100 cm, beyond which s_c and s_p assume constant "compacted" values.

The logarithmic mean of the ratio of frontal to lateral cohesive resistance at impact [Table III(c)] is 11. This is about the same as the ratio of compressive to crushing strength for brittle materials.

All experiments were conducted at vertical incidence except one with $\gamma = z = 25°$ and $w_0 = 578$ cm/sec; it gave the crater ellipticity as $\epsilon = 0.039 \pm 0.008$ (measured), to compare with a theoretical value of 0.046 according to Eq. (28).

For some of the impact experiments of Table III(c), on sites II, III, and V, the typical crater characteristics mostly averaged in the form in which they occur in Eq. (7), and are collected in Table IV. The second half of the table contains the relative diameter limits approached by throwout of the qualitatively described intensity.

F. Kinetic Efficiency and Throwout

In low velocity collisions of stone or other brittle substances, part of the kinetic energy is lost to heating, destruction, or rotation, so that the reflected translational kinetic energy is a fraction λ^2 of the original [see Eqs. (19) and (25)]. In the proposed cratering model (Fig. 1), the fraction is assumed to vary uniformly from 0 (at $x = x_0$) to λ^2 (at $x = 0$) at constant u, so that the average translational kinetic energy per gram of the ejecta at $y =$ const is $\frac{1}{4}\lambda^2 u^2$.

Experiments with stony projectiles falling on a massive stony surface from a height of 0.5–2 m and reflected from it gave $\lambda^2 = 0.23$ as an average. If an equal amount were stored in rotation, the coefficient of elastic reflectivity would be 0.46. The experiments consisted in measuring the length of flight [Eq. (45)]; on the assumption of $z = 45°$, the apparent values of λ^2 ranged from 0.08 to 0.43 ($n = 26$) with an apparent average of 0.180; allowing for a dispersion in z, a correction factor of $4/\pi$ was then applied.

The greater internal friction during cratering is likely to lead to a smaller value of λ than that of the simple two-body collision. From Table IV it can be seen that the ejecta spread over an extreme diameter of $6.8B_0$ (Experiment 26), or over a horizontal distance of $L \sim 47$ cm. The flight distance is given by

$$L = (v^2/g) \cdot 2 \sin z \cos z \qquad (45)$$

where $z = \beta$ (Fig. 1) is the zenith angle of ejection. From Eq. (4) and $s = 4.87 \times 10^4$, $\rho = 1.7$, and $w_0 = 1266$ cm/sec (as for that experiment), $u_s = 169$ cm/sec. With the condition $u < w_0$, instead of Eq. (17), Eq. (16) yields $y > y_Q = u_s/w_0 = 0.133$ in the present case. Taking $y = 2y_Q = 0.266$ as a middle value for top ejection, $u = 635$ cm/sec, $\sin \beta_0 = 0.8$ and $\sin \beta = 0.213$, $\cos \beta = 0.977$ according to Eq. (27), the top velocity of ejection according to Eq. (45) becomes $v = 333$ cm/sec. Hence $\lambda \simeq 333/635 = 0.527$ and $\lambda^2 = 0.275$ (with a considerable margin of freedom, however). It is perhaps an overestimate since for constant λ, u increases with decreasing y and the farthest throwout will come from the innermost portions, from $y \sim y_Q$. Now, taking $y = 0.15$, $u = 1126$ cm/sec, and $\sin z = 0.12$, we obtain $v = 440$ cm/sec, whence $\lambda = 0.391$ and $\lambda^2 = 0.153$. The grain size (0.4 cm, $m = 0.7$ gm/cm^2) was such that over the flight length of 47 cm or through an air mass of about 0.07 gm/cm^2, about 5% of the velocity would have been lost at the endpoint through air drag. A 2.5% increase in λ would be required, making $\lambda = 0.401$ and $\lambda^2 = 0.16$. This latter value is probably the best guess that can be made.

With the adopted cratering model [Eqs. (4), (16), (24), (25), (26), and (27)], the fraction f_b (fallback) of ejecta falling inside a radius $\frac{1}{2}B$ from the center of impact and originating along shock surface P of $u = $ const (Fig. 1) or $y = $ const is

$$f_b = \{[aby^2 + by(B/B_0) - y^{1/2}]/(1 - y^2 \sin^2 \beta_0)^{1/2}\}/(1 + aby^2). \quad (46)$$

The term $y^{1/2}$ represents conventionally the distance IE of the ejection point (Fig. 1) in units of B_0. When $f_b > 1$ is obtained, $f_b = 1$ is to be taken. Here

$$a = (8x_0 \sin \beta_0)/B_0, \quad (47)$$

so that ab accounts for the work of gravity in lifting the ejecta from the depth of the crater to its surface, and

$$b = gB_0\rho/(4\lambda^2 s \sin \beta_0) \quad (48)$$

takes care of the horizontal length of the trajectory.

The total deposition of ejecta inside $\frac{1}{2}B$ is obtained by numerical integration:

$$F_B = \int_0^1 f_b \, dy = y_Q f_Q + \int_{y_Q}^1 f_b \, dy. \quad (49)$$

Of the two parameters, b is by far the more important one. Figure 2 represents the function $\log(1 - F_B)$ as a function of $\log b$ for four selected values of a and for $\sin \beta_0 = 0.80$. The insert (a), valid for $a > 0.91$, represents $\log(1 - F_B)$ as a function of $\log(ab)$.

It may be noted that the destructive and ejective phenomena depend primarily on the properties of the rock target (ρ, s, u_S). For a crater of given dimensions, the distribution of the ejecta will be practically the same, whatever the velocity of the projectile, or whatever the origin of the crater—meteorite impact, high explosive, or nuclear blast if the charge is properly placed (not too deep and not too near the surface) so that a not too abnormal crater profile results. The difference in the origin of the blast, velocity and impacting mass, etc., would reveal itself chiefly in the central funnel Q, while over most of the crater volume this is irrelevant once the crater size is given. (The size, of course, is determined by the conditions around the center of impact.)

We chose the "Teapot" nuclear crater (Shoemaker, 1963; Nordyke, 1961), whose profile is reproduced in Fig. 1. The dimensions are: $2(I)L_3 = B_0 = 10{,}500$ cm; the ground-level bowl, $2IL = 9100$ cm; $x_0 = 0.8x_p = 3150$ cm; $x_c = 0.61x_0 = 1920$ cm. The target is "a loose sand–gravel mix with a density of 1.5–1.7 and a water content (at depth) of about 10%." We assume $\rho = 1.7$ *in situ* and $\rho = 1.5$ for the ejecta deposit, whose volume is thus to be multiplied by a factor of $1.5/1.7 = 0.883$ to reduce it to that of the parent matrix. In

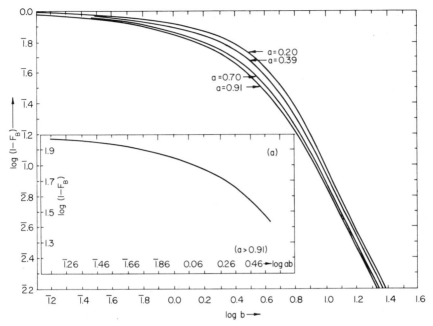

FIG. 2.   Throwout integral $1 - F_B$ (ordinates, logarithmic scale) as a function of log b for four discrete values of a (0.20, 0.39, 0.70, and 0.91). Insert (a) same as a function of log(ab), when $a > 0.91$. The parameters: $a - 8x_0 \sin \beta_0/B_0$; $b = g\rho B_0/(4\lambda^2 s \cdot \sin \beta_0)$; $\sin \beta_0 = 0.8$.

similar ground, the "Scooter" TNT explosion indicated a radial stress of 600 psi at a distance of 200 ft (Murphey, 1961) that, at the crater bowl (150 ft) and using an inverse-cube law for the stress, would correspond to lateral strength $s = 9.8 \times 10^7$ dyne/cm²; with a depth $x_c \simeq 1800$ cm, and $f_s = 0.78$ (a small frictional component of 2.1×10^6 would make little difference), $s_c = 9.6 \times 10^7$ dyne/cm² according to Eq. (11). Within $+100$ to -50%, this should hold also for the "Teapot."

Integrations, according to Eqs. (46) and (49), with $a = 1.92$, $b = 0.80$, and $\sin \beta_0 = 0.800$, gave the distribution of the ejecta as shown in Table V. The value for $B/B_0 \leqslant 1.000$ is the true fallback (F_b , Fig. 1). There is a systematic difference between the observed and calculated values that cannot be removed by using a different set of parameters. Namely, only the choice of b is to some extent free, while a and β_0 are prescribed by the crater profile. With a change in b, all the calculated values of F_B move in the same direction, so that an improvement at one end of the table will be countered by deterioration at the other. Some of the difference may be attributed to air drag, which forced the ejecta to deviate from the purely ballistic trajectories and descend at distances smaller than those of Eq. (45). The air drag on these massive ejecta does not depend on grain size but rather on the total mass m of the stream; if m_a is the traversed air mass, the loss in velocity is

$$\Delta v/v = -m_a/m = \Delta L/L.$$

The relative loss in the distance is then equal to the average loss $\Delta v^2/v^2$ over the entire trajectory, or to one-half the final loss, $\frac{1}{2}\Delta v^2/v^2$, which is $\Delta v/v$ as indicated. With this, and m determined from the thickness of the deposit (from 4 to 1 m), the ballistic distances are decreased to B'/B_0 as given in the bottom part of the table. The last two lines contain the comparison between observation and calculation with this refinement. The discrepancy is diminished but still persists. Nevertheless, for an *a priori* approach, the results are quite satisfactory.

With $b = 0.80$, $a = 1.92$, $\sin \beta_0 = 0.8$, Eqs. (47) and (4) yield

$$\lambda^2 u_s{}^2 = \lambda^2 s/\rho = 4.0 \times 10^6,$$

or, with $\rho = 1.7$,

$$\lambda^2 s = 6.8 \times 10^6$$

for the desert alluvium. With $s = 5$ to 20×10^7 as for "Scooter," one would obtain $\lambda^2 = 0.14$ to 0.034. Taking $\lambda^2 = 0.16$ as for the gravel craters, $s_c = 4.2 \times 10^7$ dyne/cm^2 can be regarded conventionally as the best estimate for the "Teapot" alluvium, the gravity friction correction amounting to a mere 2.5×10^6 dyne/cm^2.

III

PLANETARY ENCOUNTERS

The surface properties of the Moon cannot be interpreted well without considering its past history, beginning with its origin and followed by further exposure to collisions with stray interplanetary bodies. This purpose is basically served by the theory of interplanetary encounters, which in its original "linear" form (Öpik, 1951, 1963a), is concerned with very small collision cross sections as compared to the orbital dimensions. Supplemented by the consideration of acceleration in repeated gravitational "elastic" encounters (Öpik, 1966a), the theory requires essential modifications when dealing with rings of "plane-tesimals" orbiting in tightly packed, nearly circular orbits. The relative velocities are then small, the cross sections large, and, therefore, the linear approximation (i.e., treating the orbital arc segments near the points of encounter as straight lines) is no longer workable. Appropriate formulas for these cases are given for the first time further below [Eqs. (50)–(76)].

The linear approximation formulas for planetary encounters are as follows. In a Jacobian frame of the restricted three-body problem, a smaller body (planet, Moon—to be called "satellite" hereafter) of relative mass μ revolves around a central body (Sun, Earth—to be called "main body") of mass 1 ($1 - \mu$, more precisely) in a nearly circular orbit of radius 1 and period 2π, so that its orbital velocity is taken as unity, and the gravitational constant is also 1. A stray body (to be called "particle") when at distance 1, has a velocity U relative to the circular velocity of the relative mass μ in the units chosen, such that

$$U^2 = 3 - 2[A(1 - e^2)]^{1/2} \cos i - (1/A),\tag{50}$$

and a radial component U_r,

$$U_r^2 = 2 - A(1 - e^2) - (1/A),\tag{51}$$

where A is the semimajor axis, e the eccentricity, and i the inclination of the orbit of the particle relative to that of the satellite.

A particle that can pass at distance 1 "crosses" the orbit of the satellite without necessarily intersecting. Due to secular perturbations, precession of the node, and advance of the periastron of the satellite's orbit, resulting in a secular motion of the argument of periastron (perihelion, perigee) ω, a particle crossing the orbit of the satellite will intersect it twice during the period of ω, $t(\omega)$. For the Earth, in heliocentric orbit, the repetition interval is $\frac{1}{2}t(\omega) - 32{,}000$ y. The "probability" P_e (mathematical expectation) of encounter for one revolution of the particle is then

$$P_e = (\sigma_e^2/\pi \sin i)(U^2 + 0.44e_0^2)^{1/2}/(U_r^2 + 0.44e_0^2)^{1/2},\tag{52}$$

where e_0 is the eccentricity of the satellite's orbit and σ_e is the target radius or encounter (collision) parameter in units of the orbital radius of the satellite. For the inclination the average vector sum

$$\sin i = (\sin^2 i_p + \sin^2 i_0)^{1/2}\tag{53}$$

must be taken, where i_p and i_0 are the average orbital inclinations of particle and satellite, respectively, to the invariable plane of the system. However, when the encounter lifetime [see Eq. (59)] is shorter than one-half the "synodic" period of orbital precession, the "instantaneous" value of the inclination must be taken: a similar restriction holds for the term $0.44e_0^2$ in Eq. (52) with respect to the synodic period of the longitude of periastron.

For physical collision with the satellite of radius R_p (in the same relative units),

$$\sigma_e^2 = \sigma_0^2 = R_p^2[1 + (2\mu/R_p U^2)],\tag{54}$$

where

$$2\mu/R_\mathrm{p} = v_\infty{}^2 \tag{55}$$

is the square of the escape velocity from the surface of the satellite. For a complete gravitational "elastic collision," yielding a mean angular deflection of 90°, the cross-sectional radius is defined through

$$\sigma_\mathrm{e}{}^2 = \sigma_\mathrm{a}{}^2 = T_\mathrm{a}\ln[(R_\mathrm{a}{}^2 + T_\mathrm{a})/(\sigma_0{}^2 + T_\mathrm{a})], \tag{56}$$

with

$$T_\mathrm{a} = 16\mu^2/\pi^2 U^4, \tag{57}$$

where

$$R_\mathrm{a} = (\tfrac{1}{2}\mu)^{1/3} \tag{58}$$

is the radius of the "sphere of action" upon the particle of the satellite against the main body. This is not a clear-cut limit of action, but its use in logarithmic form renders this uncertainty unimportant. The average perturbation vector of the main body on the radial acceleration of the particle relative to the satellite is zero, so that there is virtually no limit of action, only a disarrangement by the perturbation.

The lifetime of the particle with respect to a given type of encounter with probability P_e is

$$t(\sigma) = 2\pi A^{1.5}/P_\mathrm{e}, \tag{59}$$

and the true probability of encounter during a time interval t is

$$\eta_t = 1 - \exp[-t/t(\sigma)]. \tag{60}$$

The validity of the linear approximation P_e is restricted to the case where the curvature of the arc of encounter is less than the target radius σ, which, for σ_0 (as for the Earth and the near-circular orbits of the particles), requires (Öpik, 1951)

$$e > 0.0063, \qquad \sin i > 0.0064, \qquad U > 0.0090 = 0.27 \quad \mathrm{km/sec}. \tag{61}$$

Further, provided that $\sin i$ and e exceed the target radius σ, the lifetime must exceed one-half the period of the argument of periastron (not the synodic period in this case), or

$$t(\sigma) > \tfrac{1}{2}t(\omega). \tag{62}$$

If this condition is not fulfilled,

$$t(\sigma) \simeq \tfrac{1}{2}t(\omega) \tag{63}$$

must be taken, unless a shorter lifetime not depending on the secular advance of ω is indicated [cf. Eqs. (69)–(73)].

The breakdown of the linear approximation leads to unreasonably high values of P_e in Eq. (52). In such a case, an upper limit P_m to P_e is set (Öpik, 1966a), still dependent upon the secular variation of ω, yet independent of the orbital elements e and i,

$$P_e \leqslant P_m = 2\sigma_e/3\pi U. \tag{64}$$

When this condition is not fulfilled, $P_e = P_m$ must be taken instead of P_e (unless superseded by another limit).

For small values of U, when $\sigma_a \gg \sigma_0$, repeated elastic encounters often bring the variable $\sin i$ and U_r values near to zero, so that sometimes Eq. (64) has to be used instead. The statistical mean probability P_e is then

$$(P_e)_{av} = P_0 = K_p \sigma^2/U, \tag{65}$$

where $K_p \simeq 3$ for heliocentric encounters with the Earth, and $K_p \simeq 2$ for those with Jupiter (Öpik, 1966a). Applying Eq. (64) to $P_e = P_0$ with $K_p = 3$, the condition for the validity of Eq. (65) becomes

$$\sigma_e < 2/9\pi = 0.0707 \quad \text{or} \quad \sigma^2 < 0.005; \tag{66}$$

otherwise P_m as in Eq. (64) must be used.

On the other hand, the target radii should not exceed the sphere of action, i.e.,

$$\sigma_e \leqslant R_a . \tag{67}$$

When any of Eqs. (54), (56), and (66) exceed this limit, $\sigma_e = R_a$ shall by convention, be taken; although action is not limited to this distance, it cannot be treated by the simple statistical model of two-body encounters; classical perturbational methods must then be used instead.

The preceding equations of encounter apply when the orbital range of the particle comes within the extent of the orbital range of the satellite, augmented by the target radius. The range of applicability is defined by the two conditions of full crossing that must be fulfilled simultaneously; when $e > e_0$, these conditions are

$$A(1 - e) < 1 - e_0 - \sigma_e$$
$$A(1 + e) > 1 + e_0 + \sigma_e + \cdots \tag{68}$$

and when $e < e_0$, the roles of particle and satellite are interchanged.

A fractional factor may sometimes be applied to the P values, allowing for partial crossing (Öpik, 1951, 1963a).

For very small values of U, such as those that would occur in preplanetary rings of planetesimals, an overall upper limit to the probability of encounter evidently is $P_e < 1$. However, two narrower limits exist, which cannot be surpassed; the average lifetime for an encounter, whatever its target radius, must be longer than the shorter of the following two: either one-half the synodic period of revolution t_s of the particle, or the time of unperturbed fall from a distance of $\sqrt{2}$ (middle of circular orbit) under the attraction of the satellite. Thus,

$$t(\sigma) > \tfrac{1}{2}t_s = \pi A^{1.5}/|\ A^{1.5} - 1\ | \tag{69}$$

or

$$t(\sigma) > t_F = \pi/(2.83\mu)^{1/2}, \tag{70}$$

where 2π is the orbital period of the satellite around the central body.

These cases occur only when the orbital semimajor axis of the particle is close to unity, or

$$A = 1 \pm \varDelta, \tag{71}$$

so that a linear approximation to Eq. (69) can be used,

$$t_s = 4\pi/3\ |\ \varDelta\ |. \tag{72}$$

The half period $\tfrac{1}{2}t_s$ is shorter than t_F [Eq. (70)] when

$$\varDelta^2 > 1.26\mu \tag{73}$$

and when $\tfrac{1}{2}t_s$ is the lower limit to $t(\sigma)$. When the condition of Eq. (73) is not fulfilled, t_F is the limit. When the lifetime, as calculated from Eq. (59), is shorter than the limit, either $\tfrac{1}{2}t_s$ or t_F must be substituted for it; the equivalent probability P_e can then be calculated from Eq. (59).

As can be seen, complications arise when U is small. A "Fermi-type" acceleration of the encounter velocity for a noncircular precessing orbit of the satellite (Öpik, 1965a, 1966a) could increase U sufficiently before a collision takes place, so that either Eq. (65) or (64) could apply. The acceleration is given by

$$d(U^2)/dt = 1.23F(\omega)(0.625e_0{}^2 + \sin^2 i_0)/t(\sigma_a) \tag{74}$$

in former notations, where $1.23 = \pi^2/8$ and $F(\omega) = 1$ when $t(\sigma_a) > \tfrac{1}{2}t(\omega)$. When the deflection lifetime is shorter than $t(\omega)$,

$$F(\omega) = 4[t(\sigma_a)/t(\omega)]^2. \tag{75}$$

While being deflected and accelerated in the elastic gravitational encounters, the particles are removed by physical collisions, so that they virtually disappear before a certain average value of U is reached. The fraction surviving is

$$\zeta(U)_2 = \exp\left[-\int_0^t (\sigma_0/\sigma_a)^2\ dt/t(\sigma_a)\right], \tag{76}$$

when U is accelerated from U_1 at $t = 0$ to U_2 at t. In this equation, $F(\omega)$ does not appear and there is no restriction depending on lifetime, although it enters implicitly through Eq. (74), from which the interval t is to be determined.

When U exceeds the critical value of $\sqrt{2} - 1 = 0.414$, additional depletion of the particle population begins through ejection from the system by way of hyperbolic orbits.

For particles encountering the Earth with low initial encounter velocities, $U_1 = 0.1$ (3 km/sec), rapid depletion by physical collisions prevents 99.9% of the particles from reaching $U > 0.3$ (9 km/sec). The average encounter velocity of such particles, when captured by the Earth or the Moon, is then $U = 0.178$ (5.3 km/sec) (Öpik, 1965a, 1966b). Of course, gravitational action will increase this value to a collision velocity of about 12 km/sec for the Earth and 5.8 km/sec for the Moon. The fraction accelerated above $U = 0.414$ is less than 10^{-13}, so that ejection is negligible, the entire population being removed in collisions. For Jupiter, however, the conditions are very different, and so are those for the Moon with respect to earthbound orbiting particles.

The preceding equations apply to free orbiting particles. In a preplanetary ring, mutual collisions and drag will reduce U to very small values and will also prevent the acceleration mechanism from working (Öpik, 1966a), conditions which must have prevailed during the origin of the Moon. Also, when the "particle" is no longer of infinitesimal dimensions as compared to the "satellite," the radius $R_p = R_1 + R_2$ and the mass $\mu = \mu_1 + \mu_2$ must be taken as the sum of the values for the two colliding or interacting bodies, i.e., the satellite and the particle.

IV

THE ORIGIN OF THE MOON

A. Theoretical and Observational Basis: The Alternatives

The events that shaped the present surface of the Moon must be traced to the very origin of our satellite as an individual body. Three principal modes of origin have been envisaged.

(a)   The fission theory proposed by Sir George Darwin, which at present has fallen into disrepute although without convincing reason;

(b)   the theory of formation from a swarm of planetesimals orbiting the Earth, simultaneously with the formation of our planet (Schmidt, 1950; Öpik, 1962a); and

(c)   the theory of capture, suggested by an extension of Darwin's calculations backward by Gerstenkorn (1955) (Öpik, 1955, 1962a), and recently sponsored by Urey (1960a) and Alfvén (1963, 1965).

As will be seen, there may be more variants of these typical hypotheses.

Hypothesis (b), originated by Schmidt (1950), has been strongly supported by Russian astrophysicists, e.g., Ruskol, Levin, and others. Levin (1966a) provides a fair survey not only of the work of Schmidt's school in this direction, but also of work done elsewhere on hypothesis (c); he rejects hypothesis (a) outright because of the impossibility "of the smooth separation of a rotating fluid mass." The objection to hypothesis (a) holds only if a ready-made Moon is supposed to be the end product. However, the products of fission, broken up into numberless fragments inside Roche's limit, could later on gather and recede, leading thus to a variant of hypothesis (b) (Öpik, 1955).

Observational data, based on the statistics of ellipticities of lunar craters and the geometry of tidal deformations (Öpik, 1961b), point with a good (though not overwhelming) probability to the craters in the lunar continents having formed at a distance of from 5 to 8 Earth radii, thus supporting hypothesis (b) as outlined by Schmidt. While the upper limit is uncertain, owing to the statistical error of sampling, the lower limit is well determined. It is pretty certain that the lunar craters in the highlands were not formed at a distance closer than 4 Earth radii.

After Gerstenkorn, retrospective calculations of the evolution of the Moon's orbit have been made by MacDonald, Slichter, and Sorokin, with very different results, depending on the assumed parameters (Levin, 1966a). All these point to a minimum distance somewhere near or inside the present Roche's limit, $2-5 \times 10^9$ years ago. However, the history of the Moon preceding this minimum distance or "zero hour" cannot be decided mathematically because not only the tidal friction parameters but even the masses of the interacting bodies themselves could have been variable and their identities unknown. [There could have existed several Moons, of which only one survived; and the Moon may never have gone through this stage at all (Öpik, 1955).] It is reasonable to assume that zero hour was some time near the beginning of the solar system, 4.5×10^9 years ago. At that time, the mass of the Earth was accumulating, and capture of the Moon could have taken place at close approach into any near-parabolic orbit (and not necessarily into a retrograde one) by nontidal trapping through increase of the Earth's mass and loss of momentum in collisions during the passage.

It must be emphasized that direct condensation of the moon from a gaseous state is a rather incredible proposition. Even if the required extremely low temperature and high density of the gas prevailed, the Earth would have profited from it first, turning into a giant planet like Jupiter. Accretion of particulate matter is the only reasonable way the Moon could have come into being. The impact velocities must not have exceeded 11 km/sec, otherwise loss of mass instead of accretion would have resulted (Öpik, 1961a) for the present lunar mass; a lower limit down to 2 km/sec and less must be set for the growth of a smaller mass [cf. Eq. (22)]. It is therefore imperative that accretion must have taken place from some kind of a ring of solid particles in which the relative velocities were small.

B. Mass Accumulation from Orbiting Debris

Even in the capture hypothesis of the Moon, it must have entered the sphere of action of the Earth on a near-parabolic relative orbit, or $U \simeq 0$. According to Eq. (50), this requires $A = 1$, $e = 0$, and $i = 0$. The Moon must have formed on the same circular orbit with the Earth inside the preplanetary ring and from the same material. Any hope to find on the Moon cosmic material of different origin than that of terrestrial material is thus not justified. Also, the time scale of the major accumulation, or depletion of the preplanetary ring, was determined by the Earth as the major body.

In the preplanetary ring, a remnant of the solar nebula, the original cosmic distribution of the elements with the predominance of hydrogen must have prevailed. Jupiter and the outer planets apparently have incorporated hydrogen, helium, and other volatiles in cosmic proportion, while the terrestrial planets consist, to 99.9%, of nonvolatile silicates and iron. If in cosmic proportion, the Earth would have captured about 100 times its mass in hydrogen, enabling it to keep, gravitationally, this and other volatiles at any imaginable temperature. Therefore, the gaseous constituents of the nebula must have been swept away somehow from terrestrial space before being sucked into the Earth, while the refractory materials gathered into a common plane, i.e., into a thin sheet similar to Saturn's rings. For a ring spread from 0.9 to 1.1 a.u., over a width of 0.2 a.u., the total mass of the Earth–Moon system would correspond to a mass load of $m_0 = 21.3$ gm/cm² over the orbital plane. A spherical planetesimal of density $\delta = 1.3$ (cometary nucleus without the ices) and radius R_c (cm) has a mass load

$$m_c = 4R\delta/3 \tag{77}$$

or $1.73R_c$ gm/cm². The damping lifetime of the relative velocity U at orbital inclination i of a particle that has to pass through the ring twice during 2π Jacobian units of time or one orbital revolution (a year) is, in the relative units chosen

$$t_z = \pi m_c \sin i / U m_0, \qquad t_z/2\pi = \tfrac{1}{2} m_c \sin i / U m_0 \quad \text{yr} \tag{78}$$

and the damped value of U after a time interval t is

$$U_2 = U_1 \exp(-t/t_z). \tag{79}$$

The orbits of the planetesimals when perturbed will rapidly become circles again while the Jacobian velocity decays on a time scale of

$$t_z/2\pi = 0.02R_c/\eta_m \quad \text{yr} \tag{80}$$

for a typical case of $\sin i/U = 0.5$. Here η_m is the fraction of the total mass in the ring that has not yet been accreted by the planet. Thus, for a typical projectile producing a crater about 10 km in diameter, $R_c \sim 1$ km and $t_z/2\pi = 2000/\eta_m$ yr, thus short by cosmogonic standards.

Damping is very much greater for $i = 0$, when the U vector is in the plane of the ring. In that case, instead of the cross sections, the linear encounter diameter $\sim 4R_c/3$ is the sweeping unit (R_c is assumed to be greater than the thickness of the ring, and slightly displaced from its plane); the linear load of the planetesimal is then

$$m_c' = \pi R_c^2 \delta \tag{81}$$

or $4.1 R_c$ gm/cm. Over a path dL it sweeps a mass $\eta_m m_0\, dL$ per centimeter. The radial displacement being $dL_r = |\, U_r\, |\, dL/U$, the radial damping length then becomes

$$L_r = |\, U_r\, |\, m_c'/U m_0 \eta_m \tag{82}$$

or typically, with $|\, U_r\, |/U_0 = 0.5$ and $m_0 = 21.3$,

$$L_r = 0.1 R_c^2/\eta_m \quad \text{cm}, \tag{83}$$

and the damping time is (independent of the ratio U_r/U)

$$t_r = 2L_r/(U \cdot 3 \times 10^6)\ (\text{sec}) = 2 \times 10^{-15} R_c^2/(U\eta_m)\ (\text{yr}). \tag{84}$$

For $R_c = 10^5$ cm as before, $L_r = 10^9/\eta_m$ cm and $t_r = 2 \times 10^{-5}/U\eta_m$ yr.

The damping would be highly efficient and, unless disturbed by the growing Earth or other centers of condensation, the particles of the ring would all move in coplanar circular orbits and mutual coagulation would stop when they were touching side-by-side, as envisaged by Jeffreys for Saturn's rings (Jeffreys, 1947b). With small particles, an almost continuous disk would thus be formed, which, from orbital friction and gravitational instability, would then break up into larger planetesimals through coagulation of neighboring regions. When their size and damping time were sufficiently large, they could be collected gravitationally by the growing planetary nucleus when planetary perturbations diverted them into its path. Also, perturbations would change the orbital elements e and i of the Earth's nucleus, thus increasing its range of heliocentric distance and sweeping ability. Encounters with other massive nuclei would also lead to changes in the orbital elements.

Disregarding damping at first, the Earth could collect the particles from the ring only when their circular orbits were perturbed so that they could cross the orbit of the Earth. An exception would be those that lay within a range from $1 + e_0$ to $1 - e_0$ heliocentric distance, where e_0 is the eccentricity of the Earth's orbit.

From Eq. (50), it can be shown that, for $A = 1 + \Delta A$, $i = 0$, and $A(1 - e) = 1$ just sufficient for orbital crossing, the encounter velocity becomes (to terms of the second order)

$$U^2 = \tfrac{1}{4}(\Delta A)^2 \quad \text{or} \quad U = \tfrac{1}{2}|\Delta A|. \tag{85}$$

Hence, when perturbations or collisions induce particles from orbit A to cross, and thus subject them to chances of collision, the U parameter will be close to that of Eq. (85).

For the envisaged ring, U values up to 0.05 would thus be expected, with an average about $0.025 = 0.75$ km/sec. In such a case, for bodies even much smaller than the Earth, with an escape velocity $v_\infty > 1.5$ km/sec, the unity term in Eq. (54) can be dropped, and the collisional cross section of the growing Earth would then become [as from Eq. (54)]

$$\pi \sigma_0^2 = \pi \cdot 2.63 \times 10^{-10}(1 - \eta_m)^{4/3}/U^2, \tag{86a}$$

and

$$\sigma_0 = 1.62 \times 10^{-5}(1 - \eta_m)^{2/3}/U. \tag{86b}$$

With the collision probability from Eq. (65), which holds, the corresponding collision lifetime from Eq. (59) would be

$$t(\sigma)/2\pi = 1.27 \times 10^9 U^3/(1 - \eta_m)^{4/3} \quad \text{yr.} \tag{87}$$

For $U = 0.025$ corresponding to $A = 1.05$ or 0.95 as the median for the ring, and $\eta_m = 0.5$, the lifetime would be 50,000 yr; at the most, this might attain 400,000 yr. The period of ω may set a lower limit of 30,000 yr.

Provided perturbations were available soon enough—which may not have been the case at all–a minimum time scale for accretion of the Earth may be set at 50,000 yr. The effective time may be several times longer.

One source of the perturbations would be the Earth itself, which would pass the particles at the close range of ΔA during a synodic period

$$t_s/2\pi = (1.5\,\Delta A)^{-1} \quad \text{yr,} \tag{88}$$

to first-order approximation. During this period the eccentricity would be excited by Earth's periodic perturbations to a value of about

$$e' \simeq 2\mu(1 - \eta_m)/\pi(\Delta A)^2, \tag{89}$$

the perihelion or the direction of the e' vector revolving with the synodic period. To reach the Earth's orbit, $e' = |\Delta A|$ is required, which yields

$$|\Delta A|' = [2\mu(1 - \eta_m)/\pi]^{1/3} = 0.0124(1 - \eta_m)^{1/3}, \tag{90}$$

or practically the radius of the sphere of action of the accumulated nucleus
[Eq. (58)]

$$R_a = 0.0115(1 - \eta_m)^{1/3}. \tag{91}$$

A secular increase of the semimajor axis of the particle's orbit with a time scale of

$$t_A = \Delta A/(dA/dt) \simeq (\Delta A)^4/24\mu^2(1 - \eta_m)^2$$
$$= 5 \times 10^9 (\Delta A)^4/(1 - \eta_m)^2 \quad \text{yr}, \tag{92}$$

would give $t_A = 3200$ yr at $\Delta A = 0.02$, 1.3×10^5 yr at $\Delta A = 0.05$, and 2×10^6 yr at $\Delta A = 0.10$. This would have the effect of moving away the outer portion of the ring and bringing nearer the inner portion.

Of course, with the distribution of masses in the solar system already settled, perturbations by the other planets would add to the effect. The time scale of secular perturbations here is of the order of 50,000 years (half period, quite sufficient except for their small amplitude, only 0.05 in the eccentricity).

To make perturbations (including acceleration) work, damping must be overcome. For the periodic perturbations, $\frac{1}{2}t_s < t_z$ or t_r is required. From Eqs. (88), (80), and (83) we thus obtain, for $U = 0.025$, $\eta_m = 0.5$, $\Delta A = 0.05$: $R_c > 170$ cm from t_z, but $R_c > 65$ km from t_r. Moderate clumping would be needed to counteract damping, provided the perturbations included inclination. The case of $i = 0$ (with the sheet of particulate matter thinner than the diameter of the planetesimal) is too extreme, and the clumping limit too high to be considered: there would always be some deviation at right angles to the plane, $i \neq 0$.

For long-period perturbations, including those in i, to be effective, we find $R_c > 12.5$ km for $t_z > 50,000$ yr. Below this value, the particles of the ring must respond to the perturbations somehow in a cooperative way.

It seems that, with a secular amplitude of about 0.05 in the eccentricity of the Earth, with a similar value of e for the larger particles above the damping limit (caused by perturbations of the major planets), and with additional perturbations by the Earth in close passages, the particles may be accreted indeed at an average encounter velocity of $U = 0.025$ and a time scale of 50,000 yr,

$$\eta_m = \exp(-t/50,000) \tag{93}$$

being the unaccreted fraction left in the ring after a lapse of t years.

C. Capture Hypothesis of the Origin of the Moon

Moon Formed Independently and Captured by Nontidal Process

The increment of mass of two bodies placed in the same medium is proportional to their collisional capture cross section $\pi\sigma_0^2$. For low velocities of encounter, the unity term in Eq. (54) can be disregarded; the rate of accretion of two independent nuclei of equal density (for the sake of simplicity) is then proportional to the 4/3 power of the mass. The differential equation of growth of two independent centers of accretion can be integrated and the result represented as a variable ratio of the masses, i.e.,

$$\mu_1/\mu_2 = (1 + C\mu_1^{1/3})^3. \tag{94}$$

With the adjustable parameter $C = 3.34$ and $\mu_1 = 1$, $\mu_1/\mu_2 = 81.5$ is obtained as for the present mass ratio of Earth to Moon. Table VI, then, represents the variation of the mass ratio as a function of the value of μ_1—the variable mass of the Earth in the course of accretion.

Thus, going backward in time during the process of accretion, the mass ratio decreases. At $\mu_1 = 10^{-3}$, when the radius of the Earth was one-tenth its present value, the mass ratio was only 2.37. The initial difference in the size of the nuclei could have been very small, just a matter of chance. Also, in the beginning there could have been many competing nuclei of comparable size.

If J_m is the mass accretion per unit of surface area and time, w the impact velocity, T_0 the original temperature of the accreting material in space, T_e the surface radiation temperature, c_1 the average specific heat of the solid, and $k_s = 5.67 \times 10^{-5}$ Stefan's radiation constant, the subsurface temperature T_s of the accreting material will more or less satisfy the equation

$$J_m[\tfrac{1}{2}w^2 - c_1(T_s - T_0)] = k_s(T_e^4 - T_0^4). \tag{95}$$

For silicate material, $c_1 = 9 \times 10^6$ erg/gm-K, also $T_0 = 300$ K can be assumed. Because of surface shielding and finite conductivity of the solid,

$$T_s > T_e > T_0. \tag{96}$$

Setting $T_e = T_s = T_s'$ in Eq. (95), a lower limit for the temperature is obtained. An upper limit T_s'', corresponding to zero radiation losses, obtains for $T_e = T_0$. However, when the temperatures reach values above $T_m = 1800$ K, the temperature of fusion, Eq. (95) does not apply. When the lower limit T_s', is below the fusion limit and $w < w_f \simeq 1.64$ km/sec, fusion cannot take place even at complete shielding, and the upper limit is then

$$T_s'' = T_0 + (w^2/2c_1). \tag{97}$$

When $w > w_f$, in the case of extreme shielding, partial fusion must take place. Let θ be the melted fraction, and let the same fraction of the surface be unshielded liquid (lava) radiating with the intensity

$$Q_0 = k_s(T_m^4 - T_0^4) = 6 \times 10^8 \quad \text{erg/cm}^2\text{-sec}, \tag{98}$$

the rest of the surface being completely shielded (e.g., by insulating dust) and at $T = T_0$. The maximum melted fraction (on the surface as well as in the subsurface) is then

$$\theta_{max} = J_m(\tfrac{1}{2}w^2 - H_0)/(Q_0 + H_f J_m) \leqslant 1, \tag{99}$$

where $H_0 = 1.35 \times 10^{10}$ erg/gm is the heat required to raise the temperature from T_0 to T_m, and $H_f = 2.7 \times 10^9$ erg/gm is the heat of fusion.

When θ exceeds 1, complete fusion takes place. The liquid is assumed to radiate to space unshielded at $T_e = T$, and probable temperature T (not a limit) is then determined by the equation

$$J_m[\tfrac{1}{2}w^2 - (H_0 + H_f) - c_2(T - T_m)] = k_s(T^4 - T_0^4), \tag{100}$$

where c_2 is the specific heat of the liquid.

Over the short time scale of accretion, conductive exchange of heat with the interior will not greatly change the results.

For a planet of density δ and radius R_p accreting on a time scale of $t(\sigma)$, the accretion is

$$J_m = \tfrac{1}{3}R_p\eta_m\delta/t(\sigma)(1 - \eta_m). \tag{101}$$

Neglecting the small role of the independently accreting Moon, at $\eta_m = 0.5$, $R_p = 5.1 \times 10^8$ cm, $\delta = 5.5$, $t(\sigma) = 50{,}000$ yr, we find $J_m(\text{Earth}) = 6.6 \times 10^{-4}$ gm/cm²-sec falling at a velocity of 8.4 km/sec upon the half-mass Earth. For the Moon at 1/50th of the Earth's mass (cf. Table VI), the accretion per unit area at constant U, $\sim\sigma_0^2/R_p^2$ [Eqs. (54) and (65)], is 1/14 that for the Earth or $J_m(\text{Moon}) = 4.7 \times 10^{-5}$ gm/cm²-sec. With $w_\infty = 2.0$ km/sec for the Moon at that epoch (mass $= 0.6$ of present Moon), and $U = 0.025 = 0.75$ km/sec, $w^2 = 4.6 \times 10^{10}$ (cm/sec)² and $w = 2.1$ km/sec is the velocity of fall.

With these data, for the independent Moon at epoch $\eta_m = 0.5$ of accretion and a time scale of 50,000 yr, Eq. (95) yields $T_s > T_s' = 404$ K, thus a low minimum value of the temperature, although heating is not negligible. The true temperature would be near this value for continuous accretion of finely divided material that does not penetrate deep into the surface.

The other extreme, e.g., conditioned by an insulating dust layer of low thermal conductivity covering every bit of a solid area, would allow heating of the bulk of the mass to nearly 1800 K. According to Eq. (99), the fraction melted as well as the fraction of exposed molten silicates would then be only

$$\theta_{max} = 1.6 \times 10^{-3}.$$

A hot solid body with some lava enclosures and exposures, just sufficient to radiate away the extra heat, could be envisaged. The lava exposures act as a thermostat, keeping the mean temperature near the melting point without complete melting.

The craters in the lunar continentes correspond to the accretion of the top fraction of about 3×10^{-5} of the lunar radius or 9×10^{-5} of the mass (Öpik, 1961b). At that stage, the collision cross sections of Earth and Moon were in a ratio of 280 to 1 [as they are now, cf. Eq. (54), with $U = 0.025$], so that there was left unaccreted in the ring a fraction

$$\eta_m = 9 \times 10^{-5} \times 281/(81.5 + 1) = 3.1 \times 10^{-4}.$$

According to Eq. (93), this would require a time interval of about 400,000 yr for the beginning of the formation of craters that have survived, and 600,000 yr for the practical termination of this primeval crater-forming epoch, as reckoned from the epoch of half-accretion ($\eta_m = 0.5$). Accretion must have been slower in the beginning, before sizeable nuclei were formed, and the total length of accretion into the Earth–Moon system may have lasted about one million years [20 times $t(\omega)$, according to a certain model (Öpik, 1961b)].

A nontidal capture of the Moon into a direct orbit could have taken place most probably when accretion was intense, thus not at the very last stage. The craters would then have been formed on a Moon in orbit around the Earth. Whatever its original distance of closest approach was, in 25,000–100,000 yr it must have receded tidally to 12–15 Earth radii. The majority of the craters could not have been formed at 5–8 Earth radii, and their tidal distortions (inversely proportional to the cube of the distance) would have been 1/10th those measured (Öpik, 1961b), or entirely negligible.

A stronger objection comes from crater statistics. Boneff and Fielder have shown that the craters are more or less evenly distributed over the Moon's surface (continentes and maria taken separately). Contrary to expectation, the western hemisphere, which is trailing behind, even carries about 10% more craters per unit area than the eastern, which is preceding in the orbital motion (Fielder, 1965, 1966). In view of the great differences in crater densities over the Moon's surface, the small excess is not very relevant and may be caused by unequal maria flooding. Now, with the craters imprinted when the Moon was at about 10 Earth radii, at an orbital velocity (full Earth mass being attained) of 2.5 km/sec and isotropically distributed hyperbolic velocity of the infalling fragments of 3.5 km/sec, strong aberration and bias toward the eastern hemisphere should have resulted. Under these circumstances, an approximate calculation, based on encounter equations, that considers the crater numbers to increase inversely as the square of the limiting diameter or, for fixed crater diameter, as the velocity [Eq. (7)], indicates that an excess of 74% is expected for the entire eastern over the entire western hemisphere of the Moon, instead of a deficiency of 10% as observed. The crater statistics are therefore incompatible with this model of formation of the Moon.

For the Earth, Eqs. (95) and (99) with $t(\sigma) = 50{,}000$ yr and a half-mass $\eta_m = 0.5$, yield

$$T_s > T_s' = 1410 \quad \text{K}, \qquad \theta_{max} = 0.372.$$

The two extremes are in this case not very different. A partially molten Earth is indicated, with oceans of lava that must have considerably influenced the tidal history of the Moon (if it was near the Earth at that time). Otherwise these figures stand irrespective of the history of the Moon; they depend only on the time scale of encounters.

D. Accretion of an Earth-Orbiting Moon from Interplanetary Material

On this model, the overall frame of accretion of the Earth–Moon mass is the same as in Sections IV,B and C, but the Moon is now supposed to have started from a nucleus already placed in orbit around the Earth. The Moon is now the "satellite," the Earth the "main body" of our model, but the particles are now entering into hyperbolic orbits with respect to the Earth–Moon system and the equations of encounter probability per revolution of the particle are no longer valid. Instead the following obvious equation, an exact equivalent of those for elliptic orbits, applies. The total accretion rate on a moving "satellite" is

$$A_p = \pi R_p{}^2 \rho v (1 + w_\infty{}^2/v^2), \qquad (102)$$

where ρ is the space density of the particles and v their (average) velocity relative to the satellite (Öpik, 1956). Also

$$J_m = A_p/4\pi R_p{}^2 = \tfrac{1}{4}\rho v(1 + w_\infty{}^2/v^2). \qquad (103)$$

For accretion by the half-mass Earth ($\eta_m = 0.5$), $\rho = \rho_0$ is the average density of matter in the ring, $v = U = 0.75$ km/sec, $w_\infty = 8.4$ km/sec (Earth). For accretion by the Earth-orbiting Moon at 10 Earth radii, with v as the vector quadratic sum of the Moon's orbital velocity (2.5 km/sec) and the velocity of escape from 10 Earth radii w_∞' (3.5 km/sec) or $v = 4.30$, $\rho = \rho_0[1 + (w_\infty')^2/U^2]^{1/2}$ (Öpik, 1965b), the new value for accretion on the Moon as "helped" by the Earth now becomes 5.8 times greater than for the "independent" Moon, $J_m = 2.73 \times 10^{-4}$ gm/cm²-sec. The impact velocity, with $w_\infty = 2.0$ km/sec for the Moon, is now $(v^2 + w_\infty{}^2)^{1/2}$ or $w_0 = 4.79$ km/sec.

With these numerical data, for the "Earth monitored" Moon at 10 Earth radii and $t(\sigma) = 50{,}000$ yr,

$$T_s > T_s' = 850 \quad \text{K} \qquad \text{and} \qquad \theta_{max} = 0.046$$

is obtained. The minimum temperature turns out to be quite high and, if its solid surface is well insulated (or thick enough), 4.6% of melting should occur on the Moon which is kept "thermostatically" close to the temperature of fusion.

Otherwise the two objections pointed out in the preceding section and based on tidal deformations of the craters and especially on crater counts, apply here, too, rendering the model highly improbable.

E. Capture into a Retrograde Orbit

Retrospective calculations of the tidal evolution of the lunar orbit, on the assumption of invariable masses of Moon, Earth, and Sun, and an absence of other relevant interacting bodies, all lead to a minimum distance close to, yet inside, Roche's limit D_r, as given by

$$D_r = 2.46R_0(\delta_0/\delta_p)^{1/3}, \tag{104}$$

where R_0 and δ_0 are radius and density of central body (Earth) and δ_p is the density of the satellite (Moon). For the Moon and the present ratio of the densities (5.52/3.34), $D_r = 2.88$ Earth radii. With the effect of solar tides, Gerstenkorn (1955) obtains 2.86, MacDonald (1964) 2.72, and Sorokin (1965) 2.40 Earth radii for the minimum distance of the Moon based on these assumptions. On the assumption of an unbroken Moon, the calculations, extended further backward (Gerstenkorn, 1955), indicate capture into a retrograde nearly parabolic orbit at a perigee of 26 Earth radii, which then decreases, the orbital eccentricity decreasing and the inclination turning from retrograde over 90° to direct (Öpik, 1955, 1962a). We can thus distinguish an incoming phase, with the Moon approaching, and the present outgoing phase, with the Moon receding.

It seems now that, if the minimum distance were inside Roche's limit, the Moon could not have existed as an integral body, and that the calculations beyond that point could not strictly apply. Yet, when a finite number of fragments had formed (see below), orbital evolution must have been slowed down without the geometry being essentially different. Through collisional damping, the fragments were forced to stay on the same orbit, and the calculations are therefore formally valid except for the time scale. Assume, therefore, that an independently accreted body of lunar mass was tidally captured by the finally accreted Earth into a retrograde orbit and went through Gerstenkorn's incoming phase until it broke up while in a circular direct orbit (as the calculations indicate). At this moment, tidal evolution was greatly slowed down (by a factor of $N_f^{1/2}$, where N_f is the number of fragments) yet did not stop completely. The reason for this would be the strength of the solid lunar body, which must have led to fragments of finite size being formed in the breakup, as visualized by Jeffreys (1947a). The upper limit of the radius R_f of the fragments, when formed at a distance $D_f < D_r$ inside Roche's limit, is given by (Öpik, 1966c)

$$R_f \leqslant [sD_f^3/\pi G \delta_p \delta_0 R_0^3]^{1/2}, \tag{105}$$

where G is the gravitational constant and s the "lateral" crushing strength as used in Eq. (7), practically equal to s_c of Eq. (11). In cgs units, with $s = 2 \times 10^8$ dyne/cm² as for sandstone, and $D_f/R_0 = 2.5$, $R_f = 2.86 \times 10^7$ cm or a diameter of 572 km for the surviving fragments, about one-sixth that of the Moon. The number of fragments, if of equal size, would then be $N_f = 224$. At the strength of granite, $s = 9 \times 10^8$, $R_f = 6.07 \times 10^7$ cm, and $N_f = 23$. We will further consider only the first case. If released in synchronous rotation from a circular orbit, the fragments will enter elliptical orbits with the encounter velocity ranging from $U = 0$ to $U = |(1 \pm R_p/D_f)^{1.5} - 1| \simeq 1.5 R_p/D_f \simeq 0.164$, according to the distance from the center of the parent body. (There is no significant tidal deformation of the brittle solid body before it yields to the ultimate stress.) Fragments released from the earthward side would reach a perigee distance of 1.2 Earth radii if the attraction of the Moon mass on the released fragments were neglected but were actually farther out, and would break up to somewhat smaller sizes; similarly, those from the far side would have their perigees there and go out in elliptical orbits to apogees of considerably less than 5.9 Earth radii, being bent inward by the attraction of the residual lunar mass.

For free orbiting fragments at 2.5 Earth radii (in notation and units of Section III) and for collisions of two equal particles, $R_p = 2R_f = 0.0360$, orbital circular velocity 1 (4.93 km/sec), orbital period 0.23 days, $w_\infty = 0.079$ (0.392 km/sec) equal to average $U = 0.079$, the collision cross section is

$$\pi \sigma_0^2 = 2.6 \times 10^{-3}\pi, \qquad \sigma_0 = 0.0510 < 0.0707;$$

therefore Eq. (65) applies with $K_p = 2$, yielding $P_0 = 0.066$, $t(\sigma)/2\pi = 15$ orbital revolutions or 3.5 days. As to $t(\omega)$, the solar perturbation is insignificant and the only important effect stems from the oblateness of the Earth, which yields (Öpik, 1958b), at a distance of a_m/R_0 Earth radii and for an orbit of small eccentricity and inclination,

$$t(\omega) = 18(a_m/R_0)^{3.5} \quad \text{(Earth rot/24}^{\text{h}})^2 \tag{106}$$

in days; it equals one-half the period of precession of the nodes. With 4.8 hr as the period of rotation of the Earth at that epoch, $t(\omega) = 18$ days $> 2t(\sigma)/2\pi$. Hence the collision lifetime of an isolated pair of fragments would equal 9 days. With 100–200 fragments present, in a matter of hours mutual collisions would completely destroy the fragments that originally survived tidal disruption.

Originally, the fragments could be imagined to be injected into a ring about 4000 km wide or thick and 10^5 km in circumference. With $N_f = 224$, this yields a number density of $N = 1.8 \times 10^{-10}$ km^{-3}. The collisional cross section $\pi \sigma_0{}^2$ is 2.1×10^6 km^2. Hence the collisional mean free path becomes $(N \cdot \pi \sigma_0{}^2)^{-1} = 2600$ km. This is of the order of the diameter of the Moon and, therefore, collisions are not restricted to particles of neighboring origin; the full variety of encounter velocities and full gravitational interaction will be realized as has been assumed.

With $w = 5 \times 10^4$ cm/sec, $s = 2 \times 10^8$, $\rho = 3.3$, and $K = 1$, Eqs. (4) and (14) yield

$$M_c/\mu = 8$$

for the relative mass of secondary fragments when the target is much larger than μ. Here, fragments of comparable dimensions are colliding; they will be destroyed completely in the first collision, and subsequent collisions will reduce the entire mass to rubble and dust, collected in a ring whose sections are orbiting separately. Let the equatorial velocity of synchronous rotation of the parent body (0.538 km/sec) be u_0; then the ultimate heating of the mass can be assumed to correspond to the average kinetic energy of rotation, $\frac{1}{5}u_0{}^2$ erg/gm which, at $c_1 = 9 \times 10^6$ erg/gm, yields only about 60 C.

With the proportions approximately those of Saturn's inner ring, extending from 2.25 to 2.75 Earth radii, or with a surface of 3.18×10^8 km^2, the average mass load per unit surface of the ring is 2.31×10^7 gm/cm^2; at average density $\delta = 2$ for the rubble, the average thickness is 115 km.

Now, even with the low cohesion such as that of sand, clumps of dimensions smaller than R_f [Eq. (105)] will be formed again. At incidental contacts, friction at the interfaces of the independently orbiting sections may force the clumps to rotate in a retrograde direction, with an angular velocity up to

$$\omega_f = \tfrac{1}{2}\omega_0, \tag{107}$$

where ω_0 is the orbital angular velocity, and

$$\omega_0{}^2 = \tfrac{4}{3}\pi G \delta_0 (R_0/D_f)^3. \tag{108}$$

The average tensile centrifugal stress in a rotating sphere of radius R_f is (Öpik, 1966c)

$$s_t = -\tfrac{1}{4}\omega_f^{2'} \delta R_f{}^2 \tag{109}$$

and, after substituting R_f and ω_f from Eqs. (105), (107), and (108),

$$s_t = s/12 \tag{110}$$

is obtained. The ratio is of the order of the tolerance of most brittle materials, whence no separate consideration of the survival of the clumps from the standpoint of tensile stresses is needed.

The ring is to stay for several hundred years at least before it is pulled outward by the weak tidal acceleration [cf. Eqs. (111)–(113)]. Its separately rotating parts will probably possess the mechanical properties *in vacuo* similar to, or slightly harder than, desert alluvium; from Section II,F we may set $s = 6 \times 10^7$ dyne/cm² and $\rho = 2$ gm/cm³ for these "orbiting sand dunes." Equation (105) yields in this case for the newly formed clumps

$$R_f \sim (s/\delta_p)^{1/2} \qquad \text{or} \qquad R_f \leqslant 286 \text{ km} \times (0.3/0.6)^{1/2} = 202 \quad \text{km.}$$

If spherical, the average thickness is $\frac{4}{3}R$ or 269 km. This is more than the estimated thickness of the ring and would lead to loss of permanent contact among its parts, a fraction of $115/269 = 0.427$ of the ring area being occupied by the projections of the fragments. This corresponds to an average spacing between the fragments $(\Delta/R_f) = (\pi/0.427)^{1/2} = 2.71$ or 547 km. The total number of fragments or minisatellites in the middle ring is then $N_r = 10^5/547 = 180$ and, over the width of $0.5R_0 = 3200$ km, there will be 6 full rings, the total number of fragments being $N_f = 1080$ in this symmetrically arranged model. Each of the six rings is orbiting independently, small perturbations of individual members being damped in mild collisions inside a ring.

Each of the 1080-odd members or moonlets raises its own tidal bulge on the rotating Earth; the instantaneous tidal bulge is the vector sum of the component bulges and, for a precisely symmetrical arrangement of the moonlets, the resultant tidal vector would be zero. However, within each of the six rings there is some freedom of motion for its members; their grouping will be ruled by the law of chance and the average absolute value of the resultant random vector will be proportional to the square root of their number. For a Poisson distribution of N equal mass points this would be exactly true; for finite size members the freedom of rearrangement is limited, but a dispersion in the masses and radii of individual members would add additional variance. It can therefore be assumed that the tidal acceleration, or the rate of tidally induced orbital change in one of the six rings is

$$(da/dt)_f = (da/dt)_0 \, N_r^{1/2}/N_f \tag{111}$$

or

$$(da/dt)_f/(da/dt)_0 = 0.0124 = 1/80,$$

where $(da/dt)_0$ denotes the rate of orbital evolution ruled by an integral lunar mass. The time scale is thus increased 80 times and, instead of some 5 yr sojourn inside Roche's limit, this would take about 400 yr.

Neighboring rings will not add to this acceleration (their tidal bulges induced on the Earth cannot stay in resonance) except through a periodic term of accidentally fluctuating amplitude of zero expectation over the synodic period (due to "regrouping" of the members of a ring). These terms work in proportion to the square root of time and their contribution is small or negligible (a calculation has been made in this respect). It can be assumed that the contributions from other rings cancel out over one synodic period (5 days or less), and that the residual tidal effect upon one of the six rings is fully accounted for by the random wanderings of members within the same ring as expressed by Eq. (111).

For the rate of tidal orbital evolution in the outgoing phase an interpolation formula can be written, satisfactorily representing Gerstenkorn's (1955) calculations at geocentric distances smaller than 12 Earth radii, giving the time of drift in years for an integral lunar mass as

$$t_a = 0.025(a_2^{5.5} - a_1^{5.5}),\qquad(112)$$

where a_2 and a_1 are the distances in Earth radii. (Between 12 and 60 Earth radii, the average power is 7.1, as compared to an "ideal" value of 6.5 for constant inclination and friction, and the time scale should be adjusted to 4.5×10^9 yr.)

Each of the six rings drifts outward at its own rate, expected to be given by Eq. (111) with

$$(da/dt)_0 = da_2/dt_a\qquad(113)$$

as defined by Eq. (112). In the case of overtaking by members inside the same ring, collisional damping will adjust the pace. The outer and inner edges of the ring, at $a_1 = 2.75$ and 2.25, respectively, according to Eq. (111), will reach Roche's limit at $a_2 = 2.86$ within 80 times the time given by Eq. (112), or within 210 and 560 yr, respectively; the interval between the extreme rings is thus 350 years, and between two successive rings 70 years.

As soon as a ring emerges from Roche's limit, its 180-odd components will be drawn together and accrete into a moonlet of one-sixth lunar mass, with a radius of 956 km (density 3.34 assumed for the compressed and heated material), and an escape velocity of $w_\infty = 1.31$ km/sec. The ring will collapse in "free fall," the time scale being given by Eq. (70), $t_F/2\pi = 6.6$ orbital periods or 1.6×10^5 sec.

The average potential energy $\frac{3}{5} \cdot \frac{1}{2}w_\infty^2 = 5.07 \times 10^9$ erg/gm does not suffice for melting. At middle accretion or $\eta_m = 0.5$, the rate of accretion as given by Eq. (101) is 500 gm/cm^2-sec. The accretion is so intense that radiation losses are negligible. The minimum and maximum temperatures from Eqs. (95) and (97) are identical and, with $T_0 = 300$ K, for the average temperature of the accreted moonlet yield

$$T_s' = T_s'' = T_s = 863 \quad K.$$

As conditioned by tidal interaction, the moonlets thus emerge at intervals of $350/5 = 70$ yr, and with inclinations to the Earth's equator decreasing from about $42°$ for the first to $27°$ for the sixth moonlet. The compacted moonlets drift outward on a time scale 14 times faster than the rings [Eq. (111) with $N_r = 1$, and $N_f = 6$, by convention] yet still six times that of Eq. (112), so that when one reaches Roche's limit, the preceding one, with its faster rate of recession, has gone far enough to escape direct contact with the newcomer. The orbits are nearly circular although of considerable inclination (specifically for the capture model), and interaction between two consecutive moonlets begins only when they approach within the gravitational target radius R_a without their orbits intersecting or crossing. This is made possible by the law of tidal evolution as expressed in Eq. (112), which brings the two moonlets separated by a time interval $\Delta t = 70$ yr closer together the farther they go ($da/dt \sim a^{-4.5}$, thus rapidly decreasing with distance). When interaction begins, Roche's limit (mutual for the two moonlets) is always reached before physical collision can take place, because

$$D_r > R_p .$$

Therefore, the two moonlets first break up into a large number of fragments, which then, while mutually colliding, accrete into a moonlet of double mass that begins drifting outward at double speed.

The time scale of this second accretion is one-half the synodic period of revolution of the two approaching moonlets and runs into a few days. The relative orbital inclination may have any value from $i_1 - i_2$ to $i_1 + i_2$, according to the position of the precessing nodes, and will not change much during the process of accretion, the period of precession being (Öpik, 1958b)

$$t(i) = 35.8 \sec i \cdot a^{3.5} \qquad \text{(Earth rot/24}^\text{h})^2, \tag{114}$$

the period of the advancing perigee

$$t(\Pi) = 35.8 a^{3.5} \text{ (Earth rot/24}^\text{h})^2/(1.5 \cos^2 i - 0.5), \tag{115}$$

and the period of the argument of the perigee

$$t(\omega) = [1/t(i) + 1/t(\Pi)]^{-1}. \tag{116}$$

For nearly circular orbits, the motion of the perigee is irrelevant and only precession of the nodes matters. The relative inclination of two orbits varies with their synodic period of precession, which runs into tens of years in the present case.

For a pair of interacting moonlets, each one-sixth the lunar mass, the sum of the radii is $956 + 956$ km or 0.300 Earth radii, and Roche's limit is about 0.40 Earth radii, each of the moonlets breaking up into $N_f = 120$ fragments of

194 km radius ($s = 6 \times 10^7$ dyne/cm²) [Eq. (105)]. This sets the order of magnitude for how close the moonlets can be before interaction begins. The appropriate distance is reached approximately 140 years after the emergence of the preceding and 70 years after the following moonlet. Table VII shows the history of accretion of the Moon according to this scheme. At $t = 140$ yr moonlets I + II, at 280 yr moonlets III + IV, and at 420 yr moonlets V + VI are assumed to merge. These pairs then may further combine at 490 yr and after, leading to a complete merger somewhere near $a = 5$ Earth radii. On account of the high power of distance in Eq. (112), this last result is quite stable for widely differing initial assumptions.

The heating of the Moon, finally accreted at 5 Earth radii, depends partly on the time scale, which, for the combination of all the considered phases of accretion, can be set at 350 yr, yielding $J_m = 0.0174$ gm/cm²-sec as an overall average [Eq. (101)]; it depends chiefly on the average encounter velocity U_m, which, from Eq. (50) for $e = 0$ and $A = 1$, is conveniently reduced to

$$U_m{}^2 = 2(1 - \cos i_e) \qquad \text{or} \qquad U_m = 2\sin(\tfrac{1}{2}i_e), \qquad (117)$$

where i_e is the average inclination of the combining orbits to the final resultant orbit. Thus, with an average of the component inclinations of 36°, $U_m = 0.62$ is an upper limit when the resultant orbit coincides with the equatorial plane and will be less for a final inclination different from zero, depending on the phase of precession. The probable value of $U_m{}^2$, calculated as the deviation from a mean of six independent vectors, is $5U_m{}^2/6$, or $U_m = 0.565$ or 2.0 km/sec at 5 Earth radii (orbital velocity $v_0 = 3.5$ km/sec). With three-fifths of the final escape energy as an average for the potential energy, $T_0 = 863$ K as from the original formation of one moonlet, the minimum average internal temperature of the Moon at formation becomes $T_s{}' = 1680$ K [Eq. (95)], and the maximum fraction of melting is $\theta_{max} = 0.838$. These are probable values; with an improbable combination of the phases of precession at the times of interaction of the six moonlets and their resultants, both $T_s{}'$ and θ_{max} may be lower, the resultant inclination remaining large in such a case. This, however, is not supported by the majority of calculations (MacDonald, 1964; Sorokin, 1965; Slichter, 1963; Darwin, 1879), which point to a low value of 10–14° at 5 Earth radii, a distance to which the retrospective calculations are more reliable. The lower limits of heating are for zero relative inclination at encounter and are identical with those calculated in Section IV,G; they hardly apply to the case of tidal capture in which component inclinations of the order of 36° (at encounter), must have been reduced to some 12° after completed accretion and thus must have led to intense conversion of kinetic energy and heating. The average encounter velocity in our model exceeds $\sqrt{2} - 1$, and ejection of some fragments into interplanetary space becomes possible (Öpik, 1963a).

The fraction ejected is

$$f_\infty = [\sigma_a^2/(\sigma_a^2 + \sigma_0^2)][(U^2 + 2U - 1)/4U]. \tag{118}$$

Consider the middle pair of moonlets III + IV (Table VII), whose merger is supposed to take place at $t = 280$ yr and $a = 3.52$ (average of 3.37 and 3.66). The orbital periods of revolution of the two before the merger are 0.363 and 0.411 days, respectively, the synodic period 3.11 days; one-half of the latter is the time scale $t(\sigma)$. The orbital precession periods are 11.0 and 14.6 yr, respectively, and the synodic period during which the relative inclination fluctuates between 3° and 72° is 45 yr. After merger, the combined double mass settles into an intermediary orbit with inclination i_m. Neglecting the small difference between the two original inclinations, we set $i = i_1 = i_2 = 35°$. From spherical geometry we have

$$\sin i_e = \sin i \cdot \sin(\tfrac{1}{2}\alpha), \tag{119}$$

$$\tan i_m = \tan i \cdot \cos(\tfrac{1}{2}\alpha), \tag{120}$$

while the relative inclination of the two original orbits is $2i_e$. Here, α is the difference in longitude of the two nodes on the equatorial plane. With the assumed inclination, $\alpha = 90$ or $270°$ divides the equator into two equal parts, one with $U > \sqrt{2} - 1$, the other with smaller U. In the first mentioned high velocity part, $U_{av} \simeq 0.52$, $\sigma_0^2 = 2.49 \times 10^{-3}$, $\sigma_a^2 = 1.92 \times 10^{-4}$, and $f_\infty = 0.0090$. Completely negligible also is the acceleration, according to Eqs. (74)–(76), where $t(i)$ is to stand for $t(\omega)$. The total colliding mass is one-third of the mass of the Moon and, thus, in the encounter of only one pair of moonlets in this orbital configuration, 0.003 of the lunar mass is expected to be ejected into interplanetary space. From there, it returns as considered in Section IV,D and, over a period of over 50,000 yr, is captured by Earth and Moon, the share of the Earth-orbiting Moon being 1/60.7 that of the Earth. Hence, from the ejected mass the Moon will receive a final contribution equal to 0.003/61.7 or 5×10^{-5} of its mass.

The craters upon which crater statistics were based, having an average diameter less than 20 km and a depth (x_p) about 3.2 km, and covering about 50% of the continentes area, correspond to a depth of erosion of 1.6 km, involving 9×10^{-4} of the lunar radius or 2.7×10^{-3} of the lunar mass. At $w_0 = 3$ km/sec, $s = 2 \times 10^8$, and $\rho = 2.6$, $k = 1$, one obtains $M_c/\mu = 34$ [Eqs. (3) and (14)]. The impinging mass that was mainly responsible for shaping the present relief of the continentes would thus equal

$$2.7 \times 10^{-3}/34 = 8 \times 10^{-5}$$

of the lunar mass. The contribution of 5×10^{-5} or 60% of it would suffice to influence the crater statistics in a manner different from that observed:

the late interplanetary projectiles would not contribute to a systematically arranged ellipticity of the craters. Hence 20% of interplanetary fragments is, perhaps, the upper limit admissible for shaping the present surface of the continentes. The rest, or all, must be of Earth-orbiting origin.

There are altogether, as in the scheme of Table VII, four merger events, each of which would suffice to obliterate the uniformity of crater numbers and the tidal deformations of the craters (imprinted 50,000 yr later at a distance where the deformations are negligible) if they happened in the high-velocity configuration. In each case the probability of the configuration is one-half; the probability that it did not take place and that, in the case of the tidal capture theory, no ejection of fragments into interplanetary space did occur, is thus

$$(\tfrac{1}{2})^4 = \tfrac{1}{16},$$

a low though not a forbidding value. With such a probability the crater statistics can be reconciled with tidal capture of the Moon and an ensuing high inclination of its orbit when at minimum distance from Earth.

F. Origin through Fission or from a Ring inside Roche's Limit

The two possibilities are indistinguishable so far as the ultimate consequences are concerned and will be treated together.

The fission theory has been doubted, even rejected (Levin, 1966a), because it is inconceivable that a mass separating from the Earth inside Roche's limit and in violent upheaval could have preserved integrity. This, however, is not needed and, with the finite cohesion and clumping mechanism, the ring of debris could slowly recede and emerge from Roche's limit to form the Moon in a manner such as that described in Section IV,E. There is one important difference: the inclination of the ring to the terrestrial equator would have been near zero in such a case. As compared with the preceding model, the sequence of events would be essentially similar, but the kinetic energy in accretion would be smaller with U the encounter velocity of the debris and the moonlets, being near zero. Ejection of fragments into interplanetary space could not then take place, and the last fragments captured from the Earth-orbiting cloud would be co-moving with the orbiting Moon, descending on it more or less isotropically from all directions. A small preference for impacts from the rear, as revealed by crater counts (Fielder, 1965, 1966), could be expected if the last fragments were accelerated in encounters with the Moon and removed into elliptic orbits with large semimajor axes, so that they were overtaking the Moon while in direct motion near their perigees. A similar excess of directly moving meteorites, periodic comets and Apollo-type asteroids is observed in the present terrestrial space of the solar system.

Another objection—that the backward calculation of the tidal evolution, based on the present masses and angular momenta of the Earth–Moon–Sun system does not lead to a solution much closer than Roche's limit—is only of "paper" value in this case, because neither the identity nor the mass and momentum distribution of the bodies or agglomerations can be considered to be known during this primitive stage.

The parent ring is assumed to be in the Earth's equatorial plane, and so will be the component six moonlets of our idealized model. Table VIII shows their calculated hypothetical history as ending in the formation of the Moon. Because of the small relative velocity, as conditioned by the small relative inclinations, they combine sooner than in the previous scheme; i.e., I + II at $t = 70$ yr, III + IV after 210 yr, and V + VI after 350 years; but before this happens, at $t = 280$ yr, the first two pairs combine into one containing two-thirds of the lunar mass. This body (I + II + III + IV), twice the mass of the remaining pair (V + VI), drifts out twice as fast and cannot easily be overtaken by the smaller companion although their separation still decreases at first (compare 3rd and 4th lines from the bottom of the table). Instead the collision target radius σ_0 [last line of the table, from Eq. (54)] rapidly increases and, when it exceeds the separation between the moonlets, final merger occurs at $t = 420$ or 490 yr, at $a = 4.5$–5 Earth radii. Before this happens, a passage through Roche's limit of the larger body destroys the smaller body (V + VI). The radius of the larger body is then $R_p = 1519$ km $= 0.243$ Earth radii, and

$$\sigma_0 = 0.243[1 + w_\infty^2/U^2]^{1/2} \tag{121}$$

in Earth radii is calculated with $w_\infty = 2.08$ km/sec for the larger body, while U is taken to be the difference of the orbital velocities of the two circular orbits (second line from the bottom of the table).

To calculate the heating limits [Eqs. (95) and (99)], we again set $J_m = 0.0174$ gm/cm²-sec as for a time scale of 350 yr, $T_0 = 863$ K, and take $\frac{1}{2}w^2 = 1.19 \times 10^{10}$ erg/gm; this is equal to three-fifths of the kinetic energy at escape velocity of the present Moon (2.38 km/sec) less 5.07×10^9 erg/gm as the potential energy of accretion of the component moonlets. The minimum average internal temperature of the accreted Moon is then $T_s' = 1260$ K, and the upper limit of the melting fraction $\theta_{max} = 0.301$. The kinetic energy of U, or the free orbital energy, is neglected; it is nearly compensated for by the overestimate in the potential energy.

G. Thermal History and Origin

Table IX contains a summary of the preceding subsections. Although based on numerical data that are inevitably rather rough, the conclusions in each case are comparatively stable and may serve as a basis for judgment that is better than a mere qualitative approach. The following summary can be made.

Hypotheses 1, 2, and 3 disagree with crater statistics and tidal deformation trends, while 4 and 5 do agree. In Hypothesis 3, the surface during cratering is too hot and too much melted to account for the regular and dense crater coverage in the continentes (lunar bright regions or highlands). Hypothesis 4 requires an unusual combination of the nodes of the component orbits (the probability is 0.06, or less if there were more than six component bodies); also, most retrospective calculations indicate a small inclination at 5 Earth radii, contrary to the requirements of this hypothesis.

Only Hypothesis 5 is free from obvious objections and will be considered as the most probable working basis. As to the consequences for the structure of the lunar surface, Hypothesis 4, although much less probable, is almost identical with Hypothesis 5.

The thermal history of the Moon has been treated by different authors, mostly on the assumption of an originally cold accreted body heated by radioactive sources. Depending on the assumed amount and radial distribution of the heat sources, opposing conclusions have been reached; either that the melting was essentially complete (Kuiper, 1954), or that there was no substantial melting except in the deep interior (Urey, 1960b, 1966). Most comprehensive calculations have been made by Majeva (1964) and Levin (1966a, containing a review of her work and that of others); radiative transfer as a component of thermal conductivity and different abundances of the radioactive elements were taken into account, as well as differentiation of a lighter sialic crust from the heavier simatic melt; for various initial parameters, the main conclusion is that at present the Moon "is solid at least to a depth of 500–700 km. But the central part, embracing 20–40% of its mass, must have been in a molten state up till the present time."

The estimates of initial heating, as originating from gravitational energy, and as for Hypothesis 5, would enhance these conclusions. Initial melting could have occurred on a large scale as the consequence of bombardment, although for the present thermal state the difference in the initial conditions would be essentially obliterated. This is partly due to the nature of the thermal decay by cooling, partly to the sialic differentiation which transports most of the radioactive elements into the granitic–basaltic crust whence the heat easily escapes to space while further heating of the melted interior stops. Thus, melting is a regulator of internal heating that automatically limits itself as soon as it starts. In a solid body of sufficient size, radioactive sources sooner or later lead to melting; this causes sialic differentiation and removes the heat sources from the interior, so that a cooling phase starts. According to Levin (1966a), after a cool start the lunar interior would have reached maximum heating and melting "1–2 billion years after its accumulation." Levin's assumptions correspond to our Hypothesis 2, but on a very much longer time scale; with the shorter time scale, as follows from the low U values, the initial average internal temperature of the Moon should have exceeded 850 K and, with sufficient shielding by a protective crust, may have reached 1800 K with about 5% melting on the surface. This no longer is a cold Moon for a start, and in Hypothesis 5 an average interior temperature between 1260 and 1800 K is indicated with up to 30% surface melting. In such a case, the melting of crater bottoms and the lava flows that covered the maria need not be relegated to some later epoch awaiting radioactive heating, but most probably were contemporaneous with the accretion itself and the last cratering. On the continentes, a solid crust of unspecified depth, 10–20 km at least, must have existed, while the maria were overflowed by lava.

On a lava sea that is able to form a solid crust, either because of differentiation of lighter minerals or because the crust is not cracked by impacts, the "bottleneck" of heat transfer is the conductivity of the solid, radiation from the surface coping with the heat flow at a very small excess of temperature over the equilibrium temperature $T_0 \sim 300$ K. With the liquid at melting temperature, conventionally 1800 K, an assumed specific heat $\rho c = 2.7 \times 10^7$ erg/cm³-deg and heat conductivity $k_t = 3.2 \times 10^5$ erg/cm-sec-deg (allowing for the radiative component), the thickness Δh of the crust increases with time (in years) closely as

$$\Delta h = 0.018 t^{1/2} \quad \text{km.} \tag{122}$$

During an upper limit of time for crater formation on the continentes, $t = 2100$ yr, during which the Moon receded from 5 to 8 Earth radii [Eq. (112)], $\Delta h = 0.82$ km only. The crust would be too thin for the craters. The process cannot be advocated for the formation of a basis for cratering.

However, as pointed out by Urey (1966), the crust will be battered and cracked by impacts at the outset; the solid fragments, being heavier than the liquid, will sink to the bottom, leaving the open liquid surface radiating to space at a rate of 6×10^8 erg/cm²-sec. At 7.3×10^9 erg/cm³ as the heat of solidification, the solidified layer at the bottom now increases linearly with time:

$$\Delta h / \Delta t = 26 \quad \text{km/yr.} \tag{123}$$

The rate is high enough to overrule all our time scales of accretion. A pressure-dependent melting point will not essentially influence the process, except by providing a "bottom" to a superficial pool. As a result, during accretion that is too rapid to be influenced by radioactive energy release, a solid almost isothermal body is rapidly formed throughout, at a temperature near the melting point, while the excess energy is radiated away from the surface of the liquid. This is exactly the condition upon which Eq. (99) was based. This makes $\theta = 0.301$ an upper limit that is close to the real value; it differs from it only in so far as the remaining 70% of the surface, being solid, does participate in radiation to space; the participation must be small indeed. Hence, θ may represent, in fact, the instantaneous surface fraction of transient liquid pools, formed by bombardment and rapidly solidifying at the bottom. The quoted value corresponds, of course, to the middle phase of intense bombardment; at the epoch of crater formation, θ must have been near zero, incidental melting occurring from the cratering impacts into the hot substratum.

With the rapidity of solidification from the bottom, no large combined lava pools could have been formed, and the melting must have been confined entirely to the surface of the Moon. A consolidated, dense, and hot body was formed in such a manner. No lava extrusion, caused by rupture of an imaginary crust, could have taken place at this stage. The maria must have been produced superficially and locally, by impacts of a few large planetesimals, soon after the intense bombardment ended but not very much later; from the number of postmare craters on them, their age cannot differ much from the 4,500 million years of the Moon itself (Öpik, 1960).

In the process of surface melting and bottom solidification in small local pools, not much differentiation could have taken place, any difference created in the pool being locally frozen in, without exchange between different depths. Iron phase could have separated into small pockets but have been prevented from concentrating in the core. (There may now be a few percent metallic iron in the core.)

After a hot solid Moon had accreted, isothermal at the surface melting temperature but about 200 K below the melting point at the central pressure, radioactive heating of the interior and conductive cooling of the outermost few hundred kilometers must have started. From curves of radioactive heating and cooling of an initially cold Moon, Levin (1966b) concludes that widespread melting, from a depth of the order of 500 km down to the center, must have occurred about 2.0×10^9 years from the start. This corresponds to a rise of central temperature by about 1600°. With the initially hot Moon, the required heating would be 1/8 less; allowing for exponential decay of the radioactive sources, the melting should have occurred in 1/10 the time. Thus, some 200 million years after accretion, a second stage in the internal evolution of the Moon must have been reached; in the molten interior, sialic differentiation must have occurred, forming a lighter intermediate layer adjacent to the outer crust. The crust itself, however, must not have been affected, retaining its original composition and cooled by radiation. The basis of the craters—the highlands or continentes—must have been preserved as it was formed and also that of the maria. At the epoch of radioactive melting, the crust was too thick for lava extrusions or for being pierced by an impacting body; the original planetesimals must have been swept absolutely clean from the surrounding space by that time [cf. Eq. (93)], while stray objects of the required size from other parts of the solar system would have been too rare to produce one mare-generating collision (not to mention several) on the Moon (Öpik, 1958a, 1960) with a reasonable probability. Eight mare impact areas can be located on the earthward hemisphere of the Moon exceeding 500 km across or requiring projectiles larger than 25 km in diameter. For the whole Earth, one such impact is expected once in 2×10^9 years (Öpik, 1958a), and for one lunar hemisphere the time scale is 6×10^{10} years, yielding an expectation of 0.075 interplanetary impacts during 4.5×10^9 years. The Poisson formula yields a probability of 2.3×10^{-14} for having eight such impacts. Clearly, it is reasonable to assume that the maria were generated as an immediate sequel of the events, and from the same source, that finally built the Moon.

An idea of how much an initial hot stage could have influenced the present thermal state of the Moon can be obtained from the calculations of Allan and Jacobs (1956), who somehow varied their radioactivity parameters more or less as they would be influenced by melting and differentiation. For a lunar sized body, Table X shows the change in average temperature over an interval of 4.5×10^9 yr for three selected cases: A, a cold start with strong radioactive sources throughout the body, the concentrations of uranium, potassium, and thorium being those for an actual chondritic meteorite; E, a cold start but with about 1/4.5 the radioactivity, a concentration assumed to hold for the Earth as a whole; and G, a hot start, but with still less radioactivity, nearly one-half of that in E and equal to that in dunite, believed to be the main constituent rock of the Earth's mantle.

Each of these assumptions has something in its favor. Case A might appear the most probable one, yet meteoritic concentration of radioactivity, which may have prevailed at the start, must have led to melting even from a cold start and to differentiation and depletion of the internal heat sources; Case G may then represent the continuation (the absolute values of temperature are not relevant; the starting temperature could be that of melting). Case E shows that, with an average concentration of the radioactive nuclides as in the Earth, an initially cold Moon may not yet have reached the melting point; however, as has been shown, gravitational heating during rapid accretion would have overruled this restriction, too, and with the higher radioactivity as compared to Case G, a molten interior would have been preserved until our time.

A possibility that the maria were formed as the result of radioactive heating during the 200 million years following accretion, by complete melting of the mantle underneath and collapse of the solid crust inward, must be rejected for another reason, besides objections from the standpoint of thermal balance; the nondifferentiated base of the continentes would have collapsed also and have become nonexistent. Also, deep melting on the maria would have led to differentiation and formation of a light sialic crust in their place, while the continentes, if somehow preserved, would have been supported by a heavier base. Isostatic equilibrium would have sunk them deeper, lifting the maria surfaces into uplands, which is the very opposite of the actual state of things. The maria are definitely depressions as shown by Baldwin's (1963) contour map, 2.52 ± 0.13 km below the average continentes (Öpik, 1962a). Although made of the same material, the top layer of the continentes may have been battered into rubble and may be lighter. Melting at impact of the relatively hot substance (cf. Section IV,I) would favor compaction, but a ratio of about 0.8 of the density of the rubble in the highlands to the solidified rocks of the maria may be a fair estimate. The thickness of the unconsolidated material in the continentes as required by the postulated isostatic equilibrium, would then be 12.5 km or eight times the estimated thickness of the layer eroded during the formation of the presently surviving craters.

As to isostatic adjustment, it must have worked on the primeval hot lunar material as it does now on Earth. With cooling of the outer mantle, some rigidity must have developed as witnessed by the earthward bulge of about $h = 1.06$ km in excess of the equilibrium tidal configuration (Öpik, 1962a) (dynamical value from physical libration). The extra load, supported by the solid mantle whose inner and outer radii are R_i and R_e, respectively, causes a compressive stress s_m in the mantle, without participation of the liquid core,

$$s_m = \tfrac{1}{2} h \rho g / [1 - (R_i^2/R_e^2)]. \tag{124}$$

With 500 km as the thickness of the mantle at the time of the last adjustment of the bulge (not necessarily now), with $\rho = 2.6$, and $R_i/R_e \sim 0.7$, $s_c = s_m = 4.6 \times 10^7$ dyne/cm² (46 atm pressure) must have been the average compressive strength of the lunar mantle. The excess bulge, not really a "fossil" tidal bulge but rather a "lagging behind" remnant of it, would indicate also the differences in lunar level, which can be supported on a large scale without isostatic adjustment.

H. *Crater Statistics and Origin*

If the relative equality of the crater densities on the eastern and western hemispheres ("astronomical" terminology of orientation) on lunar continentes, and even a slight excess in the western can be understood, in terms of the tidally directed accretion history of the Moon and the coorbiting swarm (cf. Section IV,F), a similar distribution on the maria may appear more of a puzzle. Unlike the hypothetical primitive projectiles which were bombarding the continentes, those on the maria must have belonged to the known classes of interplanetary stray bodies—comet nuclei, Apollo group "asteroids" (extinct comet nuclei), and true asteroids deflected by Mars perturbations. With respect to this external medium, whatever the distribution of velocities, the preceding hemisphere of the Moon is subjected to a greater frequency of impacts than that trailing behind, and an excess, instead of a deficiency of craters on the eastern hemisphere should have been expected.

However, the orbital velocity of the Moon is so small compared to the inter-planetary velocities that only a small effect can be expected; this could easily be masked by sampling errors, inhomogeneities in the counts [as an example, for Mare Crisium, Baldwin (1963) finds 62 craters exceeding one mile in diameter, against 40 counted by Shoemaker and Hackman], or even by systematic differences in the mechanical properties of lunar rocks in the two hemispheres which look so different. Also, a distinction between primary and secondary craters must be made when small craters are counted although, for craters exceeding 1.6 km (one mile) in diameter, the number of secondary craters in lunar maria is only 4% according to Shoemaker (1966).

From data by Shoemaker and Hackman (1963) as adjusted by Baldwin (1963), the number density of primary craters in lunar maria in the two hemispheres is as represented in Table XI.

The purely statistical probable error of sampling is indicated. For the eastern hemisphere, the Maria Imbrium, Nubium, Humorum, and Epidemiarum, and for the western, the Maria Serenitatis, Foecunditatis, Tranquillitatis, Nectaris, and Crisium were combined. The largest and easternmost area of Oceanus Procellarum is not represented. The average number density in the eastern hemisphere, both at the 1.6 km and 3.2 km crater diameter limit, is found to be markedly smaller than in the western.

For a lunar body orbiting in the ecliptical plane with a circular velocity v_c, and meeting a stream of particles, arbitrary in direction and of velocity U relative to the Earth, integration of the accretion flux, for the linear case where v_c/U is small, yields a hemispheric ratio

$$\text{eastern/western} = 1 + 1.9 v_c/U, \tag{125}$$

that is practically independent of the inclination of the U vector to the lunar orbit; at zero inclination, the coefficient is $56/3\pi^2 = 1.89$, and at $90°$ it is $6/\pi = 1.91$. The concentrating factor, represented by the second bracketed term of Eq. (54), is not taken into account; it is of the order of $(w_\infty/U)^4$ and thus negligible for the small lunar escape velocity. For the lunar orbit 2.25×10^9 years ago during the middle interval of bombardment, $a = 53$ Earth radii and $v_c = 1.08$ km/sec can be assumed. For Apollo group objects, $U = 0.660 = 19.7$ km/sec is an observed average (Öpik, 1965a), whereas for isotropically orbiting objects at heliocentric velocity v_h, the average, weighted by the square of the encounter velocity [stream velocity times cumulative number proportional to $D^2 \sim w$ according to Eq. (7)] or the square of the average impact velocity for craters of a fixed size limit, is

$$U^2 = 1 + v_h^2 + [2v_h^2/3(1 + v_h^2)]. \tag{126}$$

For parabolic comets, $v_h = \sqrt{2}$, $U = 1.856$ or 55.3 km/sec.

The two extreme types of object yield, according to Eq. (125) expected impact ratios of 1.10 and 1.04 for the two lunar hemispheres E/W, respectively. For craters of the size limit in question, the average of the two groups may be representative (Öpik, 1958a), or a ratio of 1.07. A difference from the observed values (Table XI) appears to be well established, but it would be rather far-fetched to draw conclusions as to the origin of lunar craters from such slender deviations of the ratios from unity as was done by Fielder (1965, 1966).

Here, it may be pointed out that comet nuclei, carrying a substantial proportion of volatile ices and traveling at higher velocities, will, for equal mass, produce more violent explosions than the extinct nuclei or asteroidal objects. The effect of volatiles was not considered in connection with the origin of the Moon because it may be assumed that, in the terrestrial preplanetary ring, these volatiles would not be condensable and, apparently, were not massively represented judging from the composition of the Earth.

The density of craters in the continentes is estimated to be 19 times that in an average mare, e.g., Mare Imbrium (Fielder, 1963), or 15 times according to Baldwin (1964b). It is therefore expected that 5–7% of the craters in the continentes are of postmare origin. These may be difficult to distinguish except for the ray craters, which are apparently the result of more violent impacts, perhaps by the high-velocity comet nuclei. Of the 50 ray craters in Baldwin's (1963) list, 32 are on continentes (a few just on the margin) and 18 are on maria, which more or less corresponds to the ratio of areas on the earthward hemisphere of the moon, where 60–65% of the area is occupied by continentes (more in the limb areas that, by projection, represent a smaller apparent fraction of the visible hemisphere than that occupied by their actual area). The postmare origin of these features is thus obvious. Between the hemispheres, 26 ray craters are in the western, 24 in the eastern; if polar and centrally placed "indifferent" objects are omitted, 14 are definitely western, 15 eastern. The uniformity of distribution is also apparent, however with a large statistical sampling error implied (about $\pm 15\%$), considering the smallness of the numerical sample.

Reverting to the general crater densities in the two hemispheres, the crater numbers seem to be much more influenced by throwout from a few large cratering events than it would appear from Shoemaker's (1966) estimates. In a specially investigated area of western Mare Imbrium, covering 465,000 km², a definite increase in crater numbers is revealed in the southern portion of the mare in the vicinity of Copernicus and within the reach of its rays (Öpik, 1960). At an effective limit of 1.1 km for crater diameter, the northern half shows a crater density that is uniform within the sampling error, 13.5 ± 0.5 craters/10^4 km². From the middle of the area the densities increase southward far beyond the sampling error, so that 390 craters counted over 152,500 km² yield a density of 25.6 ± 0.9. Assuming 13.5 to be the density of primary craters (a maximum value—some secondaries may be present in the northern half, too), the excess density in the southern half is to be attributed to 185 secondaries—22.8% of craters in the entire mare; Shoemaker's (1966) graph indicates only 6% at the 1.1 km limit. Moreover, about the same relative excess persists also in southern Mare Imbrium at the higher diameter limit of 2.5 km. The number of secondaries increases southward as Copernicus is approached, from $32 \pm 6\%$ in the northern third, to $42 \pm 7\%$ in the middle third and $61 \pm 6\%$ in the southern third of the southern half of the mare (Öpik, 1960).

As shown by the Ranger photographs, the rays appear to consist of tightly distributed secondary craters (Shoemaker, 1966). Crater chains belong to the same phenomenon, produced by a salvo of projectiles or by a spinning larger clump shattered by the shock that breaks up in flight and sends out fragments with different velocities at different locations along a line.

A different kind of exceptional object is the lava-filled or flooded crater. There are 42 of them in the highlands or continentes, and 20 on the marginal regions of the maria, flooded by the latter (Baldwin, 1949); thus, they all belong to the premare stage. This is in harmony with the picture of accretion of the Moon as drawn previously; i.e., the continentes base, still hot after intense accretion had subsided, was then receptive to impact melting. In the postmare period, the crust had cooled and impact melting became much less prominent.

I. Melting of a Mare

Extrusion of lava from an inner molten core to the lunar surface is as difficult to visualize as it is for the Earth's core. On Earth, lava formation and extrusion are connected with mountain building, folding, subsequent erosion, and isostatic depression, which lead to the radioactive sources being buried deep and insulated. The rocks are heated beyond melting point in subsurface lava foci. If saturated with water vapor and other gases (water drifting down from the surface), volcanoes are formed. However, more powerful lava extrusions are the plateau basalts, coming through crustal cracks and overflowing vast areas at a time, covering hundreds of thousands of square kilometers.

On the Moon the mountain building processes are absent, erosion is too slow and surface stores of water are not available. The lava pools of the period of intense accretion must have completely solidified at its conclusion. At that time, perhaps some 2000 yr after the start of final accretion, the subsurface rocks must have been hot, from a depth of some 0.5 km [using Eq. (122) for a rough estimate] below the surface. Since they were at a temperature near the melting point, the shock energy of a cratering impact then easily caused melting.

Using Eq. (20) with $\lambda_x{}^2 = 0.5$ as an upper limit and $q = 2.7 \times 10^9$ erg/gm as for the melting of a solid already heated to the melting point, the shock velocity at the fringe of complete melting becomes $u \geqslant 1.04$ km/sec. Equation (16) with $k = 2$ then yields the ratio of the melted mass to that of the projectile to be $y_i M/\mu = k w_0/u$ or 5.77 when $s_p = 2 \times 10^8$ dyne/cm² and $w_0 = 3$ km/sec are assumed as for the low velocity primeval impact into rock softened by heat. Choosing a depth of penetration of $x_0 = 100$ km, $\gamma = z = 45°$, $\delta = 1.3$ (as for a loose sandball), and $\rho = 2.6$, the cratering equations yield $p = 1.093$, and $d = 114.3$ km for the diameter of the projectile. Further, from Eq. (11), with $f_s = 0.78$, and $s_c = 1.0 \times 10^8$ (cf. Section V), $s = 2.1 \times 10^9$ dyne/cm² obtains, caused chiefly by gravity friction. Equations (7), (4), and (14) then yield a crater or mare diameter of $B_0 = 424$ km, with $u_s = 0.285$ km/sec, $M/\mu = 21.0$ (mass affected), $M_c/\mu = 14.1$ (mass crushed or melted), and $y_i = 5.77/21.0 = 0.375$ (completely melted fraction). The projectile itself is not included here; its material may be mostly melted, while vaporization, mixing, and a distinct "central funnel" do not occur. The volume of the projectile is $V = 7.8 \times 10^5$ km³, and the volume of completely melted rock becomes $(\delta/\rho) V \times 5.77 = 2.25 \times 10^6$ km³, which could cover a crater area of 1.41×10^5 km² with a layer 16 km deep. The sprayed liquid and the rock debris, ejected with velocities of 0.2–0.5 km/sec, would fall back into the crater, little being thrown over the rim.

Outside the completely melted fraction y_i, partial melting in proportion to the heat release or to $(y_i/y)^2$ will occur [Eqs. (16) and (20)]. The total melted fraction of the mass affected is then

$$f_m = y_i^2 \int_{y_i}^1 dy/y^2 + y_i = 2y_i - y_i^2 \qquad (127)$$

or $f_m = 0.609$. The unmolten rock debris will settle down, leaving a lava sea of 3.65×10^6 km³; if spread uniformly over the crater area, a liquid layer 26 km deep would result. According to Eq. (123), under bombardment, the solidification of this lava mare would take only about a year. On the other hand, if the mare were formed when intense bombardment had subsided, the formation of an unbroken solid crust would have become possible; on the linear scale contemplated, this could have happened only through differentiation of the lighter sialic rocks, which would float on the simatic melt.

A characteristic trait of the described mare-generating mechanism is the deep penetration of the impacting body—to about one-quarter of the diameter of the mare. The depth of penetration at oblique incidence (100 km) is here less than the diameter of the projectile (114 km). For Mare Imbrium, at $B_0 = 1050$ km, the other linear dimensions must be increased somewhat more than in proportion to the crater diameter. Since the lateral strength is, in this case, closely proportional to crater depth [Eq. (11)], from Eq. (9) we have

$$x_0 \sim x_p \sim B_0^{4/3} \qquad (128)$$

whence, for Mare Imbrium, at $\gamma = 45°$, we find $x_0 = 335$ km for the penetration of the front of the projectile, whose diameter would then be 384 km (density 1.3). The average depth of the molten layer would be about 87 km. All this is on the assumption that a single event was reponsible for the creation of the mare, an assumption that is difficult to refute in view of the regular, nearly circular outline of its border. A satellite which produced Sinus Iridum may have impacted nearly at the same time.

K. The Date of Closest Approach and Alfvén's Model of Lunar Capture

Mathematical attempts to retrace, backward in time, the history of the Earth–Moon system depend, in the first place, on the assumed law of tidal friction, either as it did or did not vary in the course of time. The different results obtained by different authors (cf. Section IV,E) for the minimum distance and, especially, for the time scale, depend mainly on the assumed history of friction. The relatively short time intervals, of 2–3 billion years, obtained for the time of closest distance are undoubtedly due to an overestimate of friction, which so fundamentally depends on the distribution of the oceans and conti- nentes, as well as upon the total amount of water in the hydrosphere of the Earth. The oldest dated minerals, such as the zirkons in the gneisses of Minnesota, show an U–Pb age of 3.5×10^9 yr, equal to the oldest representatives from the Central Ukraine and the Congo, and sedimentary rocks reach down to 3×10^9 yr (Cloud, 1968). The closest approach of the Moon could not have occurred later than these dates: ocean tides of up to 10 km height, accompanied by rock tides of similar amplitude [at $\Delta h = 10$ km, rocks become plastic and cohesion no longer can prevent them from following the tidal bulge of the rotating Earth (cf. Section V,B)], and tidal friction heating of the order of 9×10^9 erg/gm of the *entire Earth* would have evaporated the oceans and melted the upper crust into a lava sea that no previous extrusive or sedimentary rocks could have survived. Indeed, the heat of tidal friction must have concentrated in the upper portion of the Earth's mantle, yielding there well over the 1.6×10^{10} erg/gm required for raising the temperature to the 1800 K necessary for melting (cf. Section IV,C). The history of the Earth's present crust must have begun with a completely molten state, synchronous with the time when the Moon was closest to Earth (either captured, or emerging from inside Roche's limit as described in Sections IV,C–G). With all the uncertainty as to the absolute time scale, it is most natural to adjust it to a more definite event—the origin of the Earth itself, 4.5×10^9 years ago. From the theory of planetary encounters (Section III), a lunar body, orbiting somewhere near the Earth's orbit, could not have escaped close approaches to Earth for longer than 10^5–10^6 yr and, if tidal capture ever did take place, it must have followed the formation of the Earth with not more than such a lag in time. For this reason alone, any conjecture as to a late capture of the Moon must be rejected as so improbable that it can be termed practically impossible. Further, the geological and geochronological record renders absolutely unacceptable theories that would put the date of lunar capture at less than 10^9 years ago (Alfvén, 1965), or that would ascribe the "Cambrian–Precambrian nonconformity" in biological–geological sequences about 700 million years ago to the events of lunar capture (Olson, 1966) (instead of repeated world-wide ice ages as testified to by boulder beds at this and earlier epochs). The medicine is too strong; instead of boulder beds and interrupted organic evolution (with algae dating 2700 million years ago), a global lava sea several hundred kilometers deep would have engulfed all traces of previous history, and not simply produced a problematic "nonconformity." Under such circumstances, the critical appraisal of Olson's suggestion by Munk (1968) sounds rather mild: "Twenty years ago a hypothesis relating this unconformity to a unique event in the Earth–Moon history might have received a sympathetic reception. Somehow the problem is less urgent now. In many places the geologic record is patched across the Precambrian–Cambrian interval, and the unconformity is not so very different from others in the geologic record. With regard to the explosive biological evolution we have succeeded only too well, by destroying all existing forms of life and insisting that life start anew. The biologists won't have it."

If the Alfvén–Olson idea of a recent (late Precambrian) catastrophic event of such a magnitude is not only refuted by geological evidence, but is also contrary to the concepts of probability of planetary encounters (the probability of a primitive Moon delaying its fatal encounter with the Earth for 3×10^9 years is less than 10^{-1000}!), the mechanical variant of the capture theory proposed by Alfvén (1965, 1968) appears highly attractive, since it seems to reconcile the few critical data relating to crater ellipticities and the time of their formation (Öpik, 1961b) with the aesthetic merits of Gerstenkorn's mathematical model of tidal capture and evolution.

According to Alfvén, if the primitive Moon were nonhomogeneous, with the outer layers being of lesser density than the average, while passing close to Roche's limit as Gerstenkorn's calculations would imply, it could have lost its lighter mantle while preserving the denser core that was still outside its own Roche's limit and have been able to keep together by gravitation. In synchronous rotation, the fragments of the tidally distorted elongated mantle released earthward would have been directed inward in elliptical orbits, possibly even falling on the Earth, while those from the opposite end, turned away from Earth and possessing more angular momentum, would swing outward in elongated elliptical orbits. On a two-body approximation, neglecting the gravitational action of the Moon's mass, from the tips of a tidally deformed body, extended to double the Moon's diameter and in synchronous rotation at a mean distance of 2.71 Earth radii, the extreme inward fragments would enter elliptical orbits

between 2.16 (apogee) and 0.73 (perigee) Earth radii, thus colliding with the
Earth; and, the extreme outward fragments would be thrown into elliptical
orbits between 3.26 (perigee) and 22.0 (apogee) Earth radii. Roche's limit is
2.71 Earth radii at density 4.14 [Eq. (104)]; if half the original lunar mass were
in a core of this density, a mantle of density 2.54 comprising the other half
(and yielding 3.34 for the mean density of the Moon) could be thrown out by
tidal action, leaving the core behind. A second approximation, on the basis of
the restricted three-body problem with Earth and Moon as the principal
partners, would lead to more complicated orbits, the Jacobi integral, however,
permitting more or less the same range of geocentric distances. Things are
complicated even more by the presence of considerable diffuse masses, and by
the acceleration of the fragments in near-miss encounters with the accreting
lunar core in a noncircular, precessing orbit, drifting outward from tidal interaction. The inward fragments, partly absorbed by the Earth, and the outward
fragments will tend to coagulate into moonlets, colliding and breaking up again
and ultimately being collected by the Moon. The outward cloud, originally
swinging on an average orbit between, say 3 and 13 Earth radii, $a_e = 8$,
$e = 0.625$, will collect into moonlets and a coherent cloud of finer debris while
more or less conserving the original angular momentum, moving in near circular
orbits such that the mean distance becomes $a = a_e(1 - e^2) = 4.9$ Earth radii;
this is approximately the distance at which the craters of the continentes appear
to have been formed, judging from their systematic trend in ellipticities
(Öpik, 1961b). The outer fragments must have been rapidly collected by the
tidally advancing Moon. As to the inner fragments, those that were preserved
from falling on the Earth may have collected into a moonlet. This inner moonlet,
too small to overtake, tidally, the main body of the Moon, was perturbed and
accelerated by the latter in apogee approaches until, in apogee, a collision
(preceded by tidal breakup) with the *earthward* side of the Moon took place;
a salvo of large fragments led thus to the formation of the lunar maria.

Thus, except for the time scale, Alfvén's model of tidal capture and subsequent
marginal close passage is able to account not only for the craters in the continentes and their systematic ellipticities, but also for the later formation of the
lunar maria on the earthward hemisphere of the Moon. Quantitatively, however,
in this case one cannot put much reliance on precise calculations of tidal evolution
near, and preceding, the stage of the Moon's closest approach because the
assumptions, either of the constancy of the lunar mass, or of the limited number
of interacting bodies, cannot be upheld even approximately.

The only difficulty with this most attractive model remains in the heat
created by the impacting bodies during the last stage of crater formation on
the continentes. The expected heating would be somewhere between that of
models 3 and 4 of Table IX, and probably nearer to the former; this is a bit
too hot, and with too much melting, for the time when the highland craters

were formed. Nevertheless, with all the other circumstances taken into account, Alfvén's model of lunar capture appears to have a good degree of probability in its favor—about as much as model 5, the most favored one of Table IX.

V

STRENGTH OF LUNAR CRUSTAL ROCKS

A. Crater Profiles

The depth to diameter ratio of lunar craters is known to decrease with crater size (Baldwin, 1949, 1963). This is obviously explained by gravitation influencing fallback. The average velocity of the ejecta is mainly conditioned by the strength of the material and the kinetic elasticity; for given velocity, the altitude and distance of flight is limited by gravitation, so that a smaller percentage of the crater volume can be ejected over the rim of a large crater than over that of a small one.

The theory outlined in Section II,F can be applied to the study of lunar crater profiles. In notations of this and the preceding sections, the apparent depth of a crater x', as measured from the undisturbed ground level to the surface of the debris at or near the center of the crater (disregarding a central peak if present), can be assumed to be

$$x' = x_p(1 - F_B). \tag{129}$$

The throwout function $1 - F_B$ is represented in Fig. 2, the fallback fraction being given by Eq. (49). This is the fraction of crushed material falling back into the crater, but it may also be assumed to cover gravitational inhibition in raising a lip and in displacing the uncrushed rock of the crater bowl ($AALL_0$ in Fig. 1); this justifies the application of the fallback factor to x_p, the total depth.

Fallback mainly depends on parameter b [Eq. (48)], which approaches zero for small craters when fallback also tends to zero. In this case the gravitational friction component in lateral strength [Eq. (11)] may also become unimportant, and the crater profile, or the ratio of depth to diameter will be determined by Eq. (9) (except for erosion for very small craters). Of the parameters in this equation, the lateral strength $s = s_c$, or the product $s\delta$, is most uncertain. Nevertheless this, as well as velocity and density can be guessed fairly well for a given cosmogonic stage.

For the large craters, when parameter b is increasing with the linear scale, gravitational friction in Eq. (11) becomes important and even dominant. Fallback then depends primarily on

$$\lambda^2 u_s^2 = \lambda^2 s/\rho \sim \lambda^2 f_s = \text{const} \tag{130}$$

or on the product of kinetic elasticity and friction, and on the marginal exit angle, β_0 (Fig. 1). Setting $f_s = 0.78$, $\sin \beta_0 = 0.8$, the depth to diameter curve for the large craters can be met by a proper choice of λ^2, which, thus, is another parameter that can be empirically determined almost independently of s_c (within the margin of uncertainty of the other, more certain parameters).

The measured crater profiles used here are from Baldwin's work (1949, 1963), where the depth is reckoned from the crest of the rim. Average rim heights were therefore added to the calculated x' values; to render them comparable with the observed depths, the calculated values of x'/B_0 as referred to ground level were multiplied by an empirical factor of 1.60 for Baldwin's class 1 craters, and by 1.80 for those of classes 2, 3, and 4; the ratios do not seem to depend on crater size.

Baldwin's crater classes are meant to represent relative age, class 1 being the youngest and showing the least signs of later impacts or the least impact erosion. The classification is supposed to be uninfluenced by the depth to diameter ratio. The later or older classes are shallower, which is partly the result of erosion but may also include some subjective bias. This is brought out by the distribution of crater classes depending on size, taken from Baldwin's later work (1963) and represented in Table XII.

The smaller craters are registered predominantly as being of class 1, while among the largest craters this class is in a minority. It seems that the small craters, being less shallow for fallback reasons, tend to impress as being less eroded. Another explanation may be that, in an incomplete list, small craters are more often selected when they are sharp and neat, which makes for a preference in favor of class 1 without classification itself being systematically at fault. In both cases only the largest craters would correctly represent the relative population of the classes. Hence we may conclude that the class 1 craters, according to Table XII, are not all of postmare age, having been produced as the last 25–30% of the total population of craters; they could well belong to a later stage of premare bombardment when the lunar crust had somewhat cooled and hardened. Craters of classes 2–4 are shallower (Baldwin, 1949, 1963) and can be explained by lower values of the cohesive strength s_c as well as of the elasticity λ^2; this may be ascribed to a hotter and softer crustal material at the earlier stages of the final bombardment.

It thus appears that the crater profile data are not a homogeneous selection. For throwout theory to be applied meaningfully, a closer study of the statistical material is required.

Table XIII represents the distribution of the craters in Baldwin's list (1963) according to their surface background. Here, the selectivity is very marked, small craters being chosen chiefly when of class 1 and on the maria, apparently because they were easier to measure without interference from other craters. Of course, all postmare craters except those of class 5 are expected to be practically unaffected by later impacts, hence the virtual absence of classes 2–4 from the maria, as revealed by the table; these classes undoubtedly represent premare objects.

In class 1, all craters on the maria are, of course, of postmare origin, with ages ranging from 4.5×10^9 yr to zero. With the exclusion of the predominantly "continental" limb areas, the maria represent about 50% of the area of selection, so that the continentes should carry a number of postmare craters equal to that on the maria. Assuming this, the percentages of postmare craters on the continentes were estimated. It appears that, in class 1, the largest craters (greater than 40 mls) are predominantly of premare age and, being less affected by later impacts, must correspond to the last stage of primitive cratering, say, at a distance of some 8 Earth radii and 2000 years after the start of accretion (cf. Section IV,E and F). Craters of class 1 in the 20–40 mls group are also predominantly of premare age, although some 22% may be of postmare origin, but craters less than 10 mls in diameter must all belong to the postmare stage, including those on the continentes. This heterogeneity of class 1 must be taken into account in the interpretation of crater profiles. From the correlations of diameter with depth published by Baldwin (1949, 1963), it may appear that heterogeneity is insignificant, the curves running smoothly over a diameter range of 10^4 to 1, from the smallest terrestrial to the largest lunar craters, yet the impression is deceptive. Systematic differences amounting to a factor of 2 or 3 in the depth to diameter ratio become inconspicuous over the wide range when log absolute depth is correlated with log diameter, instead of the ratio, and an apparently smooth run of the curves for heterogeneous material (depending on diameter) can be achieved where actually there are discontinuities in the ratio.

Baldwin's (1963) class 5, the lava filled or flooded craters, actually contains two different kind of formations. Those on the continentes can be explained by local impact melting of the hot primitive crust when the crater was formed, while those in the maria appear to be flooded from outside by *lava* from the mare.

In Tables XIV and XV (in notations of, and from, the equations of Section II), are collected some theoretically calculated depth (rim to bottom) to diameter ratios, corresponding to *a priori* assumed probable parameters of impact. A median angle of incidence $\gamma = 45°$, for isotropic bombardment, and a coefficient of friction $f_s = 0.78$ are assumed throughout. Further, in Table XIV, $w_0 = 3$ km/sec for accretion during the late premare cratering phase (Hypothesis 5, Table IX), $\rho = 2.6$, and $\delta = 1.3$ gm/cm³ for a "sandball" planetesimal are assumed. Also, with the assumed constants, and the lunar acceleration of gravity (162 cm/sec²), from Eqs. (11)–(13), we have

$$s = s_0 + 160x_p \tag{131}$$

in cgs units. In Models A, B, and D–G, the compressive strength is

assumed to be the same as that for the hot and soft premare lunar crust,
$s_p = 2 \times 10^8$ dynes/cm² (cf. Section V,B); this, according to Eq. (6), yields

$$p = x_p/d = 1.093,$$

a value that is insensitive to the actual value of s_p. In Models A and B, a constant
value of the lateral strength, about one-half of s_p, is assumed. In Model C,
a high crustal strength, such as for terrestrial rocks, is assumed; this improbable
assumption is definitely refuted by the observational data, as can be seen from
Fig. 3 in which Baldwin's data for the premare craters of classes 2–4 are plotted.

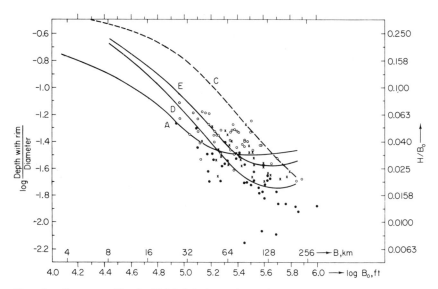

FIG. 3. Crater profiles for Baldwin's later classes (premare age). All on continentes.
Logarithmic scale. Abscissas, crater diameter (B_0, km); ordinates, depth to diameter
ratio, $(x' + H)/B_0$. Measured points: circles, Class 2; crosses, Classes 3 and 4;
dots, Class 5. The four calculated curves are those of Table XIV, with the parameters:
Curve A—$w_0 = 3$ km/sec, $\lambda^2 = 0.35$, $s_c = 1.04 \times 10^8$; Curve C—$w_0 = 3$ km/sec,
$\lambda^2 = 0.12$, $s_c = 9 \times 10^8$; Curve D—$w_0 = 3$ km/sec, $\lambda^2 = 0.25$, $s_c = 3.0–1.0 \times 10^8$;
Curve E—$\lambda^2 = 0.30$.

Of the two other models with constant s_c, Model B is completely out because
of the assumed high elasticity, while Model A with $\lambda^2 = 0.35$ leads to a better fit,
which is still bad enough as can be seen from Fig. 3. While for large craters the
curve can be adjusted by a suitable choice of λ^2, the smaller craters require an
increase in s.

It is natural to assume that, on account of cooling, the outermost crust of
the Moon acquired somewhat greater strength. Tentatively, at an age of 2000 yr
from the beginning of accretion, when the presently surviving craters in the

continentes were formed, the temperature distribution in the crust may have been about as follows [cf. Eq. (122)]:

depth, km (x_c)	0	0.4	0.80	1.2	2.0	3.2	∞
temperature, K	300	900	1250	1470	1680	1770	1800

Thus, at the depth of penetration of the smaller craters, noticeable cooling and hardening of the rocks may have taken place. In Models D and E this has been assumed, a triple value of $s_c = 3 \times 10^8$ (still only one-third that for cool terrestrial rocks) at $x_p = 1.5$ km or $x_c \sim 0.8$ km being proposed, with a corresponding softening of the material inward. The representation of the class 2–4 crater profiles (Fig. 3) is now good, the best fit being obtained at $\lambda^2 = 0.28$, an intermediate value between the two models.

Models F and G are similar to Models E and D but with more hardening of the crust, meant to represent a late stage of premare cratering, perhaps 20,000 yr after the start of accretion, when the "youngest" craters in the continentes—those of class 1—were formed. As has been pointed out above (cf. Table XIII), only the large class 1 craters of Baldwin's list are of premare age. In Fig. 4,

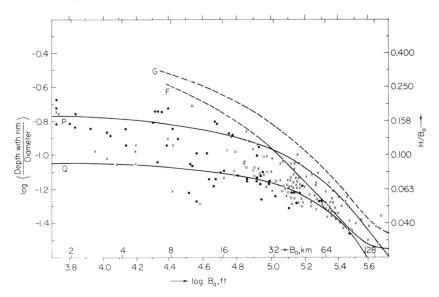

FIG. 4.   Crater profiles for Baldwin's Class 1 (postmare age). Coordinates same as in Fig. 3. Measured points: dots, on maria; crosses, on continentes; circles, ray craters. The two lower calculated curves (P and Q) are those of Table XV, and the two upper ones (G and F) are from Table XIV. The assumed parameters are: Curve F— $w_0 = 3$, $s_c = 5$–1.7×10^8, $\lambda^2 = 0.28$; Curve G— $w_0 = 3$, $s_c = 7.5$–2.5×10^8, $\lambda^2 = 0.28$; Curve P— $w_0 = 20$, $s_c = 9 \times 10^8$, $\rho = \delta = 2.6$, $\lambda^2 = 0.22$; Curve Q— $w_0 = 40$, $s_c = 9 \times 10^8$, $\rho = 2.6$, $\delta = 2.0$, $\lambda^2 = 0.28$.

the class 1 crater profiles are plotted with background (mare or continens) indicated, and it can be seen that Model F represents reasonably well the observations for the premare craters larger than 40 km, while Model G is "too strong."

The smaller craters of class 1, as well as the larger craters in the maria and all ray craters are of postmare origin. They must have been produced by high-velocity impacts of asteroidal and cometary bodies, such as those calculated in Table XV, Models P and Q. At an average age of about 2×10^9 yr, the lunar outer crust must have cooled and hardened completely, therefore, a high strength, equal to that of terrestrial granite or basalt, has been assumed. The assumption has proved a success; in Fig. 4, the observed flat run of the depth-to-diameter ratio for postmare objects (all ray craters, all craters in the maria, and all class 1 craters smaller than 32 km on the continentes) is well matched by the P and Q models, the average correlation falling between the two. From statistics of interplanetary stray bodies (Öpik, 1958a) it can be estimated that cometary impacts should account for about 40% of the cratering events at the 3 km crater diameter level, for 60% at the 10 km and 70–75% at the 40–80 km level. Accordingly, the average correlation for a mixed impacting population of asteroidal and cometary bodies should lie between Models P and Q, nearer to P for small crater diameters, and to Q for the large ones, an expectation that is in surprisingly good accord with the observations as plotted in Fig. 4. Thus, despite the heterogeneity in age and background of the class 1 crater selection, the data can be well represented as a combination of large premare craters formed at low impact velocities (3 km/sec) and of postmare asteroidal (20 km/sec) and cometary (40 km/sec) impacts. Together with the older premare craters (Fig. 3), the successful representation of the crater profiles lends some strong independent support to our concepts of lunar origin (Section IV), as well as to the quantitative theory of cratering. Worth noticing is the narrow span of the kinetic elasticity $\lambda^2 = 0.22$–0.28, corresponding to 11–14% average kinetic (throwout) efficiency of the cratering shock, higher than for sand craters and diluvium but about equal to that of hard rock at low velocities of impact (cf. Sections II,C–F).

The dispersion of the depth-to-diameter ratios for a given crater diameter is considerable, showing variation within an extreme range of about 3 to 1 (cf. Figs. 3 and 4). Yet this can be accounted for entirely by the dispersion in the angle of incidence factor $(\cos \gamma)^{1.5}$ [Eq. (9)]. Very remarkably, there is little room left for an intrinsic variation of the other relevant parameters—velocity, density, and strength of the material. This is especially true of the premare craters (Fig. 3 and class 1 craters on continentes larger than 40 km in Fig. 4), and for an understandable reason—their impact velocities must have been close to the Moon's escape velocity, and thus practically constant.

Unlike the case of the experimental "Teapot" crater (Section II,F), for the lunar craters the density of the fallback material is assumed here to be the same as that of the original "bedrock" material. For the premare craters this assumption naturally follows from the fact that the "bedrock" for new craters consists of the throwout and fallback material of their erased predecessors so that the material must be identical in all respects, including density. Besides, any significant difference in density would increase the fallback volume of large craters so much that, instead of depressions, their floor levels would appear as elevations above the original ground level, which as a rule, is not the case (with one notable exception, the large premare crater Wargentin, whose floor is 400 m above ground level). For the large postmare craters (Fig. 4), an increase in fallback volume could be countered by an increase in the elasticity coefficient λ^2, yet in this case the large depth of erosion (x_p from 5 to 12 km) would ensure pressure compaction of the partially melted rubble. However, for the smaller postmare craters, λ^2 cannot serve to balance a change in volume; instead, an increase in s_c would be required that does not appear to be plausible, the largest possible value (9×10^8 dyne/cm^2) such as that for hard rocks having been used already. It seems that, partly by "soldering" through the molten spray and partly through subsidence, helped by later impacts, the fallback must have nearly acquired the density of the original bedrocks.

Interpreted as the result of oblique impact, the premare craters in the lunar highlands (without regard to class) are found to show a rms random ellipticity of 0.070 ($n = 53$) in central regions and 0.096 ($n = 125$) in limb regions (Öpik, 1961b); the second figure is of lower weight despite being based on a greater number of craters. The values are corrected for observational error dispersion and are supposed to represent the true cosmic average of crater ellipticities; a weighted mean observed value of $\epsilon = 0.080$ can be accepted for a median diameter of about 27 km. With $p = 1.093$ and $\gamma = 45°$, Eq. (28) yields

$$\epsilon = 0.519/D, \tag{132}$$

where $D = pB_0/x_p$. This yields $\epsilon = 0.076$ for Models A and B (Table XIV) at $B_0 = 31.2$ km, and for Models D and E, at $B_0 = 28.1$ km, $\epsilon = 0.084$. The predicted values are closer to the observed ellipticity than could be expected for these *a priori* calculations based solely on physical first principles. The value depends mainly on the relative crater diameter D [Eq. (7)].

B. Orographic Relief and Strength of the Primitive Lunar Crust

The main orographic features of the Moon, including the majority of its craters, must have been formed during and immediately after the accretion phase when the crust was hot and soft, and without significant changes afterwards. From the standpoint of supporting strength, with its smaller gravity the Moon should be able to support six times greater differences in level than the Earth. Actually, the absolute differences in level on the Moon are considerably smaller than on Earth, which points to a lower strength of its crust at the time of its formation, in agreement with the conclusions drawn from crater profiles (Section V,A). The mean difference between continent and ocean levels on Earth is 4.6 km or, with the isostatic correction for the weight of sea water, the equivalent unbalanced difference amounts to 3.3 km; on the Moon this would correspond to 19.8 km while the actual mean difference between the maria and the continentes is only 2.5 km or 1/8 of that corresponding to the Earth difference (Baldwin, 1963; Öpik, 1962a).

Of course, the differences in level occurring on a large scale are isostatically balanced, and the slopes are always smaller than the angle of repose arctan f_s. Yet, when the unbalanced pressure (weight minus buoyancy) exceeds the plastic limit (compressive strength s_p), friction is unable to prevent subsidence. Differences of level Δh over short stretches or continuous slopes thus set a lower limit to s_p, which, in the extreme cases, approaches the value itself:

$$s_p \geqslant g\rho \, \Delta h. \tag{133}$$

The stress is greatest at the "foot of the mountain," i.e., at the lowest uncompensated level, from which Δh is reckoned. With depth, the stress decreases on account of isostasy, beginning from a subsurface level where the heavier rock (sima on Earth, compacted maria rock on the Moon) begins, and reaches zero at the bottom of the lighter formation, sial on Earth, battered rock on the Moon, or a depth of about 5 Δh depending on the ratio of the densities. The strength of the rock will vary with the depth, increasing on account of compaction but decreasing because of higher temperature, so that there is a certain ambiguity as to which depth Eq. (133) properly refers. It turns out that on the Moon the insulating dust layer is rather thin and that solid rock of high thermal conductivity begins soon enough below the surface [at less than 20 m (Sections VI,B and VII,C)]; the differences in temperature are therefore not large, and the lower limit of s_p would therefore correspond to the near subsurface layers, of the order of Δh, where the stress is greatest. The strength would decrease downward as the temperature rises, but this should not become significant before a depth of 10–20 km is reached.

In Table XVI, typical estimates of the compressive strength of terrestrial and lunar rocks are collected. The most prominent slopes have been chosen for the Moon from Baldwin's contour map (1963), and from the lunar limb profile as measured by C. B. Watts (1963). Of these latter data, a twofold selection is used: (a) Δh, from the average limb profile over a 10° position angle (300 km) and all librations (Öpik, 1964), in which the differences in level are partly smoothed out as it actually takes place with the highest summits and deepest troughs whose load is shared by nearby less extreme features; (b) Δh_e, from extreme differences in level over continuous slopes tabulated over a 2° position angle for zero libration (values published by Baldwin (1963) with the contour map). For terrestial features, taking isostatic compensation by water into account, 63% of the ocean depths is assumed to be the effective component of Δh, reckoned from the sea bottom plus the effective elevation over dry land and smoothed so as to eliminate extreme mountain summits. In Eq. (133), $\rho = 2.6$ for Moon and Earth alike, and the full value of Δh, or 0.8 of Δh_e, has been used to allow for sharing of load.

The largest of the limiting values of s_p should approach the true average. Hence, for the Earth, $s_p = 2 \times 10^9$ dyne/cm² seems to be indicated, a value very close to that for granite or basalt and a check on the reliability of the method. For the primitive Moon, the compressive strength is found to be 1/10 as much, $s_p = 2 \times 10^8$ dyne/cm².

C. Ray Craters and Strength of the Ejected Blocks

In Baldwin's (1963) list of 50 ray craters, 30 are on continentes and 20 on maria. This is also approximately the ratio of the respective areas occupied by continentes and maria on the earthward hemisphere of the Moon (continentes, however, prevailing in the limb areas). There is no indication of a greater density of these objects on the continentes, contrary to other types of craters. Clearly, no contribution to ray craters has come from the premare stage to which 98% of "ordinary" lunar craters, which are crowded 20 times more densely on the continentes than on the maria, belong. This is not so much a question of relative or absolute age, as that of the violence of the explosions caused by the impact of asteroids or cometary nuclei; these, at a velocity of some 15–60 km/sec, might have led to high-velocity ejecta traveling hundreds or thousands of kilometers over the Moon's surface, which the primeval impacts of planetesimals, traveling at 3 km/sec, could not have matched. The larger secondary craters (over 300 m in diameter) in the rays (Shoemaker, 1966) are of such a size that they should not have been eroded by micrometeorite bombardment even during all the 4.5×10^9 yr of exposure, while the smaller craters could last for several hundred million years (cf. Section X,D and E). And, even when the craters are eroded, the ray substance should last and maintain its lighter coloration. It is most likely that the Moon has not existed long enough for the first postmare rays of the ray craters to be erased and that the difference between ray craters and the rest lies in the original event, not in absolute age. The ages of the ray craters are expected to range uniformly over the span of 4.5×10^9 yr, with an average of 2.25×10^9 yr, the same as for the rest of the postmare craters. Besides, almost all large craters on the maria are ray craters; this supports the view that the ray craters are not exceptionally young; they were not formed during the recent one hundred million years or so—their hypothetical forerunners, whose rays were supposed to have been obliterated by age, are not there.

The secondary craters in the rays, known before but now brought to the fore by the Ranger pictures (Shoemaker, 1966), provide a means for estimating the strength of the postmare lunar crustal materials. The ejected blocks, which caused the secondary craters, have withstood high accelerations which taxed their strength to a degree that can be calculated approximately (using first principles) from the cratering formulas.

Of course, under very particular circumstances, all-sided compression may increase the strength of some blocks instead of shattering them. These, however, must be exceptional cases; in general, the ejected blocks will be representative of the strength of the parent bedrock. Clumping of pulverized debris may also produce sizeable blocks, but of inferior strength, unable to withstand high accelerations; they will not be found at great distances from the crater.

Most of the throwout leaves a crater with relatively low velocities on a short trajectory. From the central regions, the ejection velocities are higher, but most of the material is pulverized or broken up by the shock. In a high-velocity impact, leading to explosive development of gas, some blocks may be considerably accelerated and fired like missiles from a gun; the acceleration depends on favorable circumstances—position in the crater and timing of the first shock that breaks up the bedrock into large chunks. If the gas stream from the central funnel overtakes a block at the right time, it may send it out of the crater with a velocity that greatly exceeds the ordinary ejection velocity v_0 of the inelastic shock (Fig. 1). The relative mass of such high-speed ejecta may be small, yet sufficient to cause the ray phenomenon around large craters.

Without entering into details of the ejection process, the velocity w_0 (lower limit) of the ejected block can be determined from the distance of flight; its size (d_1) is then related to the diameter (B_1) of the secondary crater through Eq. (7). If ξ is the acceleration during ejection, the crushing stress (lateral strength of the material inside the primary crater) s_1 experienced by the projectile during ejection is then close to

$$s_1 = \delta \xi \, d_1 \,, \tag{134}$$

where δ is its density.

The length of path during acceleration is of the order of the depth of the primary crater (diameter B_0) or about $0.1B_0$, from which a lower limit to acceleration can be set at

$$\xi = 5w_0{}^2/B_0 \tag{135}$$

(using the equation for constant acceleration).

We assume also $\gamma = 45°$ as a median and most probable angle of ejection and impact alike; this leads to minimum velocity and minimum estimated strength (s). At this angle, the initial as well as final velocity of the projectile in its elliptical orbit is given by

$$w_0{}^2 = 2gR_\mathrm{p} \sin \psi/(\sin \psi + \cos \psi) = gL/(\sin \psi + \cos \psi) \qquad (136)$$

[$2gR_\mathrm{p} = 5.64 \times 10^{10}$ (cm/sec)^2 is the square of the velocity of escape], where R_p is the lunar radius, g the surface acceleration of gravity, 2ψ the selenocentric angular distance of flight, and $L = 2R_\mathrm{p} \sin \psi$ the linear distance of the flight measured along the chord. (For an arbitrary zenith angle γ in Eq. (136), the factor 2 is to be substituted for by $\operatorname{cosec}^2 \gamma$, and $\cos \psi \cot \gamma$ is to be taken instead of $\cos \psi$.)

The velocities are such that at impact formation of the secondary crater the stresses greatly exceed the strength of any rock, i.e., $w_0 > w_\mathrm{m}$ [Eq. (10)], wherefore Eqs. (6) and (7) with $k = 2$ should be valid. The compressive strength, which influences the result but slightly, can be set equal to

$$s_\mathrm{p} = 2.2s_\mathrm{c} \quad \text{with} \quad s_\mathrm{c} = s = \sigma_\mathrm{s} , \qquad (137)$$

gravitational friction being relatively unimportant, while σ_s denotes the lateral strength for the secondary crater. The cratering equations for the secondary crater then yield

$$D_1 = B_1/d_1 = 1.51(w_0{}^2/\delta_\mathrm{s})^{0.233}\delta^{1/4}, \qquad (138)$$

while the density ρ of the bedrock cancels out.

For $\delta = 2.6$ the numerical coefficient becomes

$$1.51\delta^{1/4} = 1.91.$$

Using this value for the density of the projectile, and setting

$$\sigma_\mathrm{s} = \eta s_1 , \qquad (139)$$

simple relations are obtained for the crushing strength (s_1) and diameter (d_1) of the ejected block, depending on the diameters of the secondary and parent craters and the velocity of ejection [Eq. (136)]:

$$s_1 = 12.1w_0{}^2\eta^{0.3}(B_1/B_0)^{1.3} \qquad (140)$$

and

$$d_1 = 0.94B_1(\eta B_1/B_0)^{0.3}. \qquad (141)$$

The secondary craters, which are considered below, are all on the maria; the depth of penetration is of the order of 100 m and more. Recent Surveyor I photographs (Newell and NASA Team, 1966; Jaffe, 1966a) show the rim of an ancient crater southwest of the spacecraft, consisting partly of rounded boulders

Fig. 5. Surveyor I photograph 66-H-834, 2 June 1966. At center left, a crater of 3 m diameter, probably of secondary origin, its rocky projectile having ricocheted out. In the foreground, a rock about 70 cm, a secondary ejectum having come to rest on the surface after ricocheting. On the horizon, a stone-walled crater 420 m in diameter, with the stone wall seen best preserved in the upper right corner; probably interplanetary primary 1500 ± 500 million years old. (Photograph by courtesy of NASA and Jet Propulsion Laboratory.)

(Figs. 5 and 6) reminiscent of a stone wall. Almost level with the general terrain, this wall can be compared to a raised lip with the top eroded. In Fig. 1, it can be compared to the lip L, bent upward and raised from level LL_1 whose original position is 0.03–0.04 crater diameters below the undisturbed surface. The crater diameter is about 420 m [estimated distance from the spacecraft is 140 m for the near side of the rim, 560 m for the far side (cf. Fig. 6)]. Hence the original depth of the rocky strata from which the lip was raised is 13–17 m. The layer of loose material in a mare must be less than this. It follows that the secondary craters that are discussed here must be the result of impact into a hard rocky substratum, not into granulated material; the strength must be of the order of that for the parent crater, so that the coefficient η [Eq. (139)] should not differ much from unity and even may exceed it, considering that s_1 is the actual stress, which the block survived and which must be smaller than the ultimate

Fig. 6. Surveyor I photograph 66-H-807, June 1966. Stone-wall detail of the crater (same as in Fig. 5) on the horizon (420 m in diameter); average blocks in its wall measure 70 cm. (Photograph by courtesy of NASA.)

strength of the parent crater interior, while σ_s is the ultimate strength of the upper crust at the point of impact.

A few typical cases of secondary craters to well-known ray craters are considered below. Although the s_1 values so calculated are inferior limits, by choosing the largest objects at a given distance, or the largest distances for a given secondary crater size, these inferior limits should come close to the actual values.

(a) In Mare Cognitum, there is a conspicuous group of secondary craters along a ray from Bullialdus as shown by Ranger VII photographs A 108, 156, 176, (NASA, 1964) and pointed out by Shoemaker (1966). Near the southeast corner of A 176, there are three large craters in a line, two of which appear to be double on closer inspection, while the middle one is single. Allowing for overlapping and scale, from a study of NASA photograph A 176, the following dimensions of the five craters have been derived:

Craters	Northernmost double	Middle single	Southernmost double	Average
		Diameter, km		
Along ray	1.61 and 1.75	2.41	2.48 and 2.05	2.06 ± 0.08
At right angles to ray	1.61 and 1.53	2.05	1.75 and 1.90	1.83 ± 0.02

The ellipticity is thus $\epsilon = (2.06 - 1.83)/2.06 = 0.112 \pm 0.040$ in the expected direction. It is, however, too uncertain for quantitative application according to Eq. (28). For the largest of the group, the middle single one, $B_1 = 2.23$ km is the average diameter. The distance from Bullialdus (south of the group) is $L = 236$ km, $2\psi = 6°48'$ (one selenocentric degree $= 30.3$ km), hence $w_0^2 = 3.58 \times 10^9$, $w_0 = 0.60$ km/sec. With $B_0 = 60$ km for Bullialdus, Eqs. (140) and (141) are transformed into

$$s_1 = 6.0 \times 10^8 \eta^{0.3} \quad \text{dyne/cm}^2,$$
$$d_1 = 0.78\eta^{0.3} \quad \text{km}.$$

For η ranging from 0.5 to 2, $s_1 = 4.9$ to 7.4×10^8 dyne/cm^2, $\sigma_s = 2.5$ to 15×10^8, $d_1 = 0.63$ to 0.96 km as the diameter of the projectile. As a lower limit, and referring to a block shattered by the blast, s_1 is found to be close to and compatible with a value of $s_c = 9 \times 10^8$, as for granite or basalt, valid for the postmare lunar rocks (in a mare) at 3–6 km below the surface.

The volume of the ejected block is about 0.25 km^3, that of the Bullialdus crater [volume crushed, Eq. (15)] about 6000 km^3, so that there is no shortage of material for these exceptional ejecta.

On the same frame A 176 (NASA, 1964) (selenographic latitude $11°23$ south, $21°44$ east) of Ranger VII, there is a group of short parallel ridges going from northwest to southeast and not in the direction of Bullialdus. They are 10–15 km long, a few hundred meters high and are also easily visible on the Earth–based Lick Observatory photographs; they appear as bright at full Moon as the continentes, in contrast to the dark mare background. They are similar in appearance to the isolated peaks in northern Mare Imbrium (Pico, Piton, and others) and are difficult to explain as ejecta from impacts. The central peak of Alphonsus (see below) belongs to the same kind. They have something to do with the melting of the mare and may be surviving relics of the premare period. O'Keefe (1964) suggests a volcanic origin for the ridges as well as for a black marking that runs in the same direction. The marking is darkest at full Moon (Lick Observatory and other photographs), reminiscent of the black spots in Alphonsus and elsewhere, and cannot have much of an elevation. Clearly, these features cannot be of direct impact origin. Secondary volcanic phenomena and lava effusion during solidification of the mare (4.5 billion years ago) can be advocated; yet there is little ground to assume "recent" volcanism (of a few hundred million years ago) as some authors would have it.

(b)   Tycho's ray (latitude $10°64$ south, longitude $20°72$ east) on Ranger VII photograph A 196 (NASA, 1964) is studded with secondaries (Shoemaker, 1966). For the largest in the group just below the middle of the frame, $B_1 = 1.02$ km, $L = 1046$ km, and $\psi = 17°52'$. With $B_0 = 88$ km and $\eta = 1$, Eqs. (140) and (141) yield $s_1 = 5.0 \times 10^8$ dyne/cm^2, $d_1 = 0.25$ km at $w_0^2 = 1.35 \times 10^{10}$,

$w_0 = 1.16$ km/sec. The blocks ejected from Tycho and originating from a continens of postmare age may be somewhat weaker than those from Bullialdus although, as a lower limit, the figure is not binding.

A crater just south of the conspicuous group but outside the ray (frame A 195) has exactly the appearance of the members of the group; if considered a secondary of Tycho with $B_1 = 1.36$ km, $s_1 = 7.5 \times 10^8$ dyne/cm^2 obtains, which makes the strength practically equal to that of the mare background of Bullialdus.

The material is not well suited for the study of crater profiles because of ambiguity in the interpretation of shadows. Also, the theory of fallback for isolated craters is not simply applicable because, in these crowded conditions, craters of an extended area mutually contribute to each other, compensating thus for the ejecta; a considerable contribution may have come from dust and rubble of the ray jet which accompanied the secondary block in flight.

The secondary craters show marked ellipticity, the study of which, however, is complicated for reasons similar to those listed above. Thus, on frame 199, the largest crater shows unusual elongation on reproductions (Shoemaker, 1966), but on the photographic original (NASA, 1964), it clearly consists of two overlapping craters, each measuring about 0.3 km in diameter. Also, it may be assumed that all the impact angles in a limited area of the ray are the same, systematically differing from the isotropic average of 45° and thus considerably influencing Eq. (28). Nevertheless, local differences can be noted even at inspection. Thus, the large ellipticities of a small group of secondaries in frame 199 (NASA, 1964) are not repeated in other groups; either the ground there is harder [smaller D, larger ϵ—Eqs. (7) and (28)], or peculiar shape and splitting of the projectiles are responsible for the deviation.

On frame 199, secondaries as small as 60 m are still visible although eroded—perhaps filled to one-half their original depth. If a diameter of 300 m is roughly the limit of erosion over 4.5×10^9 yr (Öpik, 1965c, 1966c, d), the age of a half-eroded crater one-fifth this size would be one-tenth or 4.5×10^8 years. Although one-tenth the age of the maria, this is still ten times greater than the 50 million years proposed by Shoemaker (1966).

The interior of Tycho shows on Kuiper's Atlas (Kuiper *et al.*, 1960) two craters above the limit of 2.0 km, one measuring 2.7 km on the inner eastern, the other of 3.6 km diameter on the inner western wall. For the area of 6000 km^2, Mare Imbrium carries 4.5 craters to this limit (Öpik, 1960). The age of Tycho could then be some $(2 \pm 1) \times 10^9$ years. Although uncertain, this supports the longer of the two estimates.

(c) Between Copernicus and Eratosthenes, there are magnificent crater chains produced by a salvo from the Copernicus event. A secondary of $B_1 = 6.0$ km, at a distance of $L = 150$ km (reckoned from half-way between the center and the rim of Copernicus), with $B_0 = 88$ km, $\eta = 1$, $2\psi = 4°57'$,

$w_0{}^2 = 2.33 \times 10^9$, and $w_0 = 0.48$ km/sec, yields $s_1 = 8.2 \times 10^8$ dyne/cm², $d_1 = 2.5$ km.

(d) The anomalous frequency of craters in southwestern Mare Imbrium (Öpik, 1960) suggests that secondary craters up to $B_1 = 3.0$ km have been produced by ejecta from Copernicus to a distance of $L = 590$ km. Here $2\psi = 9°44'$, $w_0{}^2 = 8.84 \times 10^9$, $w_0 = 0.94$ km/sec, $B_0 = 88$ km; with $\eta = 1$ and $s_1 = 1.3 \times 10^9$ dyne/cm², $d_1 = 1.03$ km obtains. Possibly, the value of B_1 is taken too high here, but s_1 remains within the expected range.

From the evidence presented here and in the preceding section, it appears that, from an unspecified depth (20–100 m) down to some 10 km, the strength of postmare lunar rocks is about equal to terrestrial igneous rocks.

In recent notes Kopal (1965, 1966b) expresses doubt concerning the impact origin of the "secondary" craters in Tycho's ray as revealed by Ranger VII photographs and interpreted by Shoemaker; he proposes to consider them "subsidence formations, possibly triggered by moonquakes," because the interpretation that they be secondary craters requires, according to his estimates, an unacceptably large total mass of the ejecta. From Shoemaker's (1966) crater counts in and outside the ray, and from the fact that one-fifth of the lunar surface to a distance of 1000 km around Tycho is covered with the secondaries [at an average equivalent thickness of 40 cm for the layer of ejecta (according to Kopal)], the total volume of the ejected boulders turns out to be 250 km³, which is not at all excessive for a total crater volume of 6000 km³. Kopal arrives at a much larger figure by taking a larger area of coverage, and also by overestimating by a factor of 2 the crater area densities read from Shoemaker's (1966) very primitive logarithmic graph. Besides, his use of Nordyke's empirical crater diameter–kinetic energy correlation leads to a projectile mass of 3.4×10^{13} gm to make a crater 1.04 km in diameter, while our estimate (based on first principles, especially on momentum, energy not being the proper scaling factor) yields 2.1×10^{13} gm ($d_1 = 0.25$ km, $\delta = 2.6$) or 60% of Kopal's empirical extrapolation. The disagreement is not significant, yet the smaller mass seems to be preferable. Further, the ray crater distribution is extremely patchy, and Ranger VII frames A 195–199 (NASA, 1964) on which Shoemaker's statistics mainly depend contain an exceptionally dense cluster of secondaries, which does not seem to be representative. The average coverage may be very much less. All in all, instead of Kopal's $5–9 \times 10^3$ km³, the actual secondary ray ejecta from Tycho would amount to a total volume of less than 75 km³, some 1.2% of the crushed crater volume. Subsidence craters of a regular round or elliptical shape, densely populating the area with little mutual interference, and with a "meteoritic" diameter–frequency correlation, are very difficult to understand. From the combined evidence, hardly any doubt remains concerning the secondary impact origin of the craters on Tycho's ray.

D. The Lunar Surface as an Impact Counter

On Earth, the atmosphere prevents the smaller meteoritic bodies from reaching the ground; they are not only decelerated, but also destroyed by ablation (evaporation, melting) and, in the denser atmospheric layers, through crushing and fragmentation. Iron meteorites can withstand the aerodynamical pressure to ground level up to a velocity of 55–60 km/sec, but stones, and especially the loosely bound comet nuclei (Öpik, 1966c), will be crushed at a considerable altitude and arrive as a diverging cluster of fragments. Nevertheless, when the total mass is large enough, and because the linear spread in passing through the atmosphere is more or less constant and of the order of 200 m (Öpik, 1961a), a cratering impact can take place; therefore, on Earth, meteor craters below 1 km diameter down to a few meters can only be produced by iron meteorites, while larger craters can also be produced by stony asteroidal bodies or comet nuclei. Because iron meteorites are intrinsically rare, most of the large meteorite craters on Earth must be due to the nonmetallic bodies.

The Moon, being devoid of the protective shield of an atmosphere, will register as craters the impacts of all cosmic bodies, irrespective of size. The range of craters larger than 1 km, originating from the present population of stray bodies, will be common for Moon and Earth, while smaller craters will be some 50 times more frequent on the Moon, depending on the fraction of iron meteorites among the stray body population (about 2% by mass). In addition, with insignificant erosion, the Moon has preserved all its craters of significant size and postmare age, while on Earth most of them have been erased. When the number of stray bodies of different size in terrestrial space incident on the Moon, as derived from astronomical observations, meteorite incidence, and meteor craters (with allowance for statistical selection and erosion) (Öpik, 1958a), is transformed into crater numbers with a scaling factor $D = 20$ [Eq. (7)], the number of craters in a lunar mare turns out to be in surprising agreement with the number predicted for a time interval equal to the age of the solar system, assuming a constant flux of the stray bodies (Öpik, 1960) as shown in Table XVII. The crater to projectile diameter ratio of 20 is a fair average of the two models in Table XV, $D = 15$ for asteroidal and $D = 26$ for cometary nuclei [the Apollo group "asteroids" can also be only exinct cometary nuclei (Öpik, 1965a, 1966b)], and depends on data and a priori theory unrelated to lunar crater counts, especially on the assumed high strength of the postmare lunar crust. The agreement is good and within the limits of uncertainty of the calculation; it is another link in the remarkable sequence of concordant results based on cratering theory. Archimedes, the largest crater in Mare Imbrium, is a flooded premare object of class 5 and diameter of 70.6 km; it should be excluded from the count (cf. bracketed numbers in Table XVII). In the third line of the table, the numbers, tentatively corrected for Copernican and

Eratosthenian secondaries (*see* Section IV,H), are supposed to represent primary craters only. The smaller observed number for the smallest craters could be due to incompleteness of the count, although it was considered complete by the author (Öpik, 1960). The constancy of the stray-body flux over so long an interval of time is readily explained by their transient character; their elimination lifetime is short, of the order of 10^8 yr, and they are steadily injected from two main sources that have suffered, as yet, little depletion since the beginning— chiefly from Oort's sphere of comets (Öpik, 1966a), and some few from the asteroidal belt (asteroids crossing the orbit of Mars).

Table XVIII contains a similar comparison for supposedly primary craters counted by Shoemaker and Hackman (1963) as adapted by Baldwin (1964b) over a much wider area of combined maria. The observed numbers are again smaller than the calculated ones for small craters, and definitely larger for the large craters (greater than 10 km), thus confirming the trend shown by Table XVII based on a smaller sample. The very persistence of the deviations for the two differently selected samples points toward their reality, as well as to an external cause, rather than an internal lunar factor governing the distribution.

The obvious conclusion is that all maria have been exposed to bombardment by interplanetary bodies for the same length of time, about 4.5 billion years and that, when their surfaces solidified, no significant numbers of the original swarm of planetesimals orbiting the Earth had survived.

Other comprehensive crater counts and discussions (Dodd *et al.*, 1963; Hartmann, 1965), especially the review by Hartmann (1966), which includes statistics of small craters from Ranger VII and VIII, support these conclusions.

Fielder (1963) attempted to estimate the absolute age of maria and continentes by assuming it to be proportional to the number of craters per unit area. In such a manner, assuming the continentes to be 4.5 billion years old, he ascribes to the maria an age of the order of 100 million years. This kind of reasoning is completely unfounded, even in the light of his "internal origin" hypothesis. However, relative ages can be inferred from the crater densities. The scarcity of craters on the maria rightly indicates that their surfaces solidified after the end of intense bombardment, but the time lag may be only a few thousand years. During the subsequent 4.5 billion years, about one-twentieth the number of craters per unit area were imprinted on their surfaces than were imprinted in the preceding thousand-odd years on the continentes and on their own surfaces before they were flooded.

Crater counts by Baldwin (1964b) on the flooded floors of the class 5 craters Ptolemaeus and Flammarion (in the central highlands) show intermediate crater densities between maria and continentes, about six times those in an average mare. Öpik even finds from Ranger IX photographs for Alphonsus and Ptolemaeus a density of about 20 times that in an average mare (Section V,E).

Apparently, these floors solidified at a premare stage when the remnants of the earthbound cloud of planetesimals were still there.

On the contrary, two major flooded craters around Mare Imbrium do not show excessive numbers of craterlets. On the same Mt. Wilson photograph of 15 September 1919, which was used for the Mare Imbrium counts (Öpik, 1960), there are 8 craters on the flooded floor of Archimedes (2560 km²) and six on the floor of Plato (4340 km²); down to the effective diameter limit of 1.1 km, this gives a crater density per 10^4 km² of 31 ± 8 for Archimedes, and 14 ± 4 for Plato, as compared to 13.5 for northern, and 25.6 for southern Mare Imbrium (Section IV,H) as affected by Copernican secondaries. Within the limits of the probable error of sampling (Archimedes being on the borderline between the two halves, the crater density in its strip being 20 ± 1.5), these figures seem to indicate that the floors of Plato and Archimedes were more or less contemporary with, or soon followed, the Mare Imbrium event, i.e., that no premare impacts have left their traces on them.

The excessive number of craters in Ptolemaeus and Flammarion suggested to Baldwin (1964b) the possibility that some of them might be of internal origin, blowholes on these "lava extrusions." This suggestion is unnecessary, a sufficient increase in the number of impacts being obtained by predating them a few hundred years into the premare period; the "normal" crater numbers in Plato and Archimedes weigh against the blowhole hypothesis. Why should these blowholes be present in some and absent in other lava-covered craters of comparable dimensions? As to lava "extrusions," apparently none did take place, the melting being caused *in situ* by impact heating of the already hot crustal material. Some limited volcanic events may indeed have taken place on the Moon, on the background of collisional melting, but such formations seem to be few and small, like the famous black spots in Alphonsus, which Urey (1966), on the evidence of Ranger IX photographs, considers to be caused by eruptions; they can hardly distort the cratering statistics.

Submerged "ghost craters," in which the outlines are feebly visible without any surface relief, may have a dual origin: either they are traces of normal craters caught and destroyed by the flood; or are they the result of impacts occurring when the lava had not solidified. A list of 42 more conspicuous ghost craters in the maria is given by Fielder (1962). These objects are visible because material with different reflectivity or coloration is admixed to the lava melt. The semidestroyed flooded craters, chiefly on the borders of the maria, represent a transition from normal to ghost craters; typical examples are: Fracastorius (97 km) on the southern border of Mare Nectaris, LeMonnier (53 km) on the western border of Mare Serenitatis, Kies (45 km) near to, and Campanus (48 km) on, the southern edge of Mare Nubium. Sinus Iridum is perhaps the most striking example of this type of object.

To the same category belong the extended, sharply bounded color provinces in the maria, detected through multicolor photography (Whitaker, 1966), by superimposing an infrared positive (7800 Å) on an ultraviolet negative (3800 Å). As pointed out by Kuiper (1966), they are indications of lava flows of different composition; however, one cannot agree with his comment that "the lunar maria are not covered with even 1 mm of cosmic dust, which would have obliterated the color differences" (*loc. cit.*, p. 21). There may be up to 20 cm cosmic dust material accumulated over the ages (cf. Section VII,B), but this is mixed with a much greater amount of granular material from the local bedrock, which determines the coloration.

From Ranger photographs, crater statistics have been extended down to meter sized objects by Shoemaker (1966), Hartmann (1966), and also Kuiper (1966). For the postmare craters, a very remarkable detail in the frequency curve of diameters is revealed; down from about 1.5 km diameter there is an upward surge in the crater frequencies (rate of logarithmic increase), which then is checked at about 300 m diameter where the rate of increase drops. The surge must be due to the appearance of secondary craters outnumbering the primaries, while the decline in the increment can be ascribed to erosion, which limits the lifetime of the craters and thus their number roughly in proportion to the diameter itself (Öpik, 1965c, 1966c,d). The lifetime of a 300-m crater can be set at 4.5 billion years, corresponding to an elevation (rim) of the order of 15 m being carried away and an equivalent depression filled. This more or less agrees with theoretical estimates of erosion by micrometeorite impact (Section VII,B and C, and Section X, A, D, and E).

On the Ranger IX photos, the density of craters on the flooded floor of Alphonsus is about triple that in adjacent Mare Nubium, but below a 500-m diameter the numbers in Alphonsus and in the mare become approximately equal and more or less the same as in the Ranger VII and VIII mare samples (Öpik, 1966c, from a verbal communication by Shoemaker). This implies again that the smaller premare craters have become eroded, and only postmare craters survive.

One of the requirements of the impact theory of lunar craters is the randomness of their distribution. If randomness is defined as the unpredictability of the place and time of an event, the distribution of lunar craters undoubtedly conforms to this definition.

Recently Fielder (1965), a prominent proponent of the volcanic theory, tried to prove that the craters are not distributed at random, yet he only demonstrated that the distribution is not of the elementary Poisson type—a conclusion that is obvious from even a casual inspection of the lunar map. The Poisson formula requires that all points (crater centers) be placed on the surface individually, independently, and without mutual interference. This is by no means fulfilled by the cratering phenomena; a large crater wipes out all former smaller craters within its ramparts, creating a void that would appear extremely improbable or even practically impossible in a random distribution of points.

Overlapping may not be significant in the maria, but secondary throwout craters there are very numerous; these have a tendency to group (Copernican crater chains or salvos, Tycho rays), thus rendering futile the use of the Poisson formula.

Imagine that 2500 craterlet centers are thrown at random over an area divided into 100 squares without mutual interference, so that the average number per square is 25; now let a subsequent large crater erase all the craterlets in one square, and let a cluster of 25 secondary craters be added to another square, so that the total number remains unchanged. According to the elementary Poisson formula, the mathematical expectation or, practically, the probability of having one empty square is $100 \times \exp(-25) = 1.4 \times 10^{-9}$, and the probability of having 50 objects in another is $(100 \times 25^{50}/50!) \exp(25) = 3.6 \times 10^{-4}$. The Poisson probability of both these unusual squares equals then their product or 5×10^{-13}, a practical impossibility. Yet both abnormal squares are the result of random events which are not unusual at all.

Clearly, probabilities of crater distributions calculated from the Poisson formula are meaningless, as has been pointed out by Öpik (1966e, f) and Marcus (1966a). Refined mathematical studies of the distribution of impact craters according to area density and diameter have been made by Marcus (1964, 1966a), by taking into account the formation of primary and secondary craters, overlap, and destruction by obliteration and filling, with some of the relevant parameters based on observation or experiment. The spatial distribution of the lunar craters conforms to a purely random pattern, but the observed numbers of craters less than 1 km (from Ranger VII) are much greater (10 times at 100-m diameter, 20 times at 10 m) than those predicted from experiments with terrestrial explosion craters. Marcus concludes that "If the observed excess is real, then either some primary craters produce an unusually large number of secondaries, or else many of the smaller lunar craters are of internal origin." An internal origin of the small craters is the least likely thing to assume—those which originated soon after the melting stage have been obliterated now by erosion, and recent volcanic formations are no more probable than ghosts. On the other hand, terrestrial cratering experiments have been performed on weakly cohesive media ($s_c = 6.5 \times 10^7$, desert alluvium) while the postmare lunar crust is perhaps 15 times stronger and, according to Eq. (140), would produce 8 times larger secondary craters (B_1) for a given primary (B_0). Clearly, the results of terrestrial experiments cannot be adapted directly to the lunar craters without using proper scaling procedures in which all the parameters including the strength of the bedrock should be taken into consideration.

E. Alphonsus and Its Peak

The class 5 crater Alphonsus (Fig. 7) is one of the most remarkable, yet still typical premare formations. Despite the negative conclusion regarding the

Fig. 7. Crater Alphonsus (left) and Mare Nubium (right). Ranger IX at 2 min 50 sec before impact on 24 March 1965, from an altitude of 258 miles. Dimensions of frame 121 × 109 miles. (Photograph by courtesy of the NASA Goddard Space Flight Center, Greenbelt, Maryland.)

suggested recent "volcanic eruption" from its peak, the fact of fluorescent luminescence is important in itself, and there are many features in the crater that point to some kind of plutonic activity (Urey, 1965, 1966), not recent but dating back to the premare stage. The broad features of the crater, however, can be interpreted on the collisional theory of cratering, where the original cause of melting and of the transient plutonic activity is also to be sought.

The crater is bisected by a broad band in the N–S direction, a very low uneven ridge of lighter color; this is not an indigenous feature of the crater but a "scar," a splash from the Imbrian collision, which came on top of the completely formed crater.

The simultaneous presence of a corner of Mare Nubium and Alphonsus on Ranger IX frame A36 (NASA, 1965a) (Fig. 7) lends itself readily to comparison. Table XIX contains results of crater counts by the author.

The predicted number of interplanetary impacts is calculated on the same basis (Öpik, 1958a, 1960) as in the preceding section. Contrary to the observed deficiency of small craters, as revealed in Tables XVII and XVIII, the number of craters in this part of Mare Nubium is 1.7 times the predicted number; the excess must be caused by secondaries, the region being within the reach of Tycho's rays.

The interior of Alphonsus contains 6.4 times more craters than Mare Nubium; thus, the density of craters in Alphonsus may correspond to 15–20 times that in an average mare. The frequency of these small premare craters (yet well above the erosion limit) thus exceeds the mare crater density in about the same ratio as that for large craters in the highland surroundings. This indicates that the crater floor solidified rapidly [cf. Eq. (123)] and at an early stage, to become a recipient of the premare bombardment. A superficial comparison with Ptolemaeus (also of class 5) on Ranger IX frame B17 (NASA, 1965a) indicates that its crater density, to the diameter limit 1.1 km, is approximately the same as in Alphonsus. The floors of these two craters must have solidified at about the same time, which, for similar reasons, has been suggested by Baldwin (1964b) for Ptolemaeus and Flammarion, although he obtains a systematically smaller number of small craters.

The fluctuations of crater densities over different regions in Table XIX are also much greater than their sampling errors, apparently the result of unequal occurrence of secondaries.

It has been pointed out by O'Keefe (1966a) that the absence of craters on the illuminated slope of the peak of Alphonsus must have a very particular significance; he suggests a volcanic origin of the peak. Indeed, in a search by the author, on Ranger IX frame A63 [Fig. 8 (NASA, 1965a)], as well as on several other adjacent frames, no trace of craters could be found on the main slope; there are, however, two at the southern fringe of the slope, one just at the foot where the rise begins. Four or five small shadows could be seen or suspected, but on the wrong side, indicating mounds, not cavities. This contrasts drastically with the plentitude of craterlets on the Alphonsus floor, although the brighter band of the Imbrian splash again contains fewer. Counts in sample rectangles of $4.16 \times 5.76 = 24.0$ km², equal to the illuminated area of the peak, are summarized in Table XX. In the bottom part of the table, average densities (with probable error of sampling indicated) are compared with the predicted number of interplanetary impacts, during 4.5×10^9 yr, calculated as before. The "crater density" for the peak depends upon one single entry yet, if the systematic deviation observed, or predicted, keeps the same trend as in Tables XVII and XVIII, the agreement is "perfect," the observed value being two-thirds of that predicted. Besides, many crater diameters are now near the

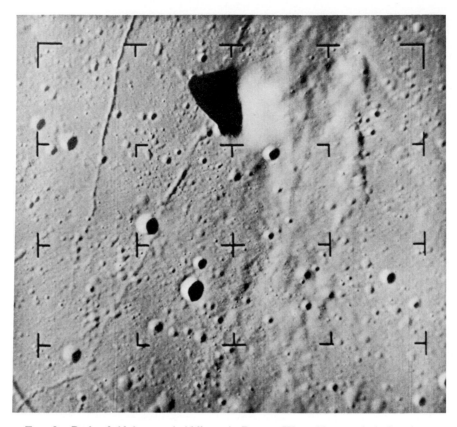

Fig. 8. Peak of Alphonsus (middle top). Ranger IX at 39 seconds before impact, from an altitude of 58 miles. Dimensions of frame 28 × 26 miles. (Photograph by courtesy of NASA Goddard Space Flight Center, Greenbelt, Maryland.)

300-m erosion limit, hence the older ones must have been strongly eroded and become unrecognizable, so that the observed number must be smaller on this account also. It is clear that the peak must be susceptible to interplanetary high-velocity collisions, but that the other kinds of impact, which account for the high crater density on the floor of Alphonsus, did not impress the peak. These are secondary impacts of low velocity; the most probable explanation is that the peak material is harder than the throwout blocks, so that they were crushed at impact without leaving a crater mark on the peak.

The altitude of the Sun over the photographed region of Fig. 8 was $10°.8$, and the slope of the peak turned toward the Sun rose another $10°.8$ from the horizon (the steepest slope was $19°.7$), so that the rays of the Sun made an angle of $21°.6$ with the slope. Shadows were shorter than on level ground, which must have made the recognition of shallow craters more difficult. This, however, cannot explain the complete absence of craters, especially because, on the sides of the peak, sunrays were falling more obliquely; not more than 20–30% of the craters could have been lost on this account. Indeed, counts in Mare Cognitum on Ranger VII photographs (NASA, 1964, 1965b), where the Sun's altitude was $22°.1$ (A) and $22°.0$ (P), respectively, showed an abundance of craters.

On Ranger VII photograph, frame A 193, in the upper left central quadrangle between the reticle marks covering $64.0 km^2$, 46 craters were counted down to an effective diameter limit of 0.24 km. Reduced to a limit of 0.270 km as in Table XX (with the inverse square of the diameter as correction factor), this yields 5700 ± 590 craters per $10^4 km^2$. The area is free from Tycho's ray (Shoemaker, 1966).

On Ranger VII, frame P_3 128, in the upper half of the square, free from clustering and probably not affected by Tycho's ray, 55 craters down to a limit of 0.260 km were counted in an area of $77.4 km^2$. Reduced to the diameter limit of 0.270 km and $10^4 km^2$, this yields a density of 6500 ± 620.

The densities of small craters in the chosen regions of Mare Cognitum are comparable to, although smaller than, those in Alphonsus (Table XX); even allowing for less favorable illumination and the choice of less crowded regions (outside conspicuous clusters), the densities probably are still lower as it should be for postmare craters if some premare craters above the erosion limit have survived in Alphonsus. At the same time it is obvious that illumination is not responsible for the absence of craters on Alphonsus' peak.

The absence or scarcity of craters on the peak is real. With an average of 38 craters found on an equal area on Alphonsus' floor, and if 20 craters are taken, allowing for less favorable illumination, the Poisson probability of finding one or none in such an area is 4×10^{-8}, too small to reconcile with an accidental avoidance.

The assumption of a recent volcanic origin of the peak, some 10^8 yr ago, may seem an easy way out. The undisturbed surface right to the foot of the peak (Fig. 8), without traces of having being disturbed by the eruption of an active volcano of this size, does not favor the suggestion. The shape is not that of a volcanic cone. Other peaks of similar chracter, like the group of angular blocks in Copernicus (shown on the much publicized picture taken on 23 November 1966, by Lunar Orbiter II), all confined to near-central regions of the respective craters, indicate close relationship to the entire buildup of the crater during impact. Further, the assumption does not help much: the crater walls of Alphonsus are also conspicuously poor, almost devoid of craterlets, while a little plateau in between the Alphonsus wall resembling a dry lake bed, according to Urey (1965), is studded with craters (Fig. 9). The crater wall cannot be explained away as being of recent origin. As was rightly pointed

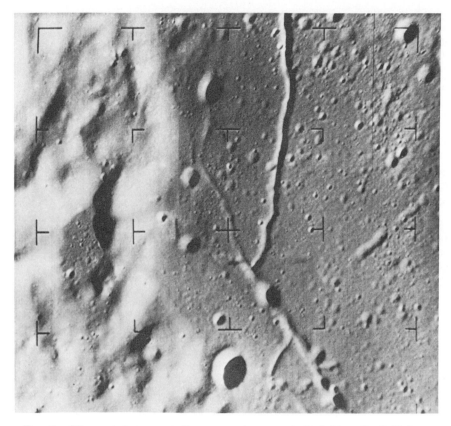

FIG. 9.   The east (astronautical) or west (astronomical) (left) wall of Alphonsus. Ranger IX at 1 min 17 sec before impact, from an altitude of 115 miles. (Photograph by courtesy of NASA Goddard Space Flight Center, Greenbelt, Maryland.)

out by Urey, the phenomenon could be explained by a harder material, possibly even nickel–iron, which is not affected by low-velocity impacts.

Another possibility suggested by Urey (1965) is that the majority of craterlets on the floor of Alphonsus are collapse features, not secondary impact craters at all. However, as shown by the counts reported above, the densities of these small craters (not the large ones) in Alphonsus are not exceptional, but are closely the same as those found in Mare Cognitum. Equality of the number of collapse features in such widely distant areas (and of different age and origin) is extremely improbable. Also, the craters at this size limit (0.2 km) are still essentially round as a rule, a strange, nay incredible regularity. There may be some collapse features (none yet proved), but their statistical importance is undoubtedly negligible.

The crevasses or rills on the floor of Alphonsus (Figs. 7–9), as elsewhere, are apparently cracks caused by solidification and cooling. The width of the strongest rill in Fig. 9 is 500–1000 m and its average depth about 75 m, with a slope of 11°—not so very steep as it looks. An impression is almost formed that the crevasses are just chains of craters and that these are just collapse features, but this is hardly true. There are so many craters on the floor that any line drawn may connect the craters like "beads" on a string, with but small deviations—and, the wiggles are actually there. Raindrops on a car window can also be seen running in almost straight lines, collecting previous drops that are distributed at random. A crater impacting near an existing crevasse will expand asymmetrically toward the void as the direction of least resistance and will thus be attracted by it. A preexisting crater will tend to collapse and join the crevasse on its nearest side.

There remains the only plausible explanation: that the Alphonsus peak and, to a slightly lesser degree, its walls consist of a hard material unaffected by the secondary impacts. The number of secondary craters on this hard rock must be reduced by a factor of at least 10–20, if not to zero. According to Shoemaker (1966), the cumulative frequency of secondary craters, both in terrestrial experiments (Sedan nuclear explosion) and on the Moon (Langrenus), varies nearly as the inverse fourth power of diameter. If, for given projectile size, the diameters are reduced to one-half on hard rock, the crater numbers will be decreased to one-sixteenth, which would suffice to explain the deficiency on peak and wall, allowing also for unfavorable illumination.

Using the suffixes a and b for the hard and soft ground parameters, respectively, the following sample calculation illustrates the point.

The unevenness of the secondary crater distribution on the floor of Alphonsus points to nearby sources of the ejecta, either inside, or near the outside of the crater. A distance of 100 km, and a velocity of $w_0 = 400$ m/sec [Eq. (45)] suggests itself. The frontal inertial ("aerodynamic") component of pressure at encounter, with $\rho = 2.6$ and $K_a = 0.75$, is then about 3×10^9 dyne/cm^2 and the total pressure higher by a value of s_p , the weaker of the two [Eq. (29)]. This is more than can be resisted by a stony material, so that Eqs. (6) and (7) would apply. With $k = 2$ for both cases, $s_p = 2 \times 10^9$ and $s_c = 9 \times 10^8$, assuming a substance such as granite for the hard substance (a), and $\delta = 2.6$, $\rho_a = 2.6$, $\rho_b = 1.3$, $s_p = 1.3 \times 10^8$, $s_c = 6 \times 10^7$ for a material such as desert alluvium for Alphonsus' floor (b) (upper 20–30 m only), $D_a/D_b = 0.447$, $(D_a/D_b)^4 = 0.04$ results, a ratio that is able to explain the virtual absence of craters on the peak, and their scarcity on the wall of Alphonsus, without recourse to exceptionally hard substances (such as iron).

So far we are mainly on a theoretical basis. If the explanation is correct, the peak should carry a great number of smaller craters, say 20, to a limit of $0.27 \times 0.447 = 0.12$ km. Unfortunately, there are no observations to confirm this, the last close view of the complete peak having been obtained on Ranger IX frame A 65; indeed, at diameter limit 0.20 km 3 craters are seen.

Better direct evidence is provided by the Alphonsus wall (Fig. 9), which also exhibits a scarcity of craters that is probably due to the same cause. On Ranger IX frame B 77, which contains a closer view of the wall, on an area of 112.7 km² the author counted 98 craters to effective diameter limit 0.141 km; this gives a density per 10^4 km² of 8700 $\pm$ 620.

For comparison, down to 0.27 km, the density on the floor of Alphonsus is 16000 $\pm$ 550 (Table XX). Another count by the author on the same frame A 63, down to diameter limit 0.54 km, gave 1440 $\pm$ 120 per 10^4 km² (six equal, marked quadrangles, excluding the two containing the peak, of 88.3 km² each, gave 17, 18, 11, 11, 9, and 10 craters each). The two counts correspond to a "population index" of 3.5 for the negative power law of cumulative crater numbers, depending on the limiting diameter. Logarithmic interpolation then yields a diameter limit of 0.322 km at a density of 8700. Assuming that at equal density, equal projectile populations were at work, the counts thus indicate that a projectile, which produced a crater of $D_b = 0.322$ km on Alphonsus floor, could only produce one of $D_a = 0.141$ km on the wall. Hence $D_a/D_b = 0.438$, as derived from the crater statistics, is in unexpectedly close agreement with the value predicted from plausible assumptions as to the mechanical properties of the surface materials. The empirical crater density ratio is then $(0.438)^{3.5} = 0.056$, essentially the same as the predicted ratio.

Combining this with crater profiles and other evidence for the mechanical properties of lunar rocks, it is evident that not only is there hard rock on the Moon under a layer of more loose material (10–30 m thick in the maria), but that crater walls and central peaks contain, or consist of, outcrops of these solid rocks, covered perhaps by a very thin insulating dust layer.

It remains to be seen how such an immense solid block could have arrived in the midst of Alphonsus (and other craters with peaks). From the shadow (Fig. 8), the summit, 970 m, is asymmetrically placed over the southwestern sector of the base, measuring 7.7 km from north to south and 6.5 km from east to west. The steepest slope is between southeast over south toward southwest, inclined 20° to the horizon, while the illuminated eastern slope is inclined 13°, and the northeastern only 10° (direction from summit to foot of the mountain) (east and west are reckoned astronomically). It could be compared to a more or less rectangular slab of butter on hot porridge. Undoubtedly, below the visible top there must be a broader extension underneath.

A tempting and most probable hypothesis is to consider the peak a direct remnant of the planetesimal which produced the crater. In the rear portion of the impacting body, the pressure is less than at the shock front in proportion to the thickness of the layer, and a certain layer may survive when the pressure is less than the plastic limit, s_p. A loose aggregate (comet nucleus) may even be compressed into a dense mineral, part of which may be destroyed by shearing, yet a part may survive. By analogy with Eqs. (134) and (135), with $x_0 \sec \gamma$ substituting for $0.1 B_0$, the average thickness h_p of a surviving hard kernel may be set equal to

$$h_p = 2s_p x_0 \sec \gamma/(w_0^2 \delta), \tag{142}$$

in former notations or, for $w_0 = 3 \times 10^5$ cm/sec as for the premare collisions, $s_p = 2 \times 10^9$ dyne/cm², $\gamma = 45°$, and $\delta = 2.6$ gm/cm³,

$$h_p = 0.024x_0 . \tag{142a}$$

Using Model D. of Table XIV, at $B_0 = 120$ km (diameter of Alphonsus), $x_p = 25$ km, $x_0 = 20$ km, and hence $h_p = 480$ m. This may be close to the average thickness of Alphonsus peak (one-third of a cone 970 m high plus 160 m underground).

It is thus possible to explain the peak as the hardened surviving kernel of the rear portion of the planetesimal, reflected back to the surface after penetration. Unlike the hot surface of the primitive Moon, easily melting at impact, the planetesimal was cold and its rear portion, suffering little compressional heating, was not melted.

The excess weight of a block of the above-mentioned dimensions, 320-m average height above ground and 160 m half-balanced by buoyancy, amounts to 1.7×10^7 dyne/cm², which is much less than can be supported by a material of the assumed strength $s_p = 1.3 \times 10^8$; the latter estimate, which successfully accounts for the scaling of cratering on the floor and walls of Alphonsus, refers to the mixed premare and postmare crater population with prevalence of the postmare stage (as follows from the comparison with Mare Cognitum) when the material had cooled and hardened. The Alphonsus event, however, belongs to the premare stage when the material was hot and soft. A minimum bearing strength of 1.7×10^7 is thus required for this stage, too. Clearly, the material could not have been liquid lava, at least not to any considerable depth, otherwise the peak would have sunk. Also, liquid lava would have solidified to hard rock, equal in strength to the peak and wall, while the cratering statistics indicate a much inferior strength for the floor. It follows that the material was not completely melted, yet was mobile enough to fill the floor to an approximately uniform level. A mechanism similar to ash flows as suggested by O'Keefe (1966a) appears to have been at work.

As shown in Section IV,I, a considerable fraction of the material must have become completely melted at impact. Where did it go? For Alphonsus, we assume Model D of Table XIV at $x_p = 25$ km, which yields the correct crater size of 120 km; the parameters are the same as those used in the mare impact model of Section IV,I except that, for the highly elastic liquid fraction, $\lambda_x^2 = 0.5$ is to be assumed, instead of $\lambda^2 = 0.25$, and $(\lambda_x^2)_{av} = 0.125$ as for the crushed

solid granular fraction with high internal friction. The diameter of the projectile
($\delta = 1.3$) is $25/1.093 = 22.9$ km and its penetration is $25 \times 0.8 = 20$ km, but,
because of flattening, the rear portion (presumably compressed from $\delta = 1.3$
to $\delta = 2.6$) will follow deep into the crater and must be reflected back to the
surface to make the peak. The melted fraction is $y_i = 0.375$, ejected with a
velocity $v = u\lambda_x = 0.74$ km/sec in a direction β (Fig. 1) such that $\sin \beta =$
$0.8y_i = 0.300$ [Eq. (27)]. The flight distance of the liquid spray is then 190 km,
starting from a melting fringe about 24 km inside the present rim. All the liquid
would have been sprayed around, far beyond the ramparts of a crater even of
Alphonsus' size. This is an intrinsic property of the mechanics of shock melting,
depending only on the linear dimension of the crater, surface gravity, and state
of preheating, and not on the velocity of impact or the strength of the material.
A cold surface would require a stronger shock for melting and would spray
the smaller liquid fraction to a greater distance. Real lava flows from meteorite
impacts can thus be caused on the Moon only on the scale of a mare. The class 5
flooded craters cannot be regarded as lava covered, but rather as filled with
the mobile "porridge" of partially molten debris, remaining in the crater
because of lower elasticity and shock velocity.

VI

THE TOP LAYER

A. *Dust and Rubble; Optical, Dielectric, and Mechanical Characteristics*

The uppermost reflecting and insulating layer on the Moon's surface has
usually been referred to as "dust." There have been objections to the term for
various reasons, partly because proponents of the dust concept have sometimes
ascribed to it extreme properties, e.g., great mobility, excessive depth, which
did not appear realistic.

The small depth to diameter ratio [Eq. (9)] of the craters on Ranger VII,
VIII, IX, Luna IX, and Surveyor I pictures definitely shows that the surface
is granular and finely divided, not pumice-like or a continuous solid. A very
convincing study in this respect by Gault *et al.* (1966) is based on cratering
experiments in fragmental media at velocities of 0.6 and 6.5 km/sec and angles
of incidence of 0 and 60°. The same follows from a consideration of the scope
of hypervelocity cratering experiments (Walker, 1967). In O'Keefe's words
(1966b), it is "a network of space with grains in it, rather than a network of
rock with space in it." "Dust" is still the best term to describe it, notwith-
standing its large rocky inclusions and its cohesive properties, the dust particles
being cemented together through contact *in vacuo*, or by deposition of vaporized
substances (from meteorite impact and solar wind sputtering). The dust possesses

potential mobility, when, by impact shock, the particles are sent flying around in small or large cratering events.

The origin of the dust is to be seen in the battering of the surface by interplanetary particles, as well as in the secondary throwout debris of cratering events. Direct accretion of micrometeoritic material is but a minor source of the dust; most of it is the product of destruction of lunar rocks by impact [cf. Eqs. (3), (4), and (14)]. For this reason the coloration of the dust must reflect the properties of the substratum from which most of the dust material is derived, whence the differences in shade not only between the maria and the continentes, but also between minor local formations such as ghost craters and color contrasts in the maria. Horizontal transport of dust on level ground is induced by micrometeorite impact; it is gravity dependent and, theoretically, limited to a few kilometers (Sections X,A,B, and C). The sharpness of some demarcation lines, such as the southern border of the eastern dark spot in Alphonsus (Fig. 7; very visible on this and original Ranger IX frames A 34–36, which show a broad dark band across the crater north of the peak, joining the eastern with a western dark spot), would limit effective migration to less than 0.5 km. This also would deny electrostatic migration, as first proposed by Gold (1955), any important role as has been already pointed out on theoretical grounds by Singer and Walker (1962); their negative conclusion is even more valid if, instead of 30 V, the photoelectric potential of the lunar surface were less than 10 V (Öpik, 1962b), the electrostatic force on a particle varying as the square of the potential. Electrostatic "hopping" of the dust would provide a means of transport almost unlimited by distance and would obliterate all sharp coloration differences on the lunar surface (except the ridges of crater walls and other elevations), which certainly is contrary to the most obvious observed facts.

A clever experiment by Gold and Hapke (1966) led to a superficially close imitation of the main features of the lunar surface microstructure down to about the 1 mm scale. By repeatedly throwing commercial cement powder (average grain about $1\,\mu$) at a layer of similar powder "until the statistical nature of the surface is no longer changed by such further treatment," a close replica of the Luna IX or Surveyor I pictures of the small-scale lunar surface near a spacecraft was obtained, including apparent "boulders" of up to 8 cm diameter. Powdered dyes were added to imitate the actual albedo and photometric properties of the lunar surface. The mineral composition of the powder is probably irrelevant, but a small particle size is essential to make it stick at impact. Little steep, almost vertical, ridges were formed, in defiance of any angle of repose. "Pebbles" were also produced, but all these formations had little strength and collapsed when touched by hand.

The experiment differs from lunar conditions in that the material was taken from outside and thrown at the surface with a relatively low velocity. On the

Moon, the material is ejected from cratering impacts, which destroy the previous structure in the crater area and which also easily blow up the false "pebbles," "boulders," or miniature ridges of low cohesion. Otherwise there is considerable similarity, and it appears that fine dust would stick even to vertical surfaces (of true boulders), somewhat protecting them from further erosion until it is shaken off by new impacts.

The density of the dust is expected to increase with depth. Experiments by Hapke (1964) on the compressivity of fine powders suggest a density–depth relationship for dunite powder on the Moon (particle size less than $10\,\mu$), if gently placed and left undisturbed under its own weight, such as that in Table XXI. The data are slightly smoothed.

On the Moon, in the absence of an atmosphere, the cohesion between grains and resistance to compression may be greater and the density smaller. On the other hand, continuous battering by meteorites (micrometeorites) should lead to tighter packing and tend to increase the density of the dust. The figures of Table XXI are thus probably minimum values, especially near the surface. Dunite is believed to be characteristic of the silicates in the Earth's mantle and more similar to undifferentiated cosmic material than granite or basalt. However, the mechanical properties of other kinds of rock powder such as basalt (now believed to represent best the composition of the lunar surface) should be similar.

Radar reflectivity provides, although not without a certain ambiguity, an observational means for estimating the density of the reflecting layer, as well as the distribution of slopes, which extend over a linear scale greater than the wavelength (Evans and Pettengill, 1963; Hagfors, 1966). In the wavelength region from 3 cm to 8 m, the lunar surface is essentially a specular reflector, which indicates that the effective roughness is less than 1 cm. From the distribution of echo ranges it is found that nearly 90% of the echo power comes from a central region of about one-tenth of the lunar radius, explained by specular reflection from a gently undulating surface, with the reflecting elements only slightly inclined to the horizon. The remaining 10% of the reflection is diffuse, continuing to the very limb, and can be ascribed to "boulders" or blocks of the order of 1 m. This description, originally proposed as a hypothesis, is now confirmed by the Luna IX and Surveyor I (also III, V, VI, VII) photographs (Figs. 5 and 6).

For long wavelengths, the dielectric constant ϵ_i equals the square of the refractive index and is determined by Fresnel's formula,

$$\epsilon_i^{1/2} = (1 + A_r^{1/2})/(1 - A_r^{1/2}), \tag{143}$$

where A_r is the reflectivity at normal incidence—which is always the case with radar. From the reflected power, A_r cannot be determined unambiguously; an assumption regarding the distribution of the reflecting elements has to be made.

The current model, confirmed by the close-up pictures, assumes reflecting elements that are small as compared to the lunar radius, with inclinations distributed at random. This leads to $\epsilon_i = 2.6$–3.0, to compare with 2.6 for dry sand, 4.8 for quartz or sandstone, 5 to 6 for most sialic rocks, 17 for olivine basalt, and 20 for meteoric material.

The interpretation in terms of bulk density (or porosity) is also somewhat ambiguous. A plausible formula by Twersky (1962) leads to

$$\rho/\rho_0 = (\epsilon_i - 1)(\epsilon_0 + 2)/(\epsilon_i + 2)(\epsilon_0 - 1), \qquad (144)$$

where ρ_0 and ϵ_0 are density and dielectric constant of the compacted parent rock, respectively, and ρ and ϵ_i those for its granular or porous derivative. For quartz sand the formula yields $\rho = 1.62$, close to the usual value. For the lunar surface, with $\epsilon_i = 2.6$, the bulk density for quartz as parent would also be 1.6, and for typical silicate rock material, $\epsilon_0 = 6$, $\rho_0 = 2.6$, and $\epsilon_i = 2.6$, the density becomes $\rho = 1.44$ with 44% porous unfilled volume. If olivine basalt is the parent, $\rho = 1.36$ with 59% porous volume would obtain. The depth to which this information pertains is of the order of a wavelength, thus from a few centimeters to 10 meters.

Another formula, by Krotikov and Troitsky (1962),

$$\rho/\rho_0 = 3(\epsilon_i - 1)\,\epsilon_0/(2\epsilon_0 + \epsilon_i)(\epsilon_0 - 1), \qquad (145)$$

yields $\rho = 1.30$ for sand, a value that is too low, and for the lunar surface the same value, or even $\rho = 0.46$ if olivine basalt is the parent rock.

On this model, the surface is a random combination of relatively smooth elements, extending perhaps for 10–1000 m and with an average inclination of 5–8° (Evans and Pettengill, 1963). According to Evans (1962), the average gradient of points spaced 68 cm apart appears to be 1 in 11.5 and of points spaced 3.6 cm apart it appears to be approximately 1 in 7.4; the radio albedo is 7.4% at meter wavelength. According to Hagfors (1966), on the scale of a meter the mean slope is 11–12° or 1 in 5, and at 3.6 cm wavelength it is about 15°.

A different model of radar reflection, proposed by Senior and Siegel (1960), by Senior (1962), and favored by Russian workers, assumes reflections from large elements comparable to the lunar radius, with corresponding radii of curvature. It requires a larger reflection area and leads thus to a smaller reflectivity; $\epsilon_i = 1.1$ and ρ of the order of 0.14 [Eq. (144)] are obtained. It is difficult to see how reflecting surfaces could retain a significant radius of curvature, or complete smoothness, on such a scale; close-up pictures of the Moon deny a reality to this model which also leads to unacceptably low values of the density.

Other models are possible, too, and there is as yet no formal way of deciding among them on the evidence of radar alone. The detailed law of the distribution of the reflecting elements leaves some freedom of adjustment. With this reservation, the Evans–Pettengill (1963) model is to be regarded as the best approach to reality. Conventionally, the density of the upper layer at decimeter to meter depth will be assumed in further calculations to be $\rho = 1.3$. This is also the probable density of comet nuclei and their dustballs after the evaporation of the ices (Öpik, 1963a, 1966a,c).

The average inclination of the reflecting elements increases with decreasing radar wavelength, which indicates increasing roughness with decreasing linear scale. This continues until, in the visible region of the electromagnetic spectrum, an extreme degree of roughness is attained, where there are only opaque reflecting grains of low albedo and much larger than the wavelength, so that diffractional backscatter is virtually nonexistent, and secondary reflections are not important. The grains are separated by cavities into which light and shadows deeply penetrate. This "fairy castle" structure explains the characteristic lunar phase effect. Near full Moon or zero phase angle (angle between incident and reflected ray), shadows are not visible and reflection is observed from the deepest interstices, which leads to the characteristic upsurge of brightness. With increasing phase angle, shadows become visible, while illuminated portions become screened and the brightness drops rapidly.

The most extended photographic measurements (on orthochromatic plates without filter) by Fedorets (1952) on 172 individual lunar points show without exception the dominance of the phase angle in the light curves; the angle of incidence, which by Lambert's photometric law is of exclusive significance, is of secondary importance, so that maximum brightness is reached not when the sun is highest but when the phase angle is near zero. The uniform brightness of the full Moon is thus accounted for, despite the rapid decrease of illumination of a horizontal surface toward the limb (Öpik, 1962a). A comprehensive review of lunar photometry has been given by Minnaert (1961). Three-color photoelectric measurements on 25 lunar features over a close range of phase angle of $\pm 28°$ were made recently by Wildey and Pohn (1964), and meticulous studies of lunar polarization have been made and discussed by Lyot and by Dollfus (1966).

The principal aim of photometric and polarimetric studies was to describe the surface structure and identify the materials. Until recently the second task proved extremely disappointing when comparisons were made with terrestrial minerals. While the geometrical buildup of the lunar surface, like a "fairy-castle" structure of opaque grains with large-scale surface undulation superimposed on it, did account qualitatively for its photometric properties, the detailed variation with phase angle and other illumination parameters, the polarization, and especially the low albedo remained unexplained until it was shown that irradiation by a proton beam changes almost all mineral powders into substances of low albedo with lunar photometric characteristics almost independent of the chemical composition or lattice structure (Hapke, 1966a). The hope for chemical or mineralogical identification of lunar materials through photometric and polarimetric methods thus vanished. It became clear instead that what we observe is the result of "radiation damage" through continuous irradiation by solar wind and cosmic rays, as modified by erosion and mixing. Despite the equalizing effect of irradiation, differences due to the parent material remain. The ambiguity as to chemical composition has now been removed by scattering experiments performed by Surveyors V (Mare Tranquillitatis), VI (Sinus Medii), and VII (continens near Tycho), which all showed a basaltic composition (Turkevich *et al.*, 1967, 1968; and NASA Reports).

Experimental and theoretical work, especially by Hapke (1966a,b; Hapke and Van Horn, 1963; Oetking, 1966; Halajian and Spagnolo, 1966; Gehrels *et al.*, 1964; Coffeen, 1965; Egan and Smith, 1965) has led to satisfactory representation or imitation of lunar photometric and polarimetric properties on the basis of the "fairy castle" model. As a result of integration of a variety of elements, agreement of the final outcome is not necessarily a proof that all the assumed details are correct. Nevertheless, the broad outlines of the photometric behavior of the lunar surface are undoubtedly explained in such a manner. Dunite powder (grain size less than 7×10^{-4} cm), after 65 coulomb/cm^2 proton irradiation, equivalent to 10^5 yr of solar wind, such as that encountered by Mariner II, closely reproduces lunar photometric and polarimetric properties (Hapke, 1966a). Hapke's improved theoretical photometric function, with a surface covered up to 90% by little steep features (about or over 45° inclination) represents lunar brightness to the very limb. These features on a subcentimeter scale, "are probably primary and secondary meteorite craters and ejecta debris..." (Hapke, 1966b).

Most remarkable is the blackening of materials under corpuscular bombardment (Wehner *et al.*, 1963a; Rosenberg and Wehner, 1964; Hapke, 1965). It is accompanied by sputtering and deposition of active silicate compounds deficient in oxygen on the rear sides of the irradiated grains. The darkening increases with decreasing grain size; coarse powders darken the least, and rough rock surfaces more than smooth surfaces (Hapke, 1966a). There are differences due to composition, but it would be difficult to extricate them from those due to grain size.

The darkening of lunar materials is akin to that of interplanetary dust whose albedo is equal to or less than that of soot and has also been explained by radiation damage (Öpik, 1956).

Sputtering by corpuscular radiation and deposition of the sputtered atoms, as well as sublimation of vaporized substances from meteorite impact, offers a means of cementation of the dustgrains. The dust will lose its mobility and become a "weak, porous matrix," according to an early statement by Whipple (1959). *In vacuo*, unimpeded by interposed air molecules, the grains may become slightly welded together by direct contact; when the contact cohesion exceeds the weight of the grain, the granular substance acquires the mechanical properties of a solid and will maintain slopes of any steepness. Clearly, the finer the grain, and the smaller the gravity, the more like a solid will the powder behave. This is the case of the experiment by Gold and Hapke (1966), and of the lunar surface as seen by Luna IX and Surveyor I (Newell and NASA Team, 1966; Jaffe, 1966a).

According to Smoluchowski (1966), cohesion forces between neighboring grains of the order of 0.5 dyne or more will be present, sufficient to counterbalance on the Moon the weight of a silicate grain 0.13 cm in diameter. At an average grain size of 0.033 cm [see Eq. (154)], cohesion between lunar dust grains would exceed 60 times their weight. For silicates in ultrahigh vacuum, Ryan (1966) also found more or less constant adhesion forces of 0.3 to 1.5 dyne at loads below 5×10^4 dynes. For higher loads, the adhesion rapidly increased, reaching 100 dynes and more at a load of 10^6 dynes; "when this type of adhesion was observed, extensive surface damage was also noted." At lunar gravity and $\rho = 1.3$, the second type of cohesion would set in at a depth of 1000 m for a grain size of 0.033 cm; the depth is inversely proportional to the square of the grain diameter. Thus, in the lunar upper layer only the first type of cohesion would be active. However, this refers to specially prepared clean samples; cementation may lead to much stronger cohesive forces.

In view of this type of cohesion, it would be wrong to treat the small elevations or craters of the dust layer from the standpoint of the angle of repose. The dust possesses no spontaneous fluidity, and the inclinations of the radar reflecting elements are not conditioned by friction. From the mechanics of cratering in a weak medium [small s_c, Eq. (9)], shallow craters and low inclinations are expected as a rule. The dust is then induced to drift downhill, whatever the slope, by micrometeorite impacts (Öpik, 1962a), so that even the smallest slopes are ultimately leveled out (cf. Section X).

As has been pointed out by Whipple (1959) and Öpik (1962a), there is no loose dust layer on the Moon, which explains also the absence of dust on Surveyor I external surfaces and the failure to record any great disturbance or raising of dust by a nitrogen jet 15 cm from the surface (Jaffe, 1966a), or the lack of a covering of dust on the Luna IX camera lens (Lipsky, 1966). This, and the firm settling of the space probes on lunar soil has even led to suggestions that the ground is not dust (Lipsky, 1966). This viewpoint has been shown to be erroneous by Hapke and Gold (1967) and was also refuted by Luna XIII (26 December 1966), which drove a rod into the lunar soil, proving that "the mechanical properties of the Moon's surface layer 20- to 30-cm deep are close to the properties of medium-density terrestrial soil" (R. N. Watts, 1967). The density of the lunar soil is estimated to be about 1.5 gm/cm^3 (Jaffe, 1966a). In what follows, we will still call this the "dust layer," with proper reservations.

The color of the Moon is reddish, its omnidirectional albedo in the optical range increasing from 0.05 in the violet to 0.073 in the visual (green–yellow) band of the spectrum. Infrared photoelectric measurements from Stratoscope II on Mare Tranquillitatis (Wattson and Danielson, 1965) showed that the increase continues in the deep infrared, the reflectivity increasing about three times between 1 and 2.3 μ. Ordinary rock powders cannot match these observations; yet an irradiation by a 2-keV proton beam equivalent to some 10^5 yr of solar wind more or less produced the desirable effect (Wattson and Hapke, 1966), except that the infrared reflectivity of the powders remained still somewhat high as compared to the Moon. The powders that responded to the treatment were from samples of basalt, tektite, and dunite. Contrary to these terrestrial samples, a powdered chondrite (Plainview meteorite) was not reddened by the proton beam, although its albedo decreased in all wavelengths. This may mean that the lunar mare material is more like the composition of the Earth's crust and not meteoritic. Other recent ground-based results point in the same direction (Binder *et al.*, 1965) and decisive evidence has come from direct Surveyor V, VI, and VII tests (Turkevich *et al.*, 1967, 1968).

B. *Thermal Properties*

Measurements of the thermal emission help to disclose some properties of the layers just below the surface. Infrared thermal emission (around 10 μ) and radio emission are used for this purpose, since it varies with the lunar day, or during eclipse, and is studied locally, as allowed by the resolving power of the instrument, or integrated over the whole disk. The effective depth for thermal emission L_e increases with the wavelength λ_e; by using different wavelengths, the thermal parameters can be studied at different depths, qualitatively at least, while absolute quantitative conclusions are less reliable in view of the many uncertainties involved in the construction of thermal models. Adapting a formula proposed by Troitsky (1962), the depth of emission from a layer of silicate rock or granulated material of bulk density ρ (gm/cm³) can be set roughly (to a factor of 2) equal to

$$L_e = 18\lambda_e/\rho, \tag{146}$$

provided the grain is small compared to the wavelength.

For $\rho = 1.3$, this becomes

$$L_e = 14\lambda_e, \tag{146a}$$

to be used for the lunar surface. The depth is much greater than for radar reflection where it is of the order of $\frac{1}{2}\lambda_e$.

The rapid variation of lunar surface temperature during eclipse led Wesselink (1948) to a calculation of the heat conductivity of lunar soil; despite an additional component from radiative conductivity that depends on grain size, it turned out to be an order of magnitude lower than for atmospheric air and to correspond to mineral dust *in vacuo* at a grain size of about 10^{-2} cm. Since then, a wealth of observational material regarding thermal emission from the Moon has accumulated. Despite elaborate models produced to account for the observations (Ingrao *et al.*, 1966; Linsky, 1966), the interpretation, in terms of realistic physical parameters, has advanced very little since Wesselink's work. One-layer and two-layer models with fixed parameters can be made to agree with one set of data, while they may fail to agree with another. In the words of one of the authors, concluding a set of critically conducted adaptations of various models solely for the Tycho region, "In the light of our present inability to decide uniquely which of several plausible models applies..., any detailed description of small-scale lunar surface structure, uncritically based upon any one kind of model yet devised, may be physically meaningless" (Ingrao *et al.*, 1966).

The "knee" in the thermal emission curve during eclipse, or the sudden change in the rate of cooling, has been interpreted to be caused by the presence of a layer of greater conductivity at a depth of a few centimeters, thus leading to the concept of a two-layer model. It seems now that solid blocks of different size strewn all over the surface, such as seen on the close-up pictures (Figs. 5, 6) (Newell and NASA Team, 1966; Jaffe, 1966a), and as suggested by the diffuse component of radar reflection (Evans and Pettengill, 1963; Hagfors, 1966), are to a considerable degree responsible for the characteristic changes in the cooling rates after the cutoff or reappearance of insolation. According to Drake (1966), "there must be a second component, not in depth, but on the surface."

This surface component, which must be responsible principally for the anomalies, through greater thermal inertia as well as through its nonhorizontal profile, has not yet been treated theoretically, except by Gear and Bastin (1962) and Bastin (1965), who considered the effect of macroscopic roughness—steep cavities and elevations—on the thermal and radiative balance of the lunar surface, as distinct from the flat surface figuring in all the usual models.

The surface component, due to stony blocks, affects certain details, yet the general run of the thermal emission curve depends on the dust layer with small inclinations, satisfactorily approximated by a horizontal outer surface. Its thermal parameters are expected to be a function primarily of depth, to some extent also of temperature which again is mainly a function of depth. By applying the equations of the one-layer model to observations relating to different depths, the effective parameters so obtained (conductivity, grain size) may yield an approximate description of their variation with depth independent of the rigid prescriptions of a two-layer model, and perhaps more realistically. In any case, the errors of such an approximate model, allowing for continuous variation of the parameters, may be smaller than those of a "rigorous" procedure based on unproved assumptions. Besides, part of the variation of the conductivity with depth, caused by mechanical compression and radiative transfer, can be estimated from first principles, so that only the grain size remains as the only depth-dependent parameter.

For thermal fluctuations of period τ_t (sec), a characteristic parameter

$$\gamma_t = (k_t \rho c_1)^{-1/2} \tag{147}$$

can be determined, to a factor of the order of unity, almost free of hypotheses. Here, k_t is the thermal conductivity (cal/cm-sec-deg), ρ the bulk density (gm/cm³), and c_1 the specific heat (cal/gm):

$$\gamma_t = (\theta_a \tau_t^{1/2} / Q_a) \times \text{const}, \tag{148}$$

where θ_a is the amplitude of temperature of the radiating surface layer, Q_a the amplitude of heat content (cal/cm² column), obtained from the observed fluctuation of radiation (insolation *minus* radiation losses). For an inside (radio) layer at effective depth x, the amplitude θ_x is given by the observations, while the heat content Q_x is itself a function of k_t or γ_t.

The amplitude decreases exponentially with depth x,

$$\theta_x = \theta_a \exp(-x/L_t), \tag{149}$$

where

$$L_t = \gamma_t k_t \tau_t^{1/2} \tag{150}$$

is the effective depth of penetration of the thermal wave, to be identified with the depth to which the mean thermal parameters apply.

With $c_1 = 0.2$ cal/gm-deg as a close value for all kinds of silicate rock, and γ_t determined observationally from Eq. (148), the ambiguity with respect to k_t rests solely with the adopted value of ρ [Eq. (147)]. Different values of γ_t, decreasing with wavelength and depth, have been obtained, indicating mainly an increase of the conductivity with depth, as could well be expected. From lunar eclipses ($\tau_t \sim 1.8 \times 10^4$ sec), Wesselink obtained $\gamma_t = 1000$ cm² deg sec$^{1/2}$/cal. In the radio range, smaller values have been found, as well as in the infrared for the lunar monthly cycle ($\tau_t = 2.5 \times 10^6$ sec). Russian authors have therefore doubted Wesselink's value. However, there is no real discrepancy, the different values being accounted for by increasing thermal conductivity with depth.

In a granular medium *in vacuo* and grain diameter d_g, the heat flow through the grain consists of three components: the conductivity flow (with a radiative component) through the grain, the radiative flow between the grains, and the contact flow by conductivity between grains pressed against each other. The relative area of contact per cm² cross section at depth x will be assumed to be

$$\sigma_p = g\rho x / s_p \tag{151}$$

in former notations. Setting the compressive strength $s_p = 2 \times 10^9$ dyne/cm², $g = 162$ cm/sec², and $\rho = 1.3$ gm/cm³, for lunar dust consisting of hard silicate grains, we obtain

$$\sigma_p = 10^{-7} x. \tag{151a}$$

If ΔT_1 is the difference of temperature along the grain and ΔT_2 the contact difference, $\Delta T = \Delta T_1 + \Delta T_2$ being the total difference over depth d_g, the effective contact gradient can be assumed to be equal to $\Delta T_2 / (\sigma_p^{1/2} d_g)$. With this, the flow of heat through the grain being equal to the contact flow plus radiative transfer as well as equal to the total flow, the double equation of heat flow can be written as

$$k_t \Delta T / d_g = k_0 \Delta T_1 / d_g = (k_0 \Delta T_2 \, \sigma_p^{1/2} / d_g) + 4 k_s T^3 \Delta T_2 ,$$

which, after elimination of the temperature differences can be reduced to

$$1/k_t = (1/k_0) + [1/(4 k_s T^3 d_g + k_0 \sigma_p^{1/2})]. \tag{152}$$

Here k_0 is the conductivity of compact rock, and k_s Stefan's constant. For our exploratory model, constant values of d_g, k_0, and T are to be used, which then can be considered as mean effective values, more or less valid at the particular depth L_t to which the observations refer.

With $T = 240$ K, $k_0 = 0.005$ cal/cm-sec-deg., and Eq. (151a), this reduces to

$$1/k_t = 200 + [10^5 / (7.5 d_g + 0.16 x^{1/2})], \tag{152a}$$

for d_g and x in centimeters.

With $\rho = 1.3$ and $c_1 = 0.2$ in Eq. (147),

$$k_t = 3.8 \gamma_t^{-2} \tag{153}$$

obtains. Further, setting $x = L_t$, as defined by Eq. (150), into Eq. (152a), d_g can be calculated. Table XXII contains some typical results. The effective value of γ_t for the lunar cycle is based on the observations of Murray and Wildey (1964) and on the calculations by Ingrao *et al.* (1966) for "temperature-independent" models.

Now instead of two or more discrete layers, a continuous increase in the conductivity caused by compression, at a more or less constant effective grain size,

$$d_g = 0.033 \quad \text{cm}, \tag{154}$$

is indicated. The effective grain size depends, of course, on the distribution of grain diameters; its constancy may indicate identical distribution at different depth, thus essentially a one-layer structure.

The low surface conductivity requires a large thermal gradient to deliver the internal flow of heat E_i, according to

$$dT/dx = E_i/k_t . \tag{155}$$

With Eqs. (152a) and (154), this can be integrated. For very different initial conditions and different content of radioactive isotopes, Levin (1966a,b) cites calculations by Majeva (1964) that give, for the present Moon, values of the thermal flux within a range of $(2.3–4.6) \times 10^{-7}$ cal/cm²-sec. Taking 4.3×10^{-7}, which is the Earth's value decreased in proportion to the radius, integration yields, for the mean temperature T at depth x,

$$T - T_0 = 8.6 \times 10^{-5}x + 0.532x^{1/2} - 0.83 \ln(1.56 + x^{1/2}) + 0.37, \tag{156}$$

where T_0 is the mean temperature of the surface. Only the first two terms are significant.

For $x = 10^4$ cm as an upper limit of validity of the model, $T - T_0 = 49°$. This is insignificant; probably a solid rocky structure begins at even a smaller depth where the conductivity will be much higher. The pressure-induced increase in conductivity is rapid enough so that the insulating capacity of the outermost dust layer has little effect on the thermal state of the Moon's interior.

As compared to the pressure effect, the increase of the radiative conductivity with depth, due to the increase of the mean temperature, is insignificant in the granular layer. In the outer layers, however, where radiative conductivity plays an important role, a curious effect arises. The diurnal fluctuations of temperature, affecting radiative conductivity, cause the daytime conductivity, when the temperature is higher, to be higher, too: Even in the absence of a net outward flux of heat from the Moon's interior, the daytime intake of solar heat by the soil requires, therefore, a smaller inward negative thermal gradient than the positive nocturnal gradient needed to restore thermal balance at the surface. The net average thermal gradient will be positive, the temperature rising inward without producing a net leakage of heat. This effect has been investigated in detail by Linsky (1966); when radiative conductivity is taken into account, the thermal gradient derived from radio data leads to a thermal flux of the order of 3.4×10^{-7} cal/cm²-sec, in agreement with theoretical limitations (Levin, 1966a), while a conductivity independent of the fluctuating temperature yields a ten times larger flux, unacceptable for various reasons. This explains also the excessive values of the lunar thermal flux, derived by Krotikov and Troitsky (1963) from the inward increase of temperature shown by radio data.

As to temperature variation, for the subsolar point, a value of 371 K (Pettit, 1961) and, for the antisolar point, 104 K (Saari, 1964) can be assumed for the typical surface. On account of low thermal inertia, the extreme afternoon maximum and the predawn minimum would differ little from these figures. These are blackbody values; a small correction for emissivity could raise these values by 1–2%. The mean equatorial near-surface temperature as determined from shortwave radio (3 mm), is near 206 K (Drake, 1966) although the scatter of individual determinations is large. In any case, the value should lie between 200 and 220 K.

If T_1 and T_2 are the temperatures of the subsolar and antisolar points, respectively, the mean arithmetical equator temperature is close to

$$T_m = \tfrac{1}{2}(T_1 + T_2) - 0.110(T_1 - T_2). \tag{157}$$

This equation is empirically adjusted to the skew radio brightness temperature at $\lambda = 3$ mm (Drake, 1966). For the thermal infrared, it yields $T_m = 208°$, which is close to, and less affected by observational error than, the radio value.

Table XXIII tentatively represents the variation of the mean subsurface temperature with depth. It is based on radio brightness temperatures as listed by Krotikov and Troitsky (1963), properly modified by Linsky (1966), with a systematic correction of $-12°$ applied to make the zero point coincide with the surface value for the infrared.

C. Thermal Anomalies

The lunar night-time temperatures are too low for accurate observations in the $10\,\mu$ window. Using the $20\,\mu$ atmospheric window, Low (1965) found a mean temperature of 90 K for the cold limb (dark near-polar or predawn). Cold spots of 70 K and lower were found, tentatively explained as those of low conductivity (γ_t about 2300 cm²-deg-sec$^{1/2}$/cal), and a hot spot of 150 K was recorded near the southeastern limb.

Hot spots, which are warmer than the normal surface during an eclipse but cooler in daytime, have been systematically observed and listed by Shorthill and Saari (1965a, 1966). Among 330 such objects, 84.5% are ray craters, craters with bright interior or bright rims at full Moon, 8.7% are bright areas of various qualifications, 0.6% are craters not bright at full Moon, the rest are not unidentified. They occur over the entire lunar surface, but somewhat more densely over the maria, being especially crowded in Mare Tranquillitatis. On a recent map (Saari and Shorthill, 1966), made from the observations of the lunar eclipse of 19 December 1964, 271 or 58.0 ± 1.7% of the hot spots are on the maria, 196 or 42.0 ± 1.7% on the continentes. Strong anomalies are even slightly more concentrated on the maria (125 out of a total of 201, or 62.2 ± 2.5%). There may be some adverse selectivity for the limb areas where continentes predominate, so that the representative areas of maria and continentes may be about equal, or slightly in favor of continentes. The excess in the maria seems thus to be real, although the distribution is quite patchy, so that the random sampling error is not representative of the actual statistical uncertainty. In any case, the anomalies are clearly postmare features. The most prominent ones are Tycho and Copernicus.

During eclipse, the interior of Tycho (diameter 88 km) retained a temperature around -70 to -60 C, with maxima of -51 and -48 C, while in the "normal" surroundings it dropped to -106 and -112 C. Just outside the crater's wall it was -82 C, at double radius around the crater -97 C, at 2.5 radii -101 C, at 3.5 radii -106 C. The resolving power of the apparatus was $10''$ of arc, corresponding to 19 km or 0.22 of the crater diameter, sufficient to observe the gradual decline in the infrared radiation around the crater. At full Moon, the crater is cooler than its surroundings ($+77$ C) by a few degrees, but this may be partly due to its greater albedo. Otherwise the anomaly is undoubtedly accounted for by greater thermal inertia, i.e., smaller γ_t , or greater conductivity and density of the material.

In addition to the spots, extended areas, occurring chiefly in the maria (Mare Humorum, Oceanus Procellarum between Aristarchus and Kepler, Northern Mare Imbrium, the continens around Tycho, and others), are warmer by about 10 C during an eclipse (Shorthill and Saari, 1965a, 1966).

From their distribution, the anomalies around craters are undoubtedly due to crater ejecta, similar to the rays but more concentrated in the vicinity of the crater. The median distance to which massive ejecta fly can be calculated from Eqs. (45), (27), (16), (4), and (19) with $s_c = 9 \times 10^8$, $\rho = 2.6$, $\lambda_x{}^2 = \frac{1}{2}\lambda^2 = 0.14$ (Table XV, Model Q), $y = 0.5$, and $\sin \beta_0 = 0.8$; this yields $v_x = 1.39 \times 10^4$ cm/sec and $L = 8.8$ km. The radius of the circle over which most of the ejecta are sprayed may be taken to be twice this value of 17 km, almost equal to the resolution of the radiometer used by Shorthill and Saari. This radius is independent of crater diameter. The effective diameter of the anomaly is then $B_0 + 34$ km, where B_0 is the crater diameter. The area cannot be less than the 34-odd km across and always well within the resolving power of the radiometer. Therefore it is expected that the measured thermal excess will not depend on the size of the crater even when its diameter is smaller than the resolving power, except for the thickness of the overlay when it drops below a certain limit.

This is exactly what Shorthill and Saari (1965b) had found, but they gave it a different interpretation. By assuming that the thermal excess is restricted to the crater itself for craters below the resolution limit, they reduced the excess to the crater area and obtained a strange increase for the smaller craters. Their "corrected signal differences" have no physical meaning; even when the meteoric theory is discarded (which no longer is possible with the present state of knowledge), a volcanic eruption on the Moon would also spread the ejecta to distances independent of crater size.

The uncorrected observed anomalies (Shorthill and Saari, 1965b) yield consistently for all craters within the diameter range from 4 to 90 km the same value of the thermal parameter $\gamma_t = 600$ from the eclipse observations. In terms of our pressure-adjusted thermal model, from Eqs. (153), (150), and (152a) with $\tau_t = 1.8 \times 10^4$ sec, the values $k_t = 1.0 \times 10^{-5}$, $L_t = 0.81$ cm, and $d_g = 0.115$ cm are obtained. The thermal anomaly of the hot spot is readily explained by assuming a coarser grain near the surface, just of the order of ordinary terrestrial sand. The upper layer is continually ground and overturned by meteorite impact and supplemented by the smoke-like products of sublimation; it is expected that, without fresh overlay, it will become more and more fine-grained with age. Also, small meteorites and especially the numerous secondary ejecta, which do not penetrate the dust and rubble layer but which are responsible for most of the overlay outside the reach of the large postmare cratering events, will eject and spread around material of a smaller grain size than large meteorites, which penetrate the top layer (craters over 0.5 km in diameter) and crush the fresh bedrock underneath. The difference in the thermal properties of the environs of large primary postmare craters, as compared to the surface at large, can thus be understood without postulating improbably short ages for them. All ages from 4.5×10^9 yr to zero would do.

In addition, there may be blocks of solid rock on the surface that, even when in small proportion, will contribute to the anomaly; in such a case the calculated effective grain size is an upper limit.

On the other hand, an exposed solid rocky surface of this size, as it figures in some interpretations, is physically inconceivable except when the ages of all these objects are assumed to be unbelievably short, i.e., some 10^6–10^7 yr. Nor would small corrugations on a centimeter scale (Gear and Bastin, 1962; Bastin, 1965) help in the interpretation—these would be leveled out by erosion and be replaced by the naturally undulating and small-scale roughness that is identical with the rest of the surface, so that no anomaly could arise on this account alone.

The young ray craters, like Tycho and Copernicus, in addition to their quality as nocturnal hot spots, have proved to be strong backscatterers of radar (Gold, 1966). As first noted by Pettengill and Henry (1962), the intensity of radar backscatter from Tycho is some 10–20 times that from the surroundings (at $\lambda_e = 70$ cm). This is due partly to greater roughness. The diffuse, non-specular general component of backscatter from the Moon (Evans and Pettengill, 1963; Hagfors, 1966) is most plausibly explained by reflections from compact rocky "boulders" [similar to those shown on Luna IX and Surveyor I pictures (Figs. 5, 6)] with dimensions larger than λ_e/π. Similar blocks in greater numbers must be present on the surface and buried among the ejecta of the ray craters.

A correction for general roughness, i.e., the more frequent occurrence of larger inclinations of the scattering elements than corresponding to normal surface undulation, has been attempted by Thompson and Dyce (1966). In this way they separated the component due to the increase in reflectivity (A_r) from that caused by roughness. In such a manner, for craters larger than the resolution limit, corrected values of the dielectric constant are suggested as follows: Aristillus, 4.5; Tycho, 5.2; and Copernicus, 6.5. Of the 25 craters that showed radar reflection enhancement, 23 would suggest dielectric constants within this range, or less, among them seven ray craters, eleven nonrayed craters of class 1, and three nonrayed class 5 craters (Atlas, Posidonius, and Vitruvius, with lesser enhancement, all within or near the borders of Mare Serenitatis and Mare Tranquillitatis). Two objects, Diophantus and Plinius, both of class 1 (nonrayed), yield extraordinary values of the dielectric constant, 15 and 35, respectively; heavy meteoritic material may be suspected in these cases.

The rest appear to have the backscatter characteristics of bare rock. However, it is not necessary to postulate a solid rock surface, either exposed or buried under a thin layer of loose material. Rocks of more than $\lambda_e/\pi = 23$ cm in diameter, buried in the upper 5–10 m of the soil, would act in a similar manner if their projected cross sections were to cover the area; for this, they need not occupy more than 10% of volume of the "boulder bed." Surveyor I pictures suggest this possibility, which, to an appreciably lesser degree, exists even in the mare landscape of Oceanus Procellarum (Figs. 5, 6).

VII

EROSION

A. Surface Modification Processes

In the following sections, a quantitative physical theory of the evolution of the lunar surface under the influence of external cosmic factors is developed. The basis for it is the theory of cratering and encounters, as well as observational data referring to features on the lunar surface and the material contents of interplanetary space. As has already been seen from the preceding sections, independent lines of evidence converge in checking and confirming the predictions. Although not precise, these predictions are supposed to be close approximations, up to 20–30% in some cases and within a factor of 2 or 3 in others when the quantitative basis is uncertain, as it is with the density of interplanetary matter. It turns out that the present state of the lunar surface can be understood completely in terms of external factors acting alone for all the 4.5 billion years of postmare existence, any signs of endogenic-volcanic or lava activity belonging to the initial short premare and mare stage.

Conclusions in the opposite sense are either based on improbable assumptions, disregard for physical realities, or qualitative judgment by terrestrial analogy not applicable to the Moon.

As an example of the latter, the very interesting article by O'Keefe *et al.* (1967), "Lunar Ring Dikes from Lunar Orbiter I" may be cited. The argument hinges on the contention that "the slopes are less than the angle of repose of dry rock; hence an explanation in terms of mass wastage is hard to support." Now, in lunar processes of erosion, the angle of repose is irrelevant. It could be decisive when the eroded chips of rock were left lying *in situ*, to roll downhill when the slope was steep enough. The only process on the Moon where this could be operative would be the destruction of rock by the extreme variation of temperature, yet, in the absence of water, it could work only to a very minor extent; otherwise the observed presence of stony blocks and extrusions would be utterly incomprehensible. Actually, however, meteoritic impact (independent of velocity) will disperse the rock fragments over a radius of about 17 km (cf. Section VI,C); Flamsteed's rocky ring is about 3 km wide (O'Keefe *et al.* 1967, Fig. 1), and the debris will disperse into the engulfing plane without any relation to the slopes of the ring. Granular material settling on the ring will be partly swept out by the impacts in a similar manner, and partly it will drift downhill under the instigation of micrometeorite bombardment at a calculable rate (Section X,B) proportional to the tangent of the slope angle, however small the angle is, independent of friction and without any relation to the angle of repose. Besides, the notion of angle of repose is inapplicable to the fine lunar dust when the cohesion between the grains exceeds their weight (grain diameter less than 0.13 cm). Also, on the photograph of Flamsteed's ring (O'Keefe *et al.*, 1967), little craterlets, witnesses of continuing erosion, are seen although in much smaller numbers than on the surrounding plain; the analogy with the peak and wall of Alphonsus is complete (cf. Section V,E). The ring may be an ancient ring dike of near-mare age although interpretation as the remnant of a raised impact crater lip is much more plausible (cf. Figs. 5, 6 and Fig. 1); however, the argument about the angle of repose is not only irrelevant, but also misleading.

The mechanical processes at work on the lunar surface have been described realistically by Whipple (1959) and Öpik (1962a). These processes are sputtering by corpuscular radiation, accretion from micrometeoric material, and destruction and transport by primary and secondary impacts.

Most of the destroyed mass, much greater than the mass of the impacting bodies, returns to the lunar surface, though not necessarily in the immediate vicinity of the impact. Except for the fallback fraction, it settles on the ground outside the crater covering previous small features with an "overlay." The overlay is subsequently disturbed by micrometeorite and small meteorite impacts, forced to leave elevations and collect into depressions. An equilibrium state between new small craters and their leveling out by erosion establishes itself, the crater profiles becoming flatter with age. This lasts until a large crater erases the traces of previous formations.

Using an ingenious method, Jaffe (1965, 1966a,b) attempted to interpret the profiles of small craters by overlay only. By sprinkling sifted sand on sharp artificial craters, imprinted in sand, the washout of laboratory crater profiles, depending on the added sand layer, was compared with the profiles of small craters (5 to 10^4 m) shown in Ranger VII, VIII, and IX photographs, and the thickness of overlay was estimated in proportion to crater diameter when the laboratory and lunar crater profiles were similar. In such a manner it was found that "at least 5 m of granular material, and probably considerably more, have been deposited on Mare Tranquillitatis, Alphonsus, and nearby highland areas, subsequent to the formation of most of the craters 55 m in diameter or larger," and similar results have been obtained for Mare Cognitum (Jaffe, 1965, 1966a,b). It is interesting to note that, despite objections to the validity of the method, Jaffe's figure comes close to the estimate of 14 m made in Section VII,C for post-mare overlay, calculated *a priori* from astronomical data and cratering theory. Walker (1966) raised some objections to the method, and others are pointed out here. The profiles of lunar craters change more from erosion (which may carry away an elevation or fill a depression of 20–30 m in 4.5 × 10^9 yr) and little if at all from overlay. The washout of experimental craters depends on friction and rolling of sandgrains (size about 0.03 cm), which in air are much more mobile than in the lunar vacuum and for which the angle of repose is decisive when there are no percussions or collisions. The lunar loose material is forced downhill by meteorite impacts as well as by the secondary spray of debris, which accounts for the overlay and which at the same time disturbs the grains and sends them downhill. Apparently, the same role was played by the laboratory sand grains falling on the artificial craters and disturbing the surface by their impacts even when the slope was less than the angle of repose. The agreement between the empirical and the theoretical estimates of overlay is thus not quite as fortuitous as it seems, a certain amount of injected overlay causing a corresponding slumping of the crater profile. It seems that a given overlay of terrestrial sand, at terrestrial gravity sprinkled from a small altitude, has the same effect as an equal amount of dust and rubble on the Moon thrown with much higher velocity on the lunar compacted dust plus direct erosion by a smaller mass of interplanetary meteoric material. As to the apparent increase of the estimated overlay depth with crater diameter suggested by Jaffe's experiments, it can be ascribed to increase in age, craters below 300 m having existed but for a fraction of the total span of 4.5 billion years and thus having received a smaller sprinkling. At a diameter of 300 m, Jaffe's figures, as plotted by Walker (1966), point to an overlay of 7 m, at 5-m diameter the overlay is about 0.07 m, while in proportion to diameter or age it is expected to be 0.12 m; the order-to-magnitude agreement is satisfactory.

B. *Sputtering by Solar Wind; Loss and Gain from Micrometeorites*

Solar-wind bombardment causes the sputtering of atoms from the silicate lattice. From a semiempirical theory of sputtering, and with a pure proton solar-wind flux of 2×10^8 ions/cm²-sec (Mariner II data), Öpik's (1962b) estimates lead to a sputtering rate of 8×10^{-9} gm/cm². year. From thorough experimental investigations of sputtering of various materials, Wehner *et al.* (1963b) arrive at a much better founded sputtering rate for a stony rough surface, equal to 0.4 Å or about 1.5×10^{-8} gm/cm²-yr; their assumed flux is 2×10^8 protons and 3×10^7 helium ions/cm²-sec with energies above the sputtering threshold, in the normal solar wind and with allowance for solar storms; this figure will be further adopted. For a pure proton flux, Wehner's figure would be 4×10^{-9} gm/cm²-yr, or one-half this author's estimate.

About two-thirds of the sputtered atoms are ejected with velocities greater than their escape velocity from the Moon. The annual loss to space from sputtering can thus be set at 1.0×10^{-8} gm/cm²; at a constant solar wind, this would amount to a loss in 4.5×10^9 yr of 45 gm/cm², equivalent to a dust layer ($\rho = 1.3$) of 35 cm or to 17 cm of solid rock. About an equal amount is sputtered inward and contributes to cementation of the dust; the continued stirring and turnover by micrometeorite impact [10^4 yr mixing time for the top 1 cm layer (Öpik, 1962a)] ensures mixing of the irradiated layer with the deeper lunar soil.

Meteorite influx may lead to gain or loss of mass, according to velocity (cf. Section IV,A), and to crushing and redistribution of the debris, which are ejected from the craters and spread as "overlay" over the surroundings or fall back onto the crater floor. At cosmic velocities of impact, the mass of overlay may be 2–3 orders of magnitude greater than that of the impacting bodies. While secondary ejecta may add to the overlay and its redistribution, the net balance of mass over the entire lunar surface depends, of course, only on the impacting extraneous mass and its velocity.

From the consideration of energy transfer and vaporization in the central portion (central funnel) of a meteor crater (Öpik, 1961a, Table 23), the author estimated that below a velocity of 10.7 km/sec, all the impacting mass will remain on the Moon at its present gravity; at 20 km/sec, a stony meteorite will cause a loss to space 17 times its own mass, and at 40 km/sec a 44-fold loss occurs.

The first case is that of the micrometeors of the zodiacal cloud (particle radius < 0.035 cm) which thus, at a space density of 2×10^{-21} gm/cm³ and relative velocity $U = 0.187 = 5.6$ km/sec (Öpik, 1956) lead to a gain of 0.01 gm in 10^6 yr (Öpik, 1962a) or 45 gm in 4.5×10^9 yr per square centimeter of the lunar surface. The gain turns out to be equal to the loss from sputtering, but, with factors of uncertainty of the order of 2–3 in the assumed rates, the balance is uncertain even as to sign.

Higher velocities are those of meteors and meteorites, which thus cause a net loss of mass. They consist of different populations, with different distributions of particle sizes. The ordinary "dustball" meteors, flaking off from comet nuclei and with masses in the range of 10^{-3} to about 10 gm [evaluated from theoretical luminous efficiency and empirically confirmed (Öpik, 1963c)] rapidly decrease in numbers with increasing size so that little mass is contained in the largest categories. The main mass (about 86%) of these "visual" meteors is contained in an "E-component" with asteroidal orbits (Öpik, 1956), and the total influx is estimated at $d\mu_0/dt = 8.0 \times 10^{-11}$ gm/cm²-yr, leading to a loss about 20 times this amount or 7 gm/cm² in 4.5×10^9 yr.

A more important component of the mass influx is represented by the "asteroids" of the Apollo group whose frequency exponent, a -2.7 power of the radius, appears to join them into one continuous group from meteorites of 25- to 400-cm radius up to bodies in the kilometer range. From combined observational data (Öpik, 1958a), the mass accretion from this group, in grams per square centimeter and year, and within the range of radii from R_1 to R_2 (cm), is estimated to be

$$d\mu_1/dt = 1.11 \times 10^{-11}(R_2^{0.3} - R_1^{0.3}) \tag{158}$$

at a density of 3.5 gm/cm³.

Comet nuclei, according to the same source, contribute (at density 2.0 gm/cm³)

$$d\mu_2/dt = 1.55 \times 10^{-14}(R_2^{0.8} - R_1^{0.8}), \tag{159}$$

and the Mars asteroids deflected by perturbations (at density 3.5 gm/cm³) give

$$d\mu_3/dt = 1.27 \times 10^{-18}(R_2^{1.4} - R_1^{1.4}). \tag{160}$$

These accretions correspond to differential number fluxes at the lunar surface per square centimeter and year and interval dR according to

$$dN/dt = CR^{-n}\, dR, \tag{161}$$

with $C_1 = 2.26 \times 10^{-13}$, $n_1 = 3.7$; $C_2 = 1.48 \times 10^{-15}$, $n_2 = 3.2$; and $C_3 = 1.22 \times 10^{-19}$, $n_3 = 2.6$. All the fluxes and accretions are insignificant at small radii (as compared to the visual meteors), so that $R_1 = 0$ can be assumed. For an upper limit of crater diameter of about 200 km, $R_2 = 5 \times 10^5$ cm, Eqs. (158)–(160) yield comparable values:

$$d\mu_1/dt = 5.6 \times 10^{-10}, \qquad d\mu_2/dt = 5.6 \times 10^{-10}, \qquad d\mu_3/dt = 1.2 \times 10^{-10},$$

and with the "visual" dustball meteor contributions equal to 8×10^{-11}, the total mass influx becomes 1.32×10^{-9} gm/cm²-yr. With a 30-fold loss ratio as the mean for impact velocities of 20 and 40 km/sec, the loss to space from these components would amount to 178 gm/cm² in 4.5×10^9 yr. These appear to be the dominant components; within the uncertainty of our estimates they represent also the net mass loss to space from the lunar surface, accretion from micrometeorites and sputtering by solar wind mutually canceling out.

Most of the loss is accounted for by large cratering events and, thus, affects crater interiors without directly influencing those portions of the surface between the craters. The loss from the average surface, undisturbed by the localized large cratering events, must be calculated to a crater diameter of 300 m or $R_2 = 7.5 \times 10^2$ cm, which is the limit of erosion or leveling out of the craters during the total age of the Moon. This yields now

$$d\mu_1/dt = 8 \times 10^{-11}, \qquad d\mu_2/dt = 3 \times 10^{-12}, \qquad d\mu_3/dt = 1.4 \times 10^{-14},$$

plus $d\mu_0/dt = 8 \times 10^{-11}$ from the total visual dustballs.

With a loss factor of 17 for μ_1 and μ_3 , and one of 44 for μ_2 and μ_0 , a total loss from the surface undisturbed by large surviving craters ($B_0 < 300$ m) becomes 5.0×10^{-9} gm/cm²-yr or 23 gm/cm² in 4.5×10^9 yr. This is practically the effective loss from meteorite impact for an outwardly "level" surface outside the boundaries of large craters if sputtering by solar wind is assumed to be balanced by the gain from micrometeorites. Whatever uncertainty is involved in this figure, it shows the order of magnitude of the very small changes in the mass load of the lunar surface that are caused by external factors. These are very much smaller than those due to redistribution of mass through cratering.

The most important component in the external mass exchange is the influx from micrometeorites, 45 gm/cm². Although an at least equivalent amount of mass, 45 gm/cm² from solar wind and 23 gm/cm² from meteorite impact, is sputtered back to space, this does not mean that the micrometeoric material is immediately lost again. Before being subjected to sputtering, it becomes mixed with 10–30 times its mass of overlay debris that was ejected from the bedrock by meteorite impacts (see Subsection C), and the material sputtered to space would contain only some 3–10% of micrometeoric material. With the figures of external mass exchange as estimated above, over a period of 4.5×10^9 yr there is a gain of 45 gm/cm² from micrometeors, and a loss of $45 + 23 = 68$ gm/cm², of which only 2–7 gm/cm² would belong to micrometeorites. In such a case, the present lunar surface should contain some 40 gm/cm² ($\pm$) of micrometeoric origin, admixed to, and diluted in, the overlay debris or the "dust" layer.

C. Overlay Depth

Much more important than intrinsic gain or loss is the material crushed and thrown about by cratering impacts; this may exceed by several hundred times the infalling mass. From the crater bowl excavated by the impact it is ejected to distances of tens or more of kilometers from the crater, where it settles as "overlay," a mixture of dust, rubble, and boulders that is subject to further modification by meteor and radiation bombardment, to form what we have called the lunar soil or "dust" layer. The average distances of ejection depend essentially on the strength of the material [Eq. (4), (16), (19), (27), and (45)] and are thus greatest for impact into bedrock, while insignificant for most of the stirred-up mass of dust. In large craters, when the distance of ejection is of the order of the crater radius, most of the debris falls back into the crater bowl where overlay may attain a thickness of several kilometers. For the postmare cratering impacts represented by Table XV, the thickness of overlay (due to the single impact) in central portions of the crater is

$$\Delta z = B_0[(p/D) - (0.625 \, H/B_0)]$$

Table XXIV shows the thickness of overlay in postmare craters. Undoubtedly, pressure compaction takes place when the thickness of the layer is great; except for its topmost layer, it cannot be regarded as just loose rubble or dust.

Outside the crater rim, massive ejecta may reach over a fringe of about 8–17 km (cf. Section VI,C); at 9 km beyond the rim, the thickness of ejecta can be estimated roughly to be one-eighth of Δz, which leads to thick overlay in the vicinity of large craters, and negligible overlay near small craters. The distribution of overlay must be extremely spotty, following the pattern of distribution of craters larger than 10 km in diameter, and with a more or less uniform "background" of area not disturbed by the vicinity of large craters (the areas being removed by more than 15–30 km from the nearest rim of a crater 20–100 km in diameter).

These considerations apply chiefly to the maria where a preexisting rocky (lava) base has been subjected to destruction by impacts. In the continentes, the crust appears to be completely formed by accretion of overlay during the premare stage, any earlier mare surfaces of the accreting Moon having been buried under the final shower of overlay.

The formation of overlay has been described in principle by Whipple (1959) and Öpik (1960, 1962a). There are two kinds of process at work: (1) All impacting bodies and radiations contribute to modification (grinding, cementation), mixing, and displacement of the existing overlay or "dust" layer; and (2) only those meteorites large enough to penetrate the layer contribute to erosion of the bedrock and are instrumental in adding new material to, and increasing the thickness of, the layer. Hence the growth of the layer with time becomes slower as its thickness and the inferior size limit of the active meteorite population increases.

In addition to primary impacts, those of secondary and higher order contribute to overlay. At first we will consider only postmare primary impacts on an initially hard surface, supposed to be solidified lava, of a strength about that of terrestrial igneous rocks (cf. Section V,C).

Let X (cm) be the thickness of overlay, p [Eq. (6)] the relative penetration in a layer of infinite thickness at $\gamma = z = 45°$ (as an average), and R the radius of the projectile. Although the velocity of the projectile decreases at penetration, its flattening at hypervelocity events increases the cross-sectional area, so that loss of momentum can be assumed to be roughly proportional to the depth of penetration. Hence the fraction of momentum retained after penetration of the dust layer is

$$\eta = 1 - (X/2pR). \tag{162}$$

The condition $\eta \geqslant 0$ yields a minimum radius for penetration that reaches the bedrock,

$$R_0 = X/2p. \tag{163}$$

Only projectiles with $R > R_0$ are capable of eroding the bedrock. An infalling mass $\Delta\mu$ at radius R produces a mass of overlay

$$\Delta M = \Delta\mu \, (M_c/\mu) \cdot \eta, \tag{164}$$

where M_c/μ is given by Eqs. (14), (3), and (7).

Micrometeorites and visual meteors are too small to penetrate the dust layer and do not contribute significantly to overlay (except at the very beginning, when incident on a bare rocky surface). For the remaining three components, $\Delta\mu$ is to be replaced by $[d(d\mu/dt)/dR]\Delta R$ with $R = R_2$, $R_1 = $ const as given in Eqs. (158)–(160), and Eq. (164) integrated from $R_1 \geqslant R_0$ [Eq. (163)] to an upper limit R_2. For the main background, unaffected by the vicinity of large craters, we assume an upper limit of crater diameter $B_0 = 2.48$ km and $R_2 = B_0/2D$ (cf. Table XV), for which the mean spacing in Mare Imbrium is $(4.65 \times 10^5/207)^{1/2} = 47$ km, about the double of the extreme flight distance of massive ejecta from hard bedrock. Ejecta from craters up to this limit will have by now produced a more or less uniform overlay X_0, with little spottiness. The ejecta from larger craters from $B_0 = 2.48$ km to $B_0 = 44$ km, or from corresponding larger projectiles (from R_2 to $R_M = 22/D$ km) will cause locally much deeper overlay that, spread uniformly over the entire area, would amount to an average layer X_1; however, for these larger projectiles, loss of momentum in Eq. (162) is to be calculated with $X = X_0$, not $X_0 + X_1$, because of the spottiness and low probability of coincidence of the major impacts. Actually, for these latter, R is so large that $\eta = 1$ can be assumed for the present and past state of the lunar surface. The upper limit $B_0 = 44$ km for major craters is that for which throwout is about 50% of the total detritus. For still larger craters, most of the overlay remains in the crater, forming an average layer $\frac{1}{5}$–$\frac{1}{3}$ of Δz as given in Table XXIV; it must be treated as purely local enhancement, and there is no point in calculating its contribution to the average depth of overlay elsewhere.

For the three components of meteorite influx, the following numerical constants have been assumed:

For Apollo group [Eq. (158)] and Mars asteroids [Eq. (160)], $\delta = 3.5$, $w_0 = 20$ km/sec, $p = 2.88$ (at $\rho = 1.3$, $s_p = 2 \times 10^8$ as for dust layer), $R_0 = X_0/5.76$, $R_2 = 8.3 \times 10^3$ cm, $R_M = 1.5 \times 10^5$ cm, $M_c/\mu = 232$; for comet nuclei [Eq. (159)], $\delta = 2.0$, $w_0 = 40$ km/sec, $p = 2.28$, $R_0 = X_0/4.56$, $R_2 = 4.66 \times 10^3$ cm, $R_M = 8.35 \times 10^4$ cm, $M_c/\mu = 637$. After integration, using these data, of Eq. (164), separately from R_0 to R_2 and from R_2 to R_M, with $\Delta X = \Delta M/\rho = \Delta M/1.3$ for the overlay rubble of assumed density 1.3, and omitting small irrelevant terms, the *a priori* calculated rate of growth of background overlay from primary meteorite impact (craters smaller than 2.5 km) in a lunar mare (solidified lava as the bedrock), in cm/yr (at density 1.3 gm/cm³), as listed separately for the three components, becomes

$$dX_0/dt = 2.97 \times 10^{-8} - 1.67 \times 10^{-9} X_0^{0.3} \text{ (Appolo group)}$$
$$+ 6.55 \times 10^{-9} - 8.96 \times 10^{-12} X_0^{0.8} \text{(comet nuclei)}$$
$$+ 7.0 \times 10^{-11} \text{ (Mars asteroids)}, \tag{165}$$

and the annual rate of smoothed-out growth from major impacts (craters from 2.5 to 44 km diameter) equals

$$dX_1/dt = 4.10 \times 10^{-8} \text{ (Apollo group)}$$
$$+ 6.57 \times 10^{-8} \text{ (Comet nuclei)}$$
$$+ 3.94 \times 10^{-9} \text{ (Mars asteroids)}. \tag{166}$$

Integration of Eq. (165) (in which the negative terms are not very significant, reducing the outcome by only 10%) yields, for $t = 4.5 \times 10^9$ yr, a background overlay of $X_0 = 147$ cm from primary impacts alone; to this is to be added a spotty overlay from the larger craters (2.5–44 km) of average thickness $X_1 = 468$ cm (184 cm from Apollo group, 266 cm from comet nuclei, and 18 cm from Mars asteroids). The total average overlay from primary impacts, calculated *a priori* from cratering theory and astronomical data, is $X_0 + X_1 = 615$ cm or 800 gm/cm² (at density 1.3); because of the spottiness, the figure has not a very definite meaning.

Secondary impacts probably contribute very little to craters over 2.5 km in diameter, and the value of X_1 should not need any correction in this respect. As for X_0, the contribution from secondary impacts by large ejecta from the larger craters must be considerable.

There exists a direct empirical method of evaluating the thickness of overlay, based on the volume actually excavated by observed craters in a mare. While the crater diameters are directly measured, for the depth to diameter ratio $p/D = x_p/B_0$, the average theoretical values (diameter range 2.5–44 km) of

0.1105 (Apollo group) and 0.0569 (comet nuclei) will be assumed, according to Table XV, weighted in a ratio of 1 to 3, so as to give a volume ratio proportional to the calculated values of overlay X_1, 184/266 = 1/1.45; in other words, it is assumed that, within the chosen diameter range, there are three cometary craters to one of the Apollo group. This gives an average ratio $x_p/B_0 = 0.0703$. Further, doubling the volume as for surface ejecta of density 1.3 originating from bedrock of density 2.6 (however for thick overlay or fallback, the doubling is not justified, as the material is compressed under its own weight), Eq. (15) yields the volume of ejecta to be

$$V_e = 0.0344B_0{}^3. \tag{167}$$

The thickness of overlay averaged over area S is then

$$X_1 = \sum (V_e/S). \tag{168}$$

Öpik's (1960) counts in western Mare Imbrium thus yield "observed" values of overlay as represented in Table XXV. All craters of the area except Archimedes are included; Archimedes as a premare crater $(B_0 = 70.6 \text{ km})$ is omitted. The cumulative number of craters in the third column is given in order from the largest (∞) down to the given limit, while the cumulative thickness of overlay in the fifth column is counted in the opposite direction from 2.48 km up. In the last column, the average separation of the craters at given cumulative number N, calculated from $(S/N)^{1/2}$, is given. This characterizes the spottiness of overlay; little will spread beyond a radius of $\frac{1}{2}B_0 + 20$ km from the center of the crater or, roughly, beyond an average separation of $B_0 + 40$ km. Thus the large contributions to X_1 beyond $B_0 > 21.4$ km are localized to a fraction less than $(61.4/241)^2 = 0.065$ of the area and are not characteristic of the background but largely depend on single major impacts. As the figures stand, for the diameter limits 2.5–44 km, the averaged observed overlay is $X_1 = 1489$ cm, to be compared with the value of $X_1 = 468$ cm as calculated above theoretically for primary impacts only. The difference may be due partly to some secondary craters larger than 2.5 km in the Copernicus and Eratosthenes rays, but mainly it is the manifestation of the excess in the true number of large craters above that calculated from the present population of interplanetary stray bodies, as persistently revealed also in crater statistics (Table XVII). Although due to a few individuals, the excess is always there, and essentially in the same proportion, such as in more extended counts on an 8 times larger area (Table XVIII); these counts, as presented by Baldwin (1964b), agree so closely with those in western Mare Imbrium that no revision of Table XXV is necessary.

In Table XXVI, averaged overlay from primary impacts, calculated from interplanetary data for different crater size limits, is compared with the observed values on the Moon derived from crater valume. It can be seen that the figures are in reasonable agreement except for craters larger than 21 km where there is a sixfold excess, seven craters being observed instead of 1.2 predicted for this range.

The discrepancy has little bearing on the general overlay background between the craters, determined by the contributions from smaller craters for which reasonable accord is expected as shown by the data of Table XXVI.

The data of Tables XXV and XXVI show what the average thickness of overlay would be if the ejecta were spread uniformly over the entire area. Actually no such uniformity can exist; most of the ejected mass comes from the sparsely distributed large craters whose separations greatly exceed the ejection range. The distribution of overlay must therefore be extremely spotty, following the distribution of large craters in whose vicinity the thickness of the layer must by orders of magnitude exceed that of the average background.

A schematic representation of the distribution of overlay can be obtained by assuming an effective radius L_m of spread of the ejecta around the crater center and by distributing uniformly the ejected mass over an area of πL_m^2. With the assumed elastic parameters of lunar rock as in Section II,F, the flight distance at $y = 0.5$ (half-mass of crater) is $8.8 + 0.35B_0$ km from crater center, and at $y = 0.25$ (quarter-mass or half-distance from center of crater), it is $18.5 + 0.25B_0$ km. An average can be assumed, i.e.,

$$L_m = 15 + 0.3B_0 \quad \text{km.} \tag{169}$$

With this and the data of Table XXV, the frequency distribution of overlay thickness in a mare has been calculated as shown in Table XXVII. The values are based on the actual crater statistics in Mare Imbrium but with an added average basic overlay of 13 m (as explained later in this Section) due to craters less than 2.48 km in diameter. Figure 10 represents this very uneven distribution graphically. The thickness values are averages over intervals of the crater statistics for Mare Imbrium (which is representative of all lunar maria, cf. Tables XVII and XVIII).

Were only primary impacts to be considered, the basic overlay would have been expected to be equal to $X_0 = 147$ cm as calculated above for craters of less than 2.5 km diameter. However, ejecta from secondary craters make a very much larger contribution in the small diameter range, increasing the thickness of the overlay and, by its protective action, decreasing the role of the very small primary impacts.

We shall use, as typical, the actual counts of small craters in Mare Cognitum, obtained from Ranger VII photographs by Shoemaker (1966); from the curve in his Fig. 2-42, *loc. cit.*, cumulative crater numbers at various diameters were taken (dots on full line). The curve, after a marked twist upward below $B_0 = 1.2$ km, interpreted as due to secondaries, bends sharply down below $B_0 = B_1 = 285$ meters. This can be attributed plausibly to erosion, B_1 being

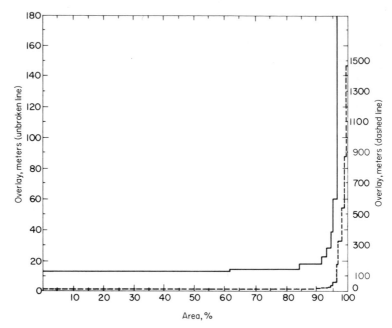

Fig. 10.   Distribution of overlay thickness in lunar maria. Abscissas, percentage area. Ordinates, thickness in meters. For unbroken line, scale to the left; for dashed line, scale to the right.

the diameter eroded in 4.5×10^9 yr and the lifetime, as well as the number of smaller craters presently surviving, being proportional to B_0/B_1. Thus, the counted numbers for $B_0 < B_1$ must be multiplied by the erosion factor

$$E_f = B_1/B_0 \tag{170}$$

to allow for eroded craters that are no longer there but whose ejecta may have contributed to the overlay.

The cumulative number of primary impacts per 10^6 km^2 and 10^9 yr is assumed according to Eq. (161), after integration and with the proper constants, with R in centimeters, to be

$$\log N_a = 11.921 - 2.7 \log R_a \tag{171}$$

for Apollo group or meteorites, and

$$\log N_c = 9.828 - 2.2 \log R_c \tag{172}$$

for the comet nuclei, while the contribution from Mars asteroids is negligible within the range considered. As in Table XV, $2R_a = B_0/14.9$ and $2R_c = B_0/26.6$ as for interplanetary stray bodies impacting on hard rock.

The adopted crater statistics are collected in Table XXVIII.

From the equations of Section II, the geometric parameters of cratering in two characteristic media—the hard bedrock and the rubble of overlay, are determined as follows.

It will be found that, in the range below 2.5 km, secondary craters are the main contributor to overlay; also, in the smallest class of craters, the Apollo-meteorite group prevails among primaries, while in the larger classes the comet-nuclei group is more prominent (cf. Table XXVIII). A simplification is therefore admissible, in assuming an equal proportion of the two groups among primaries. This gives an average of $x_p/B_0 = 0.0837$ and, instead of Eq. (167), a volume of ejecta from unprotected hard rock [cf. Eq. (15)]

$$V_e = 0.0204B_0{}^3 \tag{173a}$$

for the primaries. On the other hand, overlying rubble will prevent, partially or totally, the projectile from striking the underlying bedrock; this condition is most critical for the smallest projectiles among which the Apollo group prevails. For this group, with $w_0 = 20$ km/sec, $\gamma = 45°$, $\delta/\rho = 2$, $s_c = 6 \times 10^7$, $s_p = 2 \times 10^8$ (about as for desert alluvium), $p = 2.17$, $D = 28.0$, and $D/p = 12.9$, the penetration into an unlimited layer of rubble in a primary impact becomes

$$x_p{}' = B_0/12.9. \tag{174a}$$

For the secondaries, only those of high velocity such as in the rays of ray craters being important [the penetrating low-velocity ejecta are unable to crush the bedrock; they as well as those that are nonpenetrating are only able to produce craters in the overlay (cf. Table XXXVI (D)], we assume $w_0 = 0.75$ km/sec, $\gamma = 45°$, and $\delta = 2.6$. At impact of secondaries upon bedrock,

$$x_p''/B_0 = 0.439, \tag{175}$$

and, from Eq. (15), the volume of ejecta from unprotected hard rock becomes

$$V_e'' = 0.1070B_0{}^3. \tag{173b}$$

The relative penetration of secondaries into rubble is then

$$x_p''' = 0.352B_0 . \tag{174b}$$

With 1.3 as the density of overlay or one-half that of the bedrock, and n_i the number of impacts in 10^9 yr upon an area of $S = 10^6$ km^2 = 10^{12} m^2, the contribution to overlay thickness in 4.5 billion years from a given group of craters (B_0) equals

$$\Delta X = 2V_e \cdot 4.5n_i/S. \tag{176}$$

For the primary craters this becomes

$$\Delta X_p = 1.84 \times 10^{-11}n_pB_0{}^3, \tag{177a}$$

and for the secondaries

$$\Delta X_s = 9.63 \times 10^{-11} n_s B_0{}^3 \tag{177b}$$

in cm/4.5 $\times$ 10^9 yr when B_0 is given in meters; both equations provisionally disregard the protective layer of the overlay itself and thus represent upper limits.

In the four upper lines of Table XXIX the basic cumulative crater numbers of Table XXVIII are broken up into discrete data, interpolated for more or less comparable (not quite constant) logarithmic intervals of B_0 ; the median values of crater diameter are given in the first line, the observed differential numbers in the second, the erosion factor E_f in the third, and the rates of impact per 10^6 km² and 10^9 yr (as corrected for erosion) in the fourth line.

In the following Part (A) of Table XXIX the data are interpreted conventionally by disregarding the braking action of the overlay. While the total cratering rate n_c in the fourth line may be considered independent of this action of overlay, being based on purely empirical data, the number n_p of primary impacts (sixth line) does depend on our conventional assumption since this determines the ratio of projectile to crater diameter, thus the size and the number of projectiles. With the rubble layer, smaller projectiles will produce craters of a given size and, thus, there will be more primary impacts and, after substracting their number, the difference yields fewer secondaries. As these latter chiefly contribute to the overlay (lines 8 and 9 of the table), the overlay in Part A represents an overestimate. Even as the figures stand, new overlay cannot be produced when its thickness exceeds the imaginary "depth of penetration" into rubble (lines 11 and 12 of the table), or the crater depth in rubble of infinite thickness at crater diameter B_0 . The penetrations are given by

$$H_p = B_0/12.9 \tag{178a}$$

for the primaries and

$$H_s = 0.352 B_0 \tag{178b}$$

for the secondaries. These quantities are independent of the radius of the projectile and depend on crater diameter only. In Table XXIX, Part A, this takes place at $B_0 < 49$ m, whence a rough upper limit for overlay thickness of about 16 m follows. This compares favorably with the estimate of 13–17 m at a particular spot in Oceanus Procellarum, made in Section V,C from Surveyor I pictures of an eroded boulder wall of an ancient crater (Figs. 5 and 6).

Part (B) of Table XXIX represents a more sophisticated calculation for an assumed overlay thickness of 12 m at present. The thickness is assumed to grow uniformly with time, average values instead of differential equations being used henceforth. In Part (B), the first line gives the average age (in units of 4.5 billion years) of the presently surviving craters, calculated from

$$t_f = 1/2 E_f , \tag{179}$$

and the second line contains the average overlay thickness at the time of impact,

$$X_a = 12(1 - t_f) \quad \text{m.} \tag{180}$$

When overlay thickness exceeds "rubble penetration," i.e.,

$$X > H_{p,s}, \tag{181}$$

the bedrock is untouched and no increase of overlay takes place. For the primaries in this case $D = 28.0$ and $2R_p = B_0/28.0$. For unprotected bedrock or $X = 0$, the figures are $D = 14.9$ and $2R_p = B_0/14.9$, which also is the case in Part (A) of the Table. For a given crater diameter, the radius, and thus the predicted number of impacting projectiles, is different according to the kind of target. With the logarithmic intervals for B_0 or R in Table XXIX, the frequency index is the same as for cumulative numbers, $n - 1$ according to integration of Eq. (161). For Apollo group the index is thus 2.7, and the ratio of primary incidence in the two cases is

$$(28.0/14.9)^{2.7} = 5.50.$$

Thus, when Eq. (181) is valid, or for $B_0 < 120$ m in Part B of the table, the primary incidence will be 5.5 times that given in Part A. The incident mass, however, contains an additional factor of R^3, and thus decreases with the 0.3 power, in a ratio of

$$(14.9/28.0)^{0.3} = 0.83.$$

When Eq. (181) is not fulfilled, two-layer cratering takes place. Instead of a complicated analysis, we simply use an interpolation formula between the two extremes for $B_0 > 120$ m in Part B:

$$n_p = n_A[1 + 4.5(120/B_0)^2] \tag{182}$$

whereas for $B_0 < 120$ m, $n_p = 5.5n_A$ is to be assumed. Here n_A is the value of n_p in Part A. The calculated values of average incidence rates are given in the third and fourth lines of Part B. These are incidence rates of projectiles of the same average size that have produced the observed craters, although in the past the craters in each class—and not the projectiles—may have been smaller because of less overlay and more hard bedrock involved.

Equations (177a) and (177b) require certain additional efficiency factors η to allow for the average fraction of bedrock crushed, depending on overlay thickness and on the time t_p during which penetration to bedrock level was possible. Two cases present themselves.

For a given overlay thickness X and potential penetration H_p, the condition must be fulfilled that at $X/H_p = 1$, $\eta = 0$ and at $X/H_p = 0$, $\eta = 1$. Thus, we assume the individual efficiency in the second case as

$$\eta = 1 - (X/H_p)^2, \tag{183}$$

and the average over the entire time of existence of the mare (4.5×10^9 yr), for which $X_a = 6$ m, is

$$\eta_1 = 1 - (6/H_p)^2. \tag{183a}$$

In the first case, when the layer is thicker than the average penetration, penetration stops after a relative time interval

$$t_p = H_p/12 \tag{184}$$

(H_p being given in meters), during which the average $X_a = \frac{1}{2}H_p$. Substituting this into Eq. (183) and multiplying by t_p, the efficiency of the first case as compared to the second one becomes

$$\eta_2 = H_p/16. \tag{183b}$$

(For secondaries, use H_s for H_p.)

These efficiencies and the corresponding differential overlay accretions are given in the 5th to 8th lines of Part B of Table XXIX.

The 9th line of Part B of Table XXIX contains the cumulative accretion of new overlay, in centimeters at density 1.3 for the toal time span of 4.5×10^9 yr until the present. Extrapolation toward smaller crater diameters ($B_0 < 20$ m) will not yield much because of the rapid decrease of H_p, t_p, and η. A total extrapolated overlay thickness for Part B must be close to $X = 14$ m, while the starting assumption was $X = 12$ m. The solution is practically self-consistent. The value $X = 13$ m can be assumed as an average thickness of overlay (density 1.3) in the maria on regions removed from the vicinity of large craters, consistent with the *observed* volume and number of craters and not critically dependent on theory and interpretation. This agrees remarkably well with the estimate from the Surveyor I picture (Fig. 6) (cf. Section V,C) and is not in contradiction with more crude estimates by Jaffe (1965, 1966, 1967) which point to an overlay thickness of 5–10 m.

Besides the accretion of new overlay, from an environmental standpoint, the total influx rate of overlay material may be of interest, new material from the bedrock and old stirred up from the existing layer of overlay, i.e., the accretion rates when setting $\eta_p = \eta_s = 1$. The rates are given in the 11th and 12th lines of Table XXIX, Part B. Extrapolation to smaller crater sizes (45 m) would yield about 2000 cm/4.5×10^9 yr., but the addition consists of "soft spray," which is not relevant from most standpoints.

The preceding results refer to the maria surface, originally molten and solidified into hard rock and subsequently battered by the interplanetary population of
stray bodies, which is assumed to be the same as presently observed.

For the lunar continentes the conditions were different. We do not know what
was underneath, but the exposed top layer is battered to a great depth by
a saturation coverage of craters. The great protective depth of overlay leaves
only the larger craters to contribute to it.

Let us assume that the base of the continentes consisted of a rock surface,
possibly partly melted and solidified but still hot and soft. The impact parameters
which best suit the relevant crater range from 16 to 64 km are those of Models D
and E of Table XIV (Fig. 3). We may thus set $x_p/B_0 = p/D = 5.0/28.1 = 0.178$,
$w_0 = 3$ km/sec, $\lambda^2 = 0.25$, and $s = 2.8 \times 10^8$ dyne/cm². From Eq. (15)
the volume excavated is

$$V_e = 0.0435 B_0{}^3. \tag{184}$$

The flight range of the ejecta, proportional to the product $s\lambda^2$ at $\rho =$ const
(cf. Section II,F), is now only about 0.26 that for postmare craters. On the
other hand, the craters of relevant size on continentes are about 25 times more
numerous than on maria (Baldwin, 1964b), or there is about $\frac{1}{5}$ the distance
between them as compared with the maria: the overlapping of ejecta of neighboring craters is thus similar on the continentes and maria.

In Table XXX, Baldwin's (1964b) crater counts on the continentes as contained in his Table XV are used to estimate the thickness of *ultimate* overlay,
i.e., the layer of ejecta of density $\rho = 1.3$, produced from a rock layer of $\delta = 2.6$
by the counted visible craters of the highlands. The rubble layer thickness
so obtained is a lower limit, the original surface was rubble itself and not hard
bedrock. However, from the model of the origin of this surface as depicted in
Section IV,G, it is probable that the bedrock predating the final bombardment
was partly melted and essentially compacted, as also is supported by the evidence
of crater profiles (Table XIV and Fig. 3). The estimated overlay thickness
distribution as arrived at in Table XXX is therefore likely to be a close approximation.

In Table XXX: the 3rd column gives the crater area $S_0 = 0.75\pi B_0{}^2$; the 4th
column, the fractional crater coverage $n_i S_0/10^5$; the 5th column, the area within
a 5-km fringe over the crater rim, $S_f = 0.75\,\pi(B_0 + 10)^2$, or the effective area
covered by the ejecta, including the crater interior; the 6th column, $\tau = n_i S_f/10^5$,
the fractional coverage by crater ejecta, and the 7th, the cumulative coverage $\sum \tau$
(this may exceed unity, which means there is overlapping); the actual coverage
or fractional area is than $1 - \exp(\sum \tau) = \theta_c$. Considering that a deeper overlay
takes exclusive precedence over a shallower one, the distribution of overlay in
order of thickness is that of the distribution of the fringe areas in the order of
decreasing crater size. The percentage area in the 8th column is thus the differen-

tial of $\theta_c \times 100$, and in each group the average thickness of overlay as given in the 9th column is that over the fringe area of a crater, $X = 2V_e/S_f = 87.0B_0{}^3$ (km)/S_f (km²) (meters) according to Eq. (184), the factor of 2 allowing for the smaller density of the ejecta. Figure 11 represents this distribution of

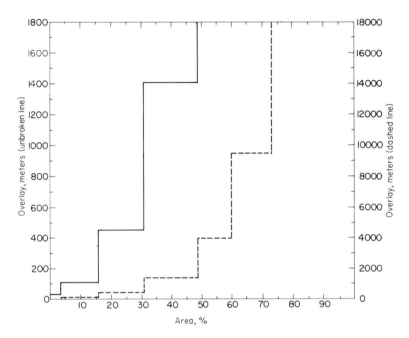

FIG. 11.   Distribution of overlay thickness in lunar highlands (continentes). (Cf. Fig. 10.)

overlay thickness in the continentes. The 10th (or last) column of Table XXX contains the differential overlay ΔX, averaged over the entire area. The total cumulative thickness exceeds 10 km, but this leveled-out average conveys an inaccurate impression of the actual nonuniform distribution featured in the 9th column and Fig. 11.

D. Overlay Particle Size Distribution

From Surveyor and Ranger pictures, as well as from an understanding of the process of fragmentation in cratering impact, it follows that the overlay rubble contains all particle sizes from microscopic dimensions up to meter-sized boulders. Let us attempt to predict the particle size distribution from the physics of cratering as outlined in Section II.

It is a well-known fact that the strength of materials increases with decreasing linear dimensions. The effect is caused mainly by imperfect cohesive coupling between the molecules of the lattice, only a small fraction of them being in full contact with each other. One of the consequences of this is the layered morphology of meteor crater debris; in the inner portions of the crater bowl, the greater shock pressure produces fine-gained rock flour, while on the outskirts the rock is fractured into sizable boulders.

Let y be the fractional crater mass as in Section II and Fig. 1. The shock pressure is proportional to y^{-2} and this can be set equal to the destruction strength s_y, i.e.,

$$s_y = s_c y^{-2}, \tag{185}$$

which results in fragments of average size R, such that

$$s_R = s_y .$$

We assume a power law for the strength dependence on size,

$$s_R = CR^{-\nu}. \tag{186}$$

The exponent ν can be estimated roughly as follows. For granite, at dimensions of the order of 20 cm as used in the building industry, $s_c = 1.2 \times 10^9$ dyne/cm^2. Its typical molecule SiO_2 has a lattice energy of 2.8 eV or 4.5×10^{-12} erg and occupies a volume of 3.7×10^{-23} cm^3, corresponding to a mean distance between lattice molecules of 3.4×10^{-8} cm. For a bond of two molecules, the energy equals one-third that of the lattice energy or 1.5×10^{-12} erg, and the cohesive force, with an inverse fifth-power law of intereaction, is

$$1.5 \times 10^{-12}/(0.2 \times 3.4 \times 10^{-8}) = 2.2 \times 10^{-4} \quad \text{dyne.}$$

Distributed over an effective contact area of $(3.4 \times 10^{-8})^2$ cm^2, this corresponds to a cohesive (tensile) strength $s = 1.9 \times 10^{11}$ dyne/cm^2, at effective dimension R (R stands here for diameter) of $3.4 \times 10^{-8} \times 2^{1/3} = 4.3 \times 10^{-8}$ cm. Applying Eq. (186) to the two extreme values of s, we find $\nu = 0.254$ as an average exponent over a relative range of 5×10^8 to 1 in the linear scale.

For a check, consider the Arizona crater with boulders up to 20 m, supposedly from the periphery, $y = 1$, and rock flour of $R \sim 10^{-3}$ cm at an effective value $y = y_f$. With the value of ν as suggested above, the ratio s_y/s_c becomes then $(2000/10^{-3})^{0.25} = 38$, whence, according to Eq. (185), $y = 0.16$ is found to be the fractional mass at which rock flour of the specified grain size is expected to be produced, a not unreasonable result.

Substituting $s_R = s_y$ from Eq. (186) into Eq. (185), we have

$$R = y^{2/\nu} \times \text{const}, \tag{187}$$

$$y = R^{\nu/2}, \qquad dy \sim R^{(\nu/2)-1} dR. \tag{187a}$$

The number of particles in the mass element dy is proportional to dy/R^3 or, with Eq. (187a), the frequency of fragments among crater debris ranging in size from R to $R + dR$ becomes

$$F(R)\, dR \sim R^{(\nu/2)-4}\, dR = R^{-n}\, dR, \tag{188}$$

where

$$n = 4 - \tfrac{1}{2}\nu \tag{189}$$

is the "frequency index" in the power law of particle diameters as in Eq. (161). With $\nu = 0.254$, we find $n = 3.87$ for the predicted frequency law of cratering fragments as counted in a volume. For comparison, Smith (1967) finds for the *surface distribution* of fragments, with $R = 2$–20 cm on the Russian Luna 9 pictures, $n - 1 = 2.9 \pm 0.2$ or $n = 3.9$. A similar value of $n = 3.77$ is found by Hapke (1968) from the Surveyor pictures. It may be relevant to note that for volcanic ejecta in Hawaii, which produced impact craters in the surroundings, Hartmann (1967) finds an empirical value of $n = 3.64 \pm 0.1$, close but not quite equal to the exponent for lunar overlay.

The agreement between the predicted and observed frequency functions of lunar surface debris is remarkable and quite unexpectedly supports the cratering theory as presented in Section II. Of course, erosion by micrometeorites and repeated turnover of the overlay by new impacts will tend to increase the number of small fragments at the expense of the larges ones, thus also increasing the value of n above that predicted. Apparently, none of these effects has been very efficient; the first, probably because the surface fragments are buried and protected from erosion sooner than they are eroded; the second because the mass fraction of old overlay in cratering ejecta is small compared to the contribution from new crushed bedrock.

The dependence of strength on size would apparently invite some revision of the cratering formulas of Section II. The size of the largest blocks, such as those formed at the crater rim, is about $1/40$ to $1/60$ of the crater diameter for the Arizona crater, and $4/450 \sim 1/110$ for the largest block seen on the far side of the stone-wall lunar crater of Surveyor I (Fig. 6). The Surveyor I bedrock seems to have been shattered before the formation of this crater and the blocks may be too small. It appears plausible to assume that geometric similarity holds, and that the characteristic value of the marginal crushing strength(s_c), determining the volume and diameter of the crater, corresponds to a particle diameter equal to $B_0/60$, so that for typical granitic or basaltic bedrock, the effective lateral strength, according to Eq. (186) with $\nu = \tfrac{1}{4}$, becomes

$$s_c = 4.0 \times 10^8 B_0^{-1/4} \tag{190}$$

dyne/cm², with B_0 in kilometers. According to Eq. (7), when $s = s_c$ without a gravity frictional component, the crater diameter varies as the 1.06 power of the projectile diameter, instead of varying in strict proportion.

The effect on penetration, amounting to the $-1/120$th power of linear dimension according to Eq. (6), is negligible. Thus, leaving the penetration parameter p unchanged, the cratering parameters in the first half of Table XV are somewhat changed through the application of Eq. (190) and are now like those given in Table XXXI. The new figures for crater diameter B_0, in the fourth line of the table, are now markedly larger than the former values (5th line), but the ratio of the two does not increase monotonously with crater size, the decrease in the cohesive lateral strength s_c being balanced by the increasing friction component. For this reason the effect remains small; the decrease of strength with increasing dimensions, although favoring a higher incidence of larger craters, is utterly inadequate to account for the observed excess in the numbers of big craters (Tables XVII and XVIII).

VIII

MECHANICAL PROPERTIES OF LUNAR TOP SOIL

Surveyor spacecraft pictures and experiments, as televised to Earth, have shown that the lunar soil is granular, with a very broad distribution of grain size from meter-sized boulders to submillimeter particles (Newell and NASA Team, 1966, 1967; Jaffe et al., 1966a,b; NASA, 1967; Christensen et al., 1967; Hapke, 1968). Hard pebbles are present, as well as clumps of coagulated finer material. Impacts of the surveyor footpads (Figs. 12 and 13) as monitored by a strain-gauge to record force data and supplemented by static penetration tests (Surveyor III), yielded experimental data similar to those described in Section II,E from which the strength parameters of lunar soil could be derived. The parameters can be defined in different ways, depending on the mechanical model used. Although the data are scarce, they are sufficient to show considerable qualitative and quantitative similarity with terrestrial natural beach gravel, especially in that the cohesive strength rapidly increases with depth. Equations (37) and (37a) appeared to be appropriate also for the frontal and lateral resistance of lunar soil. Some compressibility of the lunar soil was observed, although insignificant enough to justify the application of the penetration and cratering equations of Section II. Table XXXII contains the results.

In Part (a) of Table XXXII, static tests with Surveyor III ("Tidbinbilla") [on an inner crater slope of about 14°; crater about 230 m in diameter, in Oceanus Procellarum, $\phi = 2°9$ south, $\lambda = 23°3$ east (astronomical) or west (astro-nautical)] are listed and interpreted with Eq. (37) and three assumed values of a^2; test No. 6 is decisive and would require $a^2 = 2.4 \pm 0.8$ cm², while other tests are indifferent in this respect. It was decided to assume $a^2 = 2$ cm², the same as for terrestrial sand (Section II,E). There is not much uncertainty in S_p, which

Fig. 12.   Surveyor I, footpad 2 on lunar surface. Diameter of footpad top, 30.5 cm.
(Photograph by courtesy of NASA.)

little depends on the particular value of a^2, and for the value chosen the
logarithmic mean is

$$S_p = 3.21 \times 10^4 \quad \text{dyne/cm}^4 \qquad (+23 \quad \text{to} \quad -19\%),$$

to be compared with a value of 5.55×10^4 for similar experiments with terestrial
sand [Table III (a)].

Test No. 4, made on a trench bottom, yielded a $9.15/3.21 = 2.85$ times higher
value at a depth of 6 cm after removal of the overlying material; this compares
favorably with Experiment (2a) in Table III (a), where an 8.4-fold increase in
the bearing strength parameter was obtained at an excavated depth of 15 cm.

Fig. 13. Surveyor III, footpad 2, third touchdown, with surface sampler and a depression made by it (bottom left). (Photograph by courtesy of NASA.)

The dynamic tests are based on the impact of Surveyor footpads. The footpad has a circular top 30.5 cm in diameter (Fig. 12) and a total height of 12.8 cm; the circular bottom is narrower, 20.3 cm in diameter, and widens upward over a conical section of 45° angle, 5.1-cm thick. The footpad is not rigidly connected with the very much more massive main body, but it is linked to it by a system of shock absorbers with strain gauges attached. At the first contact, the footpad acts almost like an independent projectile, but as soon as it decelrates, the shock absorber yields and increases its pressure on the footpad, which no longer moves freely by its own inertia. The equations of motion of Sections II,D and E, which refer to a rigid projectile, do not apply therefore in this case. On the other hand, the strain gauge data provide a more direct means of evaluating the mechanical parameters of the soil and the amount of radial momentum transmitted during penetration.

The theory of impact cratering requires that the target material part laterally with a velocity determined by the preceding history of penetration, a value higher than that of the instantaneous penetration velocity. A cone of 45° like that in the footpad will not therefore, in its forward motion, be able to overtake and contact the material parting sideways. Therefore, it has been assumed here (contrary to some suggestions made by the NASA team) that, during impact, contact was maintained only with the bottom 324-cm² area of the footpad.

The shock absorber records give the time variation of the force F_a along the absorber axis making an angle α with the direction of impact from about 61° at no load to 70° at full load. The decelerating force is then $F_d = F_a \cos \alpha$. From graphical and tabular data describing the impact events (loc. cit.), a plausible approximation, $\cos \alpha = 0.487 - 0.147 F_a/C$ with $C = 7.5 \times 10^8$ dyne, was introduced. The maximum load that is reached at greatest penetration x_0 then yields

$$s_p(\text{max}) = F_d(\text{max})/\sigma$$

with $\sigma = 324$ cm² and, from Eq. (37) with $a^2 = 2$ cm²,

$$S_p = s_p(\text{max})/(x_0^2 + 2) \tag{37b}$$

is obtained directly.

With this parameter, the values of s_p derived from the strain gauge, when inserted into Eq. (37), yield a few discrete values of x and the average speed of penetration among them for successive intervals of time. The initial speed at impact w_0 and the shock entry speed w_1, as well as the initial deceleration dw_1/dt at entry (uninfluenced as yet by the shock absorber) having been estimated, a history of the forward motion of the footpad bottom surface can be reconstructed (graphically) to fit the average velocities and the boundary condition. In such a manner, for footpad 2 of Surveyor I, a reconstruction has been obtained as described in Table XXXIII.

The time variation of velocity as shown in Table XXXIII(b) is more or less empirical. It can be interpreted in the following way: during the first 0.002 sec, the deceleration is balanced through increasing coupling, by way of the shock absorber, with the main mass of the spacecraft; between $t = 0.002$ and $t = 0.009$ sec, the coupling accelerates the footpad to virtually the velocity of the spacecraft; after that, the footpad is brought almost to rest within the next 0.007 sec by increasing resistance and acts now as an effective brake on the main body during the time $t = 0.016$ to 0.114 sec, while its own penetration is slow and the kinetic energy of the spacecraft is dissipated in the three shock-absorber legs.

The radial momentum released in the lunar soil by the impact consists of two components—the shock momentum imparted to the target at first contact, and the hydrodynamic pressure integral

$$J = \left((w_0 - w_1) \cdot \tfrac{1}{2}R\rho\sigma + K_a\rho\sigma \int_0^t w^2 \, dt\right)\Big/2K_a,\qquad(191)$$

which is to be used with Eq. (42). The lateral strength parameter is then derived ultimately from Eq. (44), using the terrestrial beach average of $F = 0.118$ as the only available guess.

The data for footpad No. 2 of Surveyor I [which landed on a practically horizontal surface in Oceanus Procellarum, at $\phi = 2\overset{\circ}{.}5$ south, $\lambda = 43\overset{\circ}{.}3$ east (astron)] are the best of those quoted in Table XXXII(b). The penetration, 5.8 cm, was derived from the shadow of the top surface ($2R = 30.5$ cm) at low solar altitude; the surface was tilted 9 cm above undisturbed ground level at one end, 5 cm at the other, or a mean of 7 cm above ground level. Substracting 12.8 cm as the thickness of the footpad, we obtain $-x_0 = 7 - 12.8 = -5.8$ cm for the bottom surface, with 4.5 cm at one end and 7.1 cm at the other end of the bottom ($2R = 20.3$ cm). The difficulty of making estimates by mere inspection of the photographs is illustrated by the fact that, in preliminary reports (Newell and NASA Team, 1966; Jaffe and NASA Team, 1966a), the depth of penetration was estimated to be only 2.5 cm. The crater rim-to-rim diameter, B_0, was easier to estimate, although the darker material ejected beyond the rim may have produced the impression of a somewhat broader crater than the actual size (cf. Fig. 14, which shows more contrast than Fig. 12). Besides, because of the motion of the legs, as controlled by the shock absorber, the footpad came to rest about 5 cm inward (toward the spacecraft) from the original center of the crater, and thus assumed an asymmetric position (Fig. 12).

In the other four cases of Table XXXII(b) the parameters were more difficult to estimate. The publications (Jaffe and NASA Team, 1966b; Christensen et al., 1967; Newell and NASA Team, 1967; NASA, 1967) as well as NASA photographic prints were consulted and compared. The penetrations are probably good to ±0.5 cm, the crater diameters to ±2 cm, while the velocities and velocity histories were considered in parallel or in homology with the data of Table XXXIII, which were considered to be of better quality. The very low velocities for Surveyor III are not in accord with some statements in the NASA reports, but follow directly from the strain gauge time records (less reliable than those of Surveyor I) and are supported by the concordant values of S_c so obtained.

The dark ejecta surrounding the impact craters (Fig. 12) seem to indicate that blackening due to radiation damage is not a one-way process and that the very

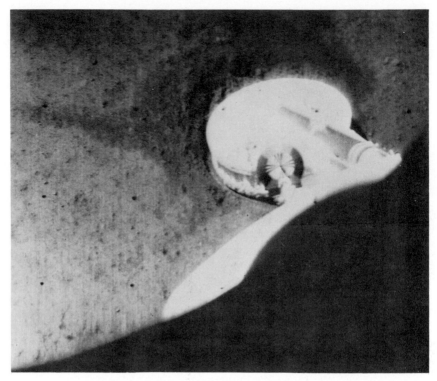

Fig. 14.   Surveyor I, footpad 2, enhanced contrast.—Courtesy of NASA.

surface, exposed to immediate radiation, becomes slightly bleached or, rather, that the material when buried and protected from direct radiation becomes spontaneously darker with time. However, as suggested by Hapke (1968), the difference in albedo may be due to different graininess and porosity, and not to physico-chemical changes in the grains. Footpad No. 2 of Surveyor III was ejected from its original crater at third touchdown and came to rest at a distance of about 30 cm from it. The bottom of the original crater (Fig. 13) (used in Table XXXII) is laid open and appears to have a higher albedo than the undisturbed surface or the ejecta—a result of compression. This seems to support the geometrical interpretation of the difference in albedo.

The dark halo of ejecta from Surveyor I (Fig. 14) shows an average outer margin at 34 cm, in some sectors reaching to 47 cm from the crater center (reckoning with the asymmetry of the footpad), and a ray reaches to a distance of at least 61 cm. The extreme, not too unusual, flight distance of the ejecta from the edge of the footpad can be set equal to $L = 47 - 10 = 37$ cm with $S_c = 1190$, $x_0 = 5.8$, and $s_c = 4.25 \times 10^4$ dyne/cm². To this a small contribution from friction, $130x_0$ [Eq. (11) and (12)] or 750 dyne/cm^2, is to be added, making $s = 4.33 \times 10^4$. With $\rho = 1.3$ gm/cm^3, u_s equals 182 cm/sec according to Eq. (4). Following the line of reasoning of Section II,E, and with $w_0 = 364$ cm/sec, from Eq. (16) we find $y_Q = u_s/w_0 = 0.5$. Further, with $\sin \beta_0 = 0.8$, $y = 0.6$ as nearest to y_Q, $u = 303$ cm/sec, $\beta = z$, $\sin z = 0.48$, $\cos z = 0.88$ [from Eq. (27)], and $g = 162$ cm/sec^2, Eq. (45) yields the ejection velocity for $L = 37$ cm to be $v = 84$ cm/sec, whence $\lambda \simeq 84/303 = 0.28$, and $\lambda^2 = 0.08$. For the conspicuous ray, $L = 51$ cm, $y = 0.5$, $u = 364$ cm/sec, $\sin z = 0.4$, $\cos z = 0.92$, $v = 106$ cm/sec, and $\lambda = 106/364 = 0.29$, $\lambda^2 = 0.08$ (or the same value). The lunar dust seems to possess higher internal friction and lower kinetic efficiency as compared to terrestrial gravel.

The two Surveyor experiments yielded very similar mechanical parameters despite the difference in terrain, Surveyor I having landed practically on level ground, Surveyor III on the inner slope of a crater wall inclined about 14° to the horizon (NASA, 1967, Part II, p. 20). Although both are on a mare surface of Oceanus Procellarum, near the lunar equator but separated by 20° in longitude or 600 km, the mechanical properties are probably representative of the upper layer of lunar soil in general and to a depth of 50–100 cm, to which depth extrapolation of Eqs. (37), (37a), and (37b) is permissible. The parameters are chiefly determined by the fine-grained matrix, of the order of 0.001–0.006 cm, as is shown by the retention of the imprints of the footpad pattern, a network of about 1 cm mesh with ridges 0.006-cm high, on the Surveyor III footpad No. 2 crater of the third touchdown (however not visible in the reproduction of Fig. 13). Lumps of coagulated grains were present, from 0.1 to 5 cm, or of about 1 cm average size; they apparently consist of loosely bound smaller grains and are easily crushed. These lumps, as well as admixed occasional hard pebbles or rock splinters, by virtue of the cooperative action of the constituent grains at inner contacts, are probably responsible for increasing the thermal conductivity and yielding a larger effective "thermal" grain size of the order of 0.03 cm (Table XXII).

Table XXXIV contains a summary of the mechanical characteristics of lunar soil (from Table XXXII) as compared with those of terrestrial natural gravel (from Table III). The notations are those of Sections II,E and F. In the fifth and eighth columns are given the surface bearing strength s_{p0} and the surface lateral strength or "cohesion" s_0, both corresponding to zero penetration $x = 0$ or $x_0 = 0$. With $s_{p0} = 6 \times 10^4$ dyne/cm^2, an astronaut with heavy equipment totaling 150 kgm but weighing only 2.4×10^7 dyne on the Moon will be supported without sinking a centimeter into the lunar surface by a 400 cm^2 contact surface—just about what would be provided by his two feet.

The minimum lateral strength of lunar soil, or its surface value of $s_0 = 2800$ dyne/cm^2 can be compared with the minimum adhesion of grains *in vacuo*, about 0.5 dynes (Smoluchowski, 1966; Ryan, 1966; cf. Section VI,A).

With an average of three contacts or 1.5 dynes per grain, it would require 1870 grains in contact per square centimeter, or an average spacing (diameter) of about 0.023 cm to account for the cohesion of the matrix. This does not differ so very much from 0.033 cm, the average "thermal" diameter (Table XXII). It may be noted that meteoric dustballs, or the grainy skeletons of cometary material that remain after ices have evaporated, have a crushing strength s_0 of about 10^4 dyne/cm^2 at average grain diameter from 0.01 to 0.1 cm (Öpik, 1958c); their strength is about that of lunar soil at $x_0 = 2$ cm, thus at an average depth of about 1 cm, although the density is less. The two kind of material seem to have much in common.

Extrapolation of Eq. (37a) with $S_c = 1400$ dyne/cm^4 would yield the strength of terrestrial alluvium $s_c = 4 \times 10^7$ dyne/cm^2 (a probable upper limit for granular material) at a depth of penetration $x_0 = 170$ cm. The overlay is much thicker than this (Section VII,C), and a constant value of cohesive strength of this order can be assumed to hold for most of the thickness of the overlay. The corresponding upper limit for the compressive strength is $s_p = 7 \times 10^8$ dyne/cm^2.

IX

THE BALLISTIC ENVIRONMENT

A. Electrostatic versus Ballistic Transport

It has been pointed out in Section VII,A that electrostatic transport of lunar dust, so ingeniously proposed by Gold (1955), does not work on the lunar surface— *de facto*, because it would have obliterated sharp transitions of contrast, which are actually observed and *de jure*, because its effect in the actual environment of the Moon cannot be significant (Singer and Walker, 1962). The ratio of electrostatic repulsion to gravity of a small particle on a planetary surface is

$$F_e/F_g = 2.66 \times 10^{-6} I^2/(gR^2 L_d \delta), \tag{192}$$

where I is the common electrostatic potential of the surface and the particle in volts, g the acceleration of gravity, R the spherical radius, and δ the density of the particle; L_d is the electrostatic plasma screening length (similar to the Debye length) for a planetary surface charge. For the Moon, $g = 162$ cm/sec^2, $\delta = 2$ gm/cm^3 for individual irregular particles, $I = +6$ V and $L_d = 100$ cm (Öpik, 1962b), whence

$$F_e/F_g = 8.20 \times 10^{-11} I^2/R^2 = 2.95 \times 10^{-9} R^{-2}. \tag{192a}$$

The effect would be noticeable for $R < 10^{-4}$ cm, $F_e/F_g > 0.3$ and of decisive importance for $R < 5.8 \times 10^{-5}$ cm, $F_e/F_g > 1$, or at submicron sizes. When disturbed by meteor impact, these small particles would float in sunlight within the screening-length range, about 100 cm from the surface, their charge sustained by the photoelectric effect, until entering a shadow when they would become neutralized and fall back. The virtual absence of any trace of detail blurring (which should be caused by particles of so high a mobility) indicates that these small particles cannot play any significant role on the lunar surface. The thermal conductivity and the cohesion of lunar soil also indicate that the relevant average particle size of lunar dust is at least 100 times greater than that at which efficient eletrostatic transport begins.

   Therefore only the ballistic transport of dust on the lunar surface, caused by meteorite impact, is of relevance.

B. *Impact Fluxes and Cratering in Overlay*

   As distinct from accretion and loss, two main sources cause the mobility of the dust: direct meteoritic impact into the dust layer, and the impact of debris from secondary ejecta broken off the bedrock and accumulating as overlay. Although the latter source signifies also a kind of transport, from the standpoint of the local material balance of a "normal" plain, undisturbed by large-scale cratering, its effect amounts to virtual accretion, while the "abnormal" area of a new crater from which the material is taken begins a new history and is atypical. Of course, the factors of dust transport begin to work at once on the surface of the newly formed crater, but some of the starting conditions, such as the thickness of overlay, are different.

   The quantitative importance of the different sources in stirring the dust can be measured provisionally by the radial momentum $J = k\mu w_0$, imparted to the overlay by nonpenetrating projectiles where μ is the impinging mass rate per unit of time—say, 4.5×10^9 yr—and the area (cm^2). For micrometeors of the zodiacal dust, $w_0 = 6.0$ km/sec (including acceleration by the Moon), $k = 2$, $\mu = 47$ gm (Table XXXVI, Part A and Section VII,B), whence $J_M = 564$ in the units chosen. For the "visible" meteors, chiefly belonging to the "E-component" and which are all nonpenetrating, $\mu = 0.36$ gm, $w_0 = 18$ km/sec (Öpik, 1965a), $k = 3.1$, and $J_0 = 20$. For the Apollo-meteorite group, $R_2 = 200$ cm(nonpenetrating, the larger members lead to basic cratering and produce new overlay from the bedrock) and from Eq. (158), $\mu = 0.25$, $w_0 = 20$ km/sec (Öpik, 1965a), and $k = 3.3$, $J_1 = 16$. For comet nuclei, Eq. (159), $R_2 = 300$ cm, $\mu = 0.0067$, $w_0 = 40$ km/sec, $k = 4.4$, and $J_2 = 1.2$. For Mars asteroids, $R_2 = 200$ cm, $\mu = 9.5 \times 10^{-6}$, $w_0 = 9$ km/sec (Öpik, 1965a), $k = 2.3$, and $J_3 = 2 \times 10^{-4}$, which is utterly negligible. As to secondary ejecta, only the hardspray component is of importance here, which component originates from the bedrock and thus is representative of the *new* overlay; its rate at present may be close to 12 m or

$\mu = 1500$ gm/cm^2 in 4.5×10^9 yr as given by the thickness of overlay (Table XXIX, Part B); with $s = 6 \times 10^8$ dyne/cm^2, $\rho = 2.6$, and $\lambda^2 = 0.22$ for the parent bedrock at $y = 0.5$, the median mass, and $x/x_0 = 0.5$, Eqs. (4), (16), (24), and (25) suggest an average ejection or secondary impact velocity of $w_0 = 0.10$ km/sec, $k = 0.63$, and an additional factor $s/s_a = V/V_d = 1.1$ for "static work" (cf. Table XXXVI, Part D); the relative radial momentum becomes $J_e = 0.63 \times 1.1 \times 1560 \times 0.10 = 108$. Hence the relative stirring power of the different components obtains as given in Table XXXV.

Because the velocities and flight distances of massive ejecta depend solely on the mechanical properties of the target, the figures of the table must approximately represent the relative mixing efficiency of the separate sources with respect to the fine granular component of overlay. There is, however, a qualitative difference depending on the statistical character of the different populations. Components J_M and J_0 are concentrated in small particle sizes and sweep the surface without much penetration and with shallow cratering, while J_1 prevails in large projectile sizes that penetrate and produce craters through the entire thickness of overlay. J_e, despite the prevalence of large sizes, possesses a low velocity and does not penetrate deeply enough to stir the entire layer. Thus, despite the lesser mechanical sweeping power, J_1 and J_e are mainly responsible for cratering in the surface layer, and, in addition, J_e provides sizable boulders; the role of J_M and J_0 then consists of leveling out the craters and craterlets, and of ablating (grinding) the boulders, or of "polishing"—smoothing out the surface roughness continually produced by the two other components and by their own action. The actual state of the surface is then determined by an equilibrium between the two opposing processes. The role of components J_2 and J_3 with respect to the overlay layer is negligible and need not be considered further in this context.

In Table XXXVI, theoretically predicted and empirically supported flux and cratering data in overlay are given for the four relevant sources of particulate flux, and for a typical level mare surface. The surface sample is supposed to be remote from large craters; it should correspond to a "normal" overlay thickness of 13 m, which, according to Table XXVII, may be representative of about 62% of the total mare surface (and, with some indulgence, even of $61.7 + 22.6 + 7.1 = 91\%$). For Parts A, B, and C of Table XXXVI, the flux rates and velocities are based on astronomical data in the author's interpretation (loc. cit.; Section VII,B and C; V, D and E, et alia), which he believes give a well-balanced account of the observations and which he is reluctant to exchange for data from other sources; the cratering parameters, equations, and notations are those of Sections II,B, C and F, while n is the frequency index of radii according to Eq. (161); the cohesive strength data for the overlay are those of Section VIII, where R is the equivalent "spherical" radius of the impacting meteoroid, B_0 the rim to rim crater diameter, x_0 the penetration, and x' the apparent crater depth below the undisturbed surface (assumed to have been flat) as corrected for fallback.

Part D of Table XXXVI contains the flux and cratering data for the ejecta that contribute to the overlay. The total mass influx is assumed to correspond to a present accretion rate of 12 m or 1560 gm/cm² in 4.5×10^9 yr, which gives 3.47×10^{-7} gm of debris per square centimeter and year. The rate is slightly less than the average arrived at in Section VII.3 and would correspond to that of the present time and to a greater protective layer than the average in the past. The figure is essentially an empirical value, since it is based on the actual volume ejected from observed craters (Table XXIX). With another empirical datum, linking the largest projectile size to the diameter of the crater (Section VII,D), the effective radius of the largest fragment will be assumed to be

$$r_{max} = B_0/120,$$

where B_0 is the diameter of the crater in the bedrock from which the fragments were ejected. With the maximum flight distance of the largest fragments set at about 3 km [5th line of Table XXXVI, Part D(a)] or a source area of 27 km², and 400 million years as their survival life on the lunar surface (cf. Section X,A), one cratering event per 27 km² and 400×10^6 yr would correspond to 9×10^4 events per 10^6 km² and 10^9 yr; in Table XXIX, this corresponds to $B_0 = 240$ m and $r_{max} = 200$ cm as an effective upper limit of debris sizes. Of course, several hundred such blocks could be ejected in one cratering event and, in the case of a large crater, the blocks could be larger such as in a strewn field in Mare Tranquillitatis (Fig. 15), where, on a lunar Orbiter II photograph, blocks up to 9-m diameter are discernible. The Surveyor fields, however, seem to agree with the expected average conditions, with blocks up to only meter size visible (Figs. 5 and 6). Setting $r_{max} = 200$ cm, $n = 3.875$ [Eq. (161) and Section VII,D], and the total mass flux being given, the cumulative number and mass infall rates as given in Part D of Table XXXVI have been calculated with the aid of well-known integral formulas (Öpik, 1956).

The velocities and angles of ejection from the parent crater, independent of the parent velocity when $y > y_Q$ and solely dependent on crushing strength and density of parent target rock, were calculated according to the formulas of Sections II,B,C, and F, with $\lambda^2 = 0.22$ (Table XV) and $s = 5.7 \times 10^8$ dyne/cm² at $B_0 = 2.4 \times 10^4$ cm as the assumed typical parent crater diameter according to Eq. (190). With $\rho = 2.6$ gm/cm³, this gives $u_s^2 = 2.19 \times 10^8$ (cm/sec)² and a maximum velocity of ejection

$$w_{max}^2 = (\lambda u_s/y)^2, \qquad w_{max} = 6950/y \quad \text{cm/sec,}$$

where y is the cumulative relative mass as given in the third line of the table, identical with the fractional crater volume of Section II,B. The average velocity of ejection is assumed equal to $w_0 = w_{max}/\sqrt{2}$, although $\frac{2}{3}w_{max}$ could also be

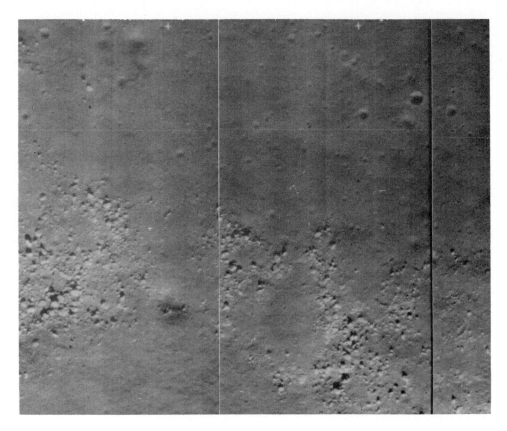

FIG. 15. Orbiter II photograph (19 November 1966) of an area 360×450 m² in Mare Tranquillitatis, showing strewn fields of blocks ranging up to 9 m in diameter. (Photograph by courtesy of NASA.)

used: the arbitrary span of the model is much greater, anyway. The upper part of Table XXXVI, Part D(a) (lines 1 to 5) contains these source data of overlay—flux, angle, velocity, and range L of the fragments.

The ejecta land at the same velocity and angle as those of ejection. This is the low-velocity problem of impact into a granular target, solved with the aid of the equations of Sections II,E and F. The lower part of the table contains the calculated cratering data, especially B_0, the crater diameter, x_0, the penetration, and x', the apparent crater depth as corrected for fallback.

We note that, in our schematically regular model, the ejecta radii are assumed to be a unique function of y, which defines the position and shock pressure inside the crater during ejection. This is assumed to be matched in a unique manner by the increasing cohesive strength as the particle size decreases, an assumption that led to a successful prediction of overlay particle size distribution (Section VII,D). In nature, there will be, of course, considerable statistical fluctuation around the average relationships. Also, those high-velocity ejection phenomena, which are connected with ray craters, are not taken into account here. Our model is meant to represent the bulk of the ejection processes, while the exclusive ray-forming processes are not quantitatively prominent enough to modify essentially our conclusions (cf. concluding paragraph of Section V,C).

Table XXXVI purports to describe quantitatively the cratering events at impact into the *granular* target of overlay. Yet when the projectile happens to hit a fragment considerably larger than itself, the projectile will react solely with this fragment; the impact will then be virtually like that onto hard rock, and not of the granular type. The last line in each part of the table contains a probability factor G_g, derived as subsequently described and indicating the fraction of impacts that are of the granular type, while the remainder, a fraction of $1 - G_g$, are limited to impacts into single large grains or blocks and are thus of the hard-target type as dealt with in Section II,B.

For a hypervelocity projectile (Groups J_M, J_0, and J_1 of the Table XXXVI, a small grain, although larger than the projectile itself, may be demolished completely and the radial momentum transmitted to other grains. The ultimate result will not differ essentially from a truly granular cratering where the grains are all smaller than the projectile. The *blocking effect* of large grains will be felt only when the shock wave from the collision does not transcend, partially at least, the boundaries of the target grain; in other words, when the virtual crater diameter B_R, produced by the impact into the hard substance of the target grain by a projectile of radius R, will be of the order of the target-grain diameter $2r$ (R stands for projectile radius and r for target grain radius). On a model of a circular target-grain cross section, the blocking effect, measured by the product of target-grain area πr^2 and the blocking efficiency, was roughly evaluated as follows (blocking efficiency is measured by the azimuthal angle of shielding by the grain).

(1) For $r \gg B_R$, the blocking effect is given by πr^2.

(2) For $r = B_R$ and $\varDelta$ equal to the distance between the grain center and the impact center, the rough estimate by zones of $\varDelta$ yields:

$\varDelta$	0	0.5r	r	1.5r
blocking angle			150°	
blocking efficiency	1	1	5/12	0
aver. block. efficiency	1	17/24	5/24	
area/πr^2	0.25	0.75	1.25	
blocking effect/πr^2	0.25	51/96	25/96 total	$(25/24)\pi r^2 \pm$

(3) For $r = \frac{1}{2}B_R$, total blocking effect $= (5/12)\,\pi r^2\pm$.

(4) For $r = \frac{1}{4}B_R$, the surface of the grain (not necessarily its root) is destroyed completely at contact and the blocking effect is zero.

We conclude therefore that, for hypervelocity (destructive) impact, the blocking effect of large target grains can be represented satisfactorily by the grain area πr^2 when $r > r_b = \frac{1}{2}B_R$, and can be taken equal to zero when $r < r_b$; r_b can be called the granular blocking limit. From the equations of Section II,B, with $\rho = 2.6$ gm/cm^3, $s_p = 2 \times 10^9$, and $s_c = 9 \times 10^8$ dyne/cm^2 for the hard target material, we find for the micrometeorities (J_M), $p = 2.01$, $D = 7.89$, $B_R = 15.8R$, and conveniently $r_b = 8R$. For the visual dustballs (J_0), p $= 0.934$, $D = 10.7$, $B_R = 21.4R$, and the blocking limit $r_b = 11R$. For the Apollo-meteorite group (J_1), $D = 14.9$ (Table XV), $B_R = 29.8R$, and $r_b = 15R$. All these limits are quite high, because of the destructive efficiency of the high-velocity impact.

The case is different with the secondary ejecta (J_e); at their low velocities, they are reflected without destruction from a larger target grain. Considering only head-on collisions, a fragment of velocity w_0 and mass μ_R will impart to a grain of mass μ_r a forward velocity of

$$w_r = \mu_R w_0(1 + \lambda)/(\mu_r + \mu_R)$$

and will itself acquire a reflected velocity of

$$w_R = w_0(\mu_R - \lambda\mu_r)/(\mu_r + \mu_R), \tag{193}$$

which is negative when the projectile is bouncing back. In the limiting case of $\mu_R/\mu_r \to 0$, $w_R = -\lambda w_0$, where $\lambda \sim 0.5$, or $(0.23)^{1/2}$ according to experiments mentioned in Section II,F, is the linear kinetic elasticity, comparatively high for this case of a single collision of rocky particles. The target grain that was hit proceeds farther as an independent projectile, but its penetration $x_0 = x_r$ into the granular substratum will be less than the normal penetration of the projectile $x_0 = x_R$. With the equations of Section II,E, the degradation of the target, measured by the ratio of the penetrations, was found to be as follows for r/R equal to 1 and 2, respectively:

μ_r/μ_R	1	8
x_r/x_R ($\lambda = 0$)	0.625	0.294
x_r/x_R ($\lambda = 0.5$)	0.825	0.382

It appears that a blocking limit of $r_b = 2R$ can be assumed. On account of the slow variation of the cumulative mass of the overlay ejecta with radius [Table XXXVI, Part D(a) 1st line], the exact limit is irrelevant. The blocking effect is thus equal to the relative cumulative cross section $\sigma_B = \sum \pi r^2$ of the overlay particles with $r > r_b$. The surface frequency exponent of particle sizes is obviously $n - 1$, where n is the volume frequency exponent in Eq. (161). (Each particle lying on the surface occupies a volume $dV = \frac{4}{3}r$ cm^{-2}, whence an additional r-factor.) The cumulative cross-sectional area is then (to a constant factor)

$$\int r^{-n+1} \cdot r^2 \, dr = |\, r_2^{4-n} - r_1^{4-n} \,|,$$

and this is exactly the same as the expression for cumulative volume, or mass reckoned per volume. Hence the blocking effect $1 - G_g$ equals the cumulative mass of the fragments for $r > r_b$ and G_g, the fraction of granular target impacts, thus equals the cumulative mass to $r \leqslant r_b$ as tabulated in the 1st line of Table XXXVI, Part D(a).

As a consequence of the broad frequency distribution of overlay particle sizes, quite a considerable proportion $(1 - G_g)$ of the impacts are nongranular in character, the proportion increasing with decreasing size. The mean values (weighted by mass) of the granular impact fractions can be assumed to be: micrometeorites (J_M), $\bar{G}_g = 0.40$; visual dustballs (J_0), $\bar{G}_g = 0.53$; Apollo meteorites $(J_1$, nonpenetrating), $\bar{G}_g = 0.88$; and the secondary ejecta (J_e), $\bar{G}_g = 0.55$. Thus, the granular target model alone cannot serve even as a first approximation. Of course, "blocked" impacts into large grains or blocks will not produce craters observable in overlay but only small craterlets or pockmarks on the rocky targets. All the craters in overlay, recognizable as such on Surveyor pictures, must therefore be produced in the granular impact process; the factor G_g gives their number relative to the total, and can be called the "overlay cratering fraction."

C. The Astronautical Hazard

The astronaut on the lunar surface is exposed to the bombardment by flying secondary debris from cratering impacts elsewhere on the moon, although mostly from his immediate vicinity, in addition to direct bombardment by interplanetary particles. The total mass of the secondary fragments exceeds by 30 times that of the incoming meteoritic mass; although its momentum, on account of the low velocity, is only one-sixth that of the meteorites (Table XXXV), the hazard from this source may appear serious. Thus, from the cumulative numbers of Table XXXVI, the number of hits per 100 m^2 and 10 yr would be for R or r equal to or greater than 0.02, 0.2, and 2 (cm), respectively:

J_M (micrometeorites)	180		
J_0 (dustballs)		0.0029	1.5×10^{-8}
J_1 (Apollo meteorites)	0.032	0.00006	1.3×10^{-7}
J_e (secondary ejecta)	550	0.73	9.5×10^{-4}

Of course, among the small particles, the micrometeorite impacts prevail over the ejecta because of their much higher velocity, despite the fact that their mass is only one-third of the mass of the ejecta. Among the larger particles the ejecta appear to dominate.

However, unlike the direct meteoritic components which appear as a flux of statistically independent individuals, the ejecta come in bursts from large and rare cratering events in the vicinity. They are spaced by long intervals of time during which no ejecta fall. The total frequency of the parent cratering events (primary meteorites and secondary ray-crater ejecta), given by the cumulative sum in the 4th line of Table XXIX, is 2.3×10^8 per 10^6 km² and 10^9 yr. The maximum flight distance of fragments with $r > 0.2$ is 10.6 km, so that spray of this size can reach a given point from a surrounding area of only about 350 km², which corresponds to an expectation of one event in 12,000 years. Here, 90% of the spray comes from $B_0 > 49$ m [Table XXIX, Part B, 9th line], with an expectation of one event in 6×10^5 yr. For comparison, the expectation that one will be killed in a car accident in the U.S.A. is one in 5000 yr and of being injured, one in 200 yr. Clearly, with all the other sources of accidents on Earth—earthquakes, hurricanes, fires, and warring hostilities—the Moon is a much safer place to be; in any case, the hazard from flying secondary debris of component J_e can be disregarded altogether, not only because of the low velocities involved but also because of the wide spacing in time.

There remains the hazard from direct hits by individual interplanetary meteorites, which may be more dangerous on account of the higher velocities involved. The shielding by the lunar body reduces the hazzard precisely to one-half that in interplanetary space. Table XXXVII contains the relevant expectations based on data from another publication (Öpik, 1961a).

On account of the micrometeorites, the hazard in the case of weak protection is quite considerable, since astronaut with 1-mm magnesium sheet metal armor runs the risk of being badly hit once in 5 yr of exposure. With 4 mm protection, the risk drops to one hit in 70,000 yr. Thus, from this standpoint also, the Moon may easily be made a much safer place to stay on than our Earth.

D. Observability of Shallow Craters and Ricocheting

From the B_0/x' ratios in Table XXXVI, Part D, it appears that the craters produced by secondary fragments (J_e) are deep and must be easily observable when not degraded by erosion. On the contrary, the craters produced by the meteoritic components are shallow and practically nonobservable even when fresh. From a study of the Surveyor I pictures and the crater counts on them by the NASA team at the Sun's altitudes of 20° and 8° (Jaffe and NASA Team, 1966b, pp. 18–25), it appears that those with profile ratios of $B_0/x' < 20$ were certainly detectable (unless covered by shadows inside larger craters); the detection of those with profile ratios from 20 to 50 is dubious, and those with $B_0/x' > 50$ are certainly missed. (In Fig. 3, the largest ratio is 120 for craters observed on the Moon at large; however, it seems that a ratio of 80 is an upper limit for recognition by repeated observations on the Moon, and for the Surveyor conditions, 50 appears to be a generous upper limit.) Taking 50 as limit, we can say that the craters produced by micrometeorites and dustballs (Parts A and B of Table XXXVI) are unobservable even when fresh, although the volume disturbed can be large and may be the chief contributor to the migration of dust; these components do not contribute to the roughness of the surface but cause only a smoothing or polishing and sweeping effect. In the Apollo Group (Part C of Table XXXVI), craters larger than 15 m in diameter ($R > 10$ cm), that are formed in overlay have observable profiles when fresh, but are still relatively shallow. On the contrary, secondary ejecta (Part D of Table XXXVI) produce, in overlay, deep, easily observable craters, the profile ratio decreasing with size. Therefore practically all craters of less than 15-m diameter that are observable on the lunar surface must be produced by the secondary ejecta.

A beautiful example of such a feature is rimmed Crater No. 5 of the Surveyor I pictures (Jaffe and NASA Team, 1966a,b; Newell and NASA Team, 1966). It is placed about 11 m to the southeast (astronautical) from the spacecraft, and can be seen to the left of the midddle on Fig. 5, and on Figs. 16 and 17 at different illumination, with the Sun at a low angle on the latter. Its diameter is 3.3 m and the depth is stated to be $\frac{2}{3}$ m. From a study of the pictures, I find a smaller depth, $x' = 34$ cm, as the depth below the undisturbed surface, which gives $B_0/x' = 9.7$. The closest description is for $r = 50$ cm as the secondary fragment radius in Part D of Table XXXVI, which gives $B_0 = 339$ cm, $x' = 44$ cm, and $B_0/x' = 7.8$. The observed crater (Figs. 5, 16, 17) is somewhat shallower, possibly due to some erosion or a different angle of impact and velocity (if the figures are taken literally). The boulder which produced this crater, however, is missing from its interior and from those of other similar craters. It must have ricocheted out, possibly even breaking up into a few large pieces, and the somewhat eroded boulder visible in the right-hand corner of the picture (Fig. 16) (measuring $59 \times 26 \times 15$ cm above ground), or another in the south-west (astronautical) ($50 \times 25 \times 15$ cm above ground), or both, could be (improbably, however, since they are too near the crater) portions of the original projectile. The remarkable feature of the small and large, often angular boulders lying on top of the lunar soil (cf. Surveyor photographs), without definite traces of cratering around them, can be explained by multiple ricocheting of the impacting fragments, something similar to what happened to Surveyor III ("Tidbinbilla") (NASA, 1967) although in this case the vernier rockets were mainly responsible. If λ^2 is the kinetic efficiency of ricocheting (in the sense of Section II,F), in repeated jumps, the velocity would decrease by a factor equal to λ, each time making smaller craters, so that in the last jump no visible crater

FIG. 16. Surveyor I photograph 66-H-589, 2 June 1966. Crater No. 5 (3 m diameter) and rock, the same as in Fig. 5. (Photograph by courtesy of NASA.)

would be produced, the fragment finally coming to rest at a depth of not more than a few centimeters as did the surveyor footpads. With $\lambda^2 = 0.09$, $\lambda = 0.3$, and the initial velocity $w_0 = 60$ m/sec as in Table XXXVI, the ricocheting velocities and distances will be ($\gamma = 45°$):

ricochet	0 (impact)	1	2	3	4
w (m/sec)	60	18	5.4	1.62	0.49
L, jumping distance (m)	2200	200	18	1.62	0.15

The Surveyor experiments permit an estimate of the ricocheting elastic efficiency $\lambda = \lambda_r$. Oscillations of the spacecraft on hard ground had a frequency of 8.0 sec^{-1}, while on the lunar surface the frequency was 6.5 sec^{-1} (NASA, 1967,

Fig. 17. Surveyor I photograph 66-H-814, 15 June 1966. Crater No. 5 and block (see Figs. 5 and 16) at low sun illumination. (Photograph by courtesy of NASA.)

Part II, p. 146). For harmonic oscillations, this means that, at equal peak load, the amplitude A_0 on hard ground was increased in the ratio of $(A_0 + A_s)/A_0 = 1.51$ on the Moon, yielding $A_s = 0.51 A_0$—the lunar surface responding with nearly one-half the amplitude of the spacecraft (which was about 0.2 cm). From computer-simulated strain guage data of landing on a hard surface (Jaffe and NASA Team, 1966b, p. 73), a very low value for the shock absorber, $\lambda_0 = 0.114t$, results, while on the lunar surface, the velocity decay ratio for footpad 2 of Surveyor I was 0.20 (the velocities being given by $\frac{1}{2}gt$, where $g = 162$ cm/sec² on the Moon, and t is the time of free flight between two touchdowns). Hence $(\lambda_0{}^2 A_0 \mid \lambda_s{}^2 A_s)/(A_0 + A_s) = (0.20)^2$ and, with the ratio of A_s to A_0 given, $\lambda_0{}^2 + 0.51\lambda_s{}^2 = 0.06$ and $\lambda_s{}^2 = 0.092$, which is, accidentally, almost exactly the value (0.08) estimated in Section VIII for the kinetic efficiency of lunar soil in a cratering process. The value of $\lambda^2 = \lambda_s^2 = 0.09$ ($\lambda = 0.3$) seems to be well justified for all impact processes in the lunar soil, and is to be compared with a value of about 0.5 for hard rock.

As a consequence of ricocheting, a rock fragment impinging onto the lunar surface with a moderately low velocity will produce several craters in successive leaps, the velocity decreasing with a damping factor equal to λ. The upper limit of the initial velocity for survival of the fragment at impact is given by Eq. (10); with $\rho = 1.3$, $s_p = 2 \times 10^9$, it is of the order of 0.5 km/sec. Hence cratering from component J_e is not completely exhausted by the data for primary impacts as contained in Table XXXVI, Part D(a). Table XXXVIII describes some crater chains produced by ricocheting, schematically calculated by assuming a constant angle γ throughout the ricocheting sequence. After a ricochet from a granular surface (probability G_g), the velocity is assumed to decrease by a factor of $\lambda = 0.3$, with a crater imprint left behind; a ricochet from a hard target (large grain, $r_b \geqslant 2r$) does not make a crater but the damping factor is larger, $\lambda = 0.5$ being assumed. The notations are those of Table XXXVI and of Section II,E and F.

The 9th column of Table XXXVIII gives $B_0{}^2$, the relative cratering area; the 10th gives $x_0 B_0{}^2$ or the total volume excavated in units of 0.363 cm^3; the 11th gives $x' B_0{}^2$, the volume ejected beyond the crater rim, in same units; the 12th column contains $k w_0$, or the average radial momentum imparted to 1 gm of the "volume affected" [Eqs. (2) and (36)]. The chain is terminated either when $L < \frac{1}{2} B_0$ or when the altitude of the rebound is less than x_0, so that the projectile falls back into its last crater. The last line in each section of the table shows the "amplification ratio" or a factor by which each of the items is increased in the sum total of the chain, as compared to a first and direct impact into granular target. Except for the total cratering area, these factors are all within the order of unity.

While G_g is the probability of impact into a granular target, a ricocheting chain may have all combinations of hard and granular impacts according to the binomial law of probability. To calculate all these combinations would mean stretching our numerical analysis too far; the calculations are very approximate anyway, although certainly better than a mere qualitative appraisal.

For $r = 0.5$ cm, $G_g = 0.52$, three typical cases have been considered: (1) an alternating chain of hard and "soft" impacts, starting with a soft one; (2) a similar chain, starting with a hard impact; and (3) a "soft" chain throughout. The true statistical mean for $G_g = 0.50$ should not differ essentially from the average of the first two cases, and a comparison with the third case could then show the error of neglecting the hard impacts altogether at the given value of $G_g = 0.50$. The comparison is made in Table XXXIX. The last line gives the ratio of the first two lines, or the correction factor in a transition from $G_g = 1$

to $G_g = 0.5$. Within the uncertainties of the model, the factor is the same for all four parameters, its average value of 0.621 showing the result of the difference in the elastic constant ($\lambda = 0.3$ and 0.5, respectively) between the two cases. For equal λ, the true average should equal $G_g = 0.5$ exactly, but because elasticity in hard impacts is higher, there is partial compensation for the chain loss; as compared to the "all soft" chain ($G_g = 1$), the correction factor of the amplification ratio can be assumed to be $G_g + \frac{1}{2}G_g(1 - G_g) = 1.5G_g - 0.5G_g^2$, which gives $1.5G_g$ at $G_g \to 0$, 0.625 at $G_g = \frac{1}{2}$, and 1 at $G_g = 1$, which very closely describes the true factor at a "soft" amplification ratio of 1.5 and is a good approximation for other ratios.

The amplification ratios A_0 calculated at $G_g = 1$ do not differ very much in the sample cases (3)–(5) of Table XXXVIII, so that averages can be taken: $A_0 = A_a = 2.17$ for cratering area, $A_0 = A_v = 1.44$ for total volume excavated and $A_0 = A_e = 1.35$ for volume ejected. The chain amplification ratio for rock fragments or hard grains impacting with moderate velocity ($\ll 500$ m/sec) and ricocheting on lunar overlay is then

$$A_q = G_g A_0 (1.5 - 0.5G_g), \tag{194}$$

with the proper value of A_0 corresponding to the particular parameter q (area, volume, etc.) to be used. The sum total for the ricocheting chain is obtained by applying the factor A_q to the area, volume, etc. of a direct "soft" first impact. This kind of amplification of cratering by ricocheting can take place only when the projectile is not destroyed, i.e., in the case of component J_e. Another kind of amplification, caused by the granular ejecta themselves, is common to all types of impact. Because of the smallness of the grain, small k and G_g values, and low velocities of ejection, cratering proper in overlay by the secondary ejecta from the overlay itself can be discounted; but, as a factor of mobility of the "dust," this has to be considered. Only transmission of radial momentum is of importance here.

Of the ejecta, only those with $w > u_s$ (all notations are those of Section II) are to be considered as a factor causing further mobility in the target. This limits the active mass to a fraction of $\lambda x'/x_p$ of the total mass affected; on the other hand, the velocity of ejection increases toward the inner portions of the crater [Eqs. (16)–(26)], which partly balances the limitation of mass. For hypervelocity impact, the ratio of radial momentum transmitted by the ejecta into the surroundings to radial momentum of the primary cratering event is found to be

$$J/J_1 = 4\lambda k[\ln(\lambda k_0 w_0/25u_s) + (25u_s/\lambda k_0 w_0)^3 - \tfrac{1}{3}]$$
$$\cdot x'(1.5G_g - 0.5G_g^2)/9x_p ,$$

where k_0 and w_0 are radial momentum factor and velocity, respectively, for the primary event, and k is the radial momentum coefficient in the secondary shower. For micrometeorite impact, $w_0 = 6.0 \times 10^5$ cm/sec, $u_s = 200$ cm/sec, and $k_0 = 2$; with $\lambda = 0.3$ and $k = 0.2$ (cf. Table XXXVIII, $r = 0.5$ cm at $w_0 = 200$ cm/sec), and from Table XXXVI, Part A, $Gg = 0.4$ and $x'/x_p = 0.75$, the factor in square brackets (accounting for increased velocity of ejection from the interior) becomes 4.0 and $J/J_1 = 0.04$—an utterly insignificant increase. For nondestructive impact (J_e), the gain in total momentum from ejecta is still smaller, and can be neglected completely. Thus, only ricocheting is of significance in amplifying the action of primary impacts on overlay, while the contribution from secondary ejecta is too small to be taken into account.

The amplification factor in Eq. (194) consists of two distinct factors: Gg, the straightforward probability of "soft" cratering, which thus rules the number of successful cratering events in each ricocheting step and is the same as for the primary impacts; and the product $A_0(1.5 - 0.5G_g)$, which measures the total quantitative gain in the parameter (sum of the area, volume, etc., of craters) for one primary impact. We may assume the crater parameter (area, volume) to decrease in geometrical progression with each ricochet (which is an idealization of a more complex process) (cf. Table XXXVIII); if $\varDelta$ is the common ratio of the progression (assumed infinite), evidently

$$\varDelta = [A_0(1.5 - 0.5G_g) - 1]/[A_0(1.5 - 0.5G_g]. \tag{195}$$

At $G_g = 1$, as for the "test case" of a completely granular traget,

$$\varDelta_0 = (A_0 - 1)/A_0 \quad \text{or} \quad A_0 = 1/(1 - \varDelta_0). \tag{196}$$

For crater area ($B_0{}^2$), $A_0 = 2.17$ and $\varDelta_0 = 0.539$, whence for crater diameter (B_0), $\varDelta_0 = (0.539)^{1/2} = 0.734$ and $A_0 = 3.76$. For total crater volume ($B_0{}^2x_0$), $A_0 = 1.44$ and $\varDelta_0 = 0.306$; this is the product of the common ratios for $B_0{}^2$ and x_0, whence the ratio for crater depth or penetration (x_0) becomes $\varDelta_0 = 0.306/0.539 = 0.566$ with $A_0 = 2.30$. Similarly, for the apparent depth (x'), $\varDelta_0 = 0.480$ and $A_0 = 1.92$. This of course is an oversimplification, as can be seen from Table XXXVIII, and is meant only to convey an overall idea of the ricocheting process, which is too complicated to be represented by a uniformly decreasing geometrical progression. Nevertheless, for the sake of simplicity, some of the ricocheting chain parameters can be expressed through such a progression of a constant common ratio with an error of only a few percent.

The cratering area of a ricocheting chain, according to Eq. (194), is amplified by a factor of 2.17 as compared to the parent crater when $G_g = 1$. However, the ricocheting craters are smaller, and when the sum total (cumulative number, area, and volume) to a fixed limit of crater diameter is taken, the ricocheting members arise from larger and less numerous crater sizes so that the relative contribution at the fixed limit is less than 2.17. The actual contribution depends on the frequency function of the primary diameters. Similarly, the contribution to crater numbers is also a decreasing progression. With an emprical value of $n = 3.27$ representing the cumulative primary crater numbers B^{-n} between $B_0 = 339$ and 7.0 cm [between $r = 50$ and 0.5 cm, last line of Table XXXVI, Part D(a)], and with the progression ratios as quoted above, simplified expressions for the ricocheting chain parameters were adopted as follows. For the cumulative crater number, primary plus ricochets, to the same limit B_0,

$$dN/dt = 1.57 \sum (1.5 - 0.5\, G_g)(G_g\, \Delta N) \qquad (197)$$

was assumed, where ΔN is the differential frequency of primary impacts, or $(G_g\, \Delta N)$ the differential frequency of "soft" (granular) primary impacts [last line of (a), Table XXXVI, Part D].

The cumulative crater coverage to the limit B_0 (area per cm^2-yr, or fractional area per year) is then

$$\sigma_B = 0.785 \sum B_1\, B_2\, \Delta N, \qquad (198)$$

where ΔN is the differential frequency of crater numbers (primary and ricochets) for the interval from B_1 to B_2 as can be obtained from Eq. (197) or from the 2nd line in Table XXXVI, Part D(b).

Through admixture of degraded shallower ricochets, crater depth is decreased. The depth-to-diameter ratio in a chain forms a progression with a common ratio of $0.566/0.734 = 0.772$ for x_0/B_0, and one of $0.480/0.734 = 0.653$ for x'/B_0. With $0.734^{+n} = 0.365$ as the "degradation ratio" of crater numbers (ratio of ricochet crater numbers to primary crater numbers at $B_0 = $ const) for a ricocheting diameter decrement of 0.734, the average penetration x_0, at constant crater diameter, requires a correction factor of

$$(1 - 0.365)(1 - 0.365 \times 0.772) = 0.884,$$

and the apparent depth x' must be multiplied similarly by a factor of 0.834. It turns out that, despite multiple ricocheting, neither the crater numbers nor their total areas and volumes are changed very much, as compared to the primary impacts, when statistics are made to constant crater diameter limits. The degraded chain members join the more numerous groups of smaller craters where their numbers are relatively small and affect the average very little.

In Table XXXVI, Part D(b), the ricocheting chain data, with primary and secondary members counted to the same limit of B_0, are given.

E. Overlapping and Survival of Craters

Table XXXVI contains the predicted rates of crater formation from the main sources. The actual crater numbers depend on the balance of formation and removal.

Two main processes of removal of craters by extraneous agents can be discerned: through superposition or overlapping of a later larger crater; and through erosion by smaller cratering impacts. Erosion works gradually, exponentially with time, and cannot erase a crater completely although the crater may become too shallow for recognition; this will be discussed in a subsequent section. Overlapping changes the terrain completely and no trace of a small crater can be expected to remain when it happens to fall within the bounds of a later, sufficiently large crater. Quantitative estimates of overlapping can be made on the basis of Table XXXVI.

Only a very schematic approach to the problem can be justified. The line of demarcation between "small" and "large" craters cannot be sharp. Nevertheless, by ignoring intermediate transitional cases, we can choose a conventionally sharp margin of crater size for which deletion by overlapping would occur, and from some rough estimates, this should lead to more or less the same statistical result as for mathematical adaptation with gradual transition.

Let B_0 and x' be the diameter and apparent depth, respectively, of the earlier crater, and B_a and $x_0 = x_a$ the diameter and depth of penetration of the later crater. Several conditions for which removal would occur can be set up to be applied in different cases.

The overall condition of removal or erasure is set by an effective minimum ratio of diameters, which we find must be close to

$$B_a \geqslant 2B_0 ; \tag{199}$$

also, the center of B_0 must fall within the boundary of B_a. This condition is sufficient only when the larger crater is of sufficient depth, namely when

$$x_a > \tfrac{1}{2}x' . \tag{200}$$

In such a case the rate of removals ν_B is evidently equal to the cumulative coverage by craters larger than $2B_0$, i.e.,

$$\nu_B = \sigma_{2B} . \tag{201}$$

When Eq. (200) is not fulfilled, or when the larger crater is much shallower than the small one, partial filling to a depth of $2x_a$ after one overlap is assumed, and the rate of removals becomes

$$\nu_B = 2x_a\sigma_{2B}/x' . \tag{202}$$

A variant consists of selecting $B_a \geqslant cB_0$, with $c \geqslant 2$ so that Eq. (200) is fulfilled, and of setting

$$\nu_B = \sigma_{cB} . \tag{203}$$

Of the alternatives presented by Eqs. (202) and (203), we shall choose that which yields the larger rate of removal.

Table XXXVI contains only those components of impacting flux that do not penetrate the overlay at present. A fourth component, represented in Table XXIX (in so far as it does not overlap with those of Table XXXVI) must be added although it is important only in the larger crater classes. The "primaries" of this component are essentially an extension of the Apollo-meteorite group of Table XXXVI and shall be entered only beginning with average values of B_0 greater than 129 m ($B_0 > 114$ m). The "secondaries" in Table XXIX are primaries from the standpoint of Table XXXVI; they are probably energetic ejecta from ray craters with respect to which Part D (component J_e) of Table XXXVI is secondary. An upper limit to crater size is also set at $B_{0(av)} < 283$ m ($B_0 < 325$ m), in conformity with our assumption that our region is "normal" and beyond the reach of large craters. In such a manner the area covered by the additional "component J_c" of the large craters, calculated from Table XXIX, is given in Table XL.

In Table XLI, a summary of crater formation and subsequent removal by overlapping is given. Only craters with a profile ratio of $B_0/x' < 50$ at the moment of formation are included. This restricts the small-crater statistics chiefly to components J_e and part of J_1 (Table XXXVI, Parts D and C), in the larger sizes supplemented by component J_c (Tables XL and XXIX). Micrometeorites and dustball meteors (Table XXXVI, Parts A and B) produce flat unrecognizable craters that are not included in the counts although their ability to delete smaller craters by overlapping must be reckoned with.

The introduction to Table XLI gives the necessary explanations. F_i (or F_0), in the 5th (Parts A and B), 4th and 10th (Part C) columns of each subsection, is the theoretically calculated differential rate of cratering (on the basis of observed interplanetary populations and for J_e, from the rate of growth of overlay, which is itself based in turn on observed excavated crater volume), with allowance for the factor G_g or the proportion of granular impacts; the cumulative rates of cratering are given in the 1st line of Table XXXVI, Part D(b).

If we provisionally ignore erosion, the time variation of crater area density n_i, subject to creation rate F_i and deletion rate v, is determined from the differential equation

$$dn_i/dt = F_i - v n_i$$

When integrated from $t = 0$, $n_i = 0$ to $t = t_0$, $n = n_i$, this yields

$$n_i = (F_i/v)(1 - e^{-v t_0}). \tag{204}$$

For $v t_0 \to \infty$, the equilibrium density F_i/v is reached. When $v t_0$ is small, $n \to F_i t_0$. In the 6th or 5th column of each Part of Table XLI the calculated crater density n_0 corresponding to $t_0 = 4.5 \times 10^9$ yr of *uneroded* existence is given.

When constructing Table XL on the basis of Shoemaker's counts, erosion was assumed to delete a crater after an erosion lifetime of

$$t' = 1.58 \times 10^5 B_0 \quad \text{yr} \tag{205}$$

when B_0 is given in centimeters. The equation is based on a discontinuity in the observed gradient dN/dB_0 explained as an erosional removal of craters with $B_0 = 286$ m in a time interval of 4.5×10^9 yr. The provisional erosion time scale as given by Eq. (205) is shown in the 8th or 7th columns of Table XLI.

When $t' > \tau_0$, erosion is too slow and craters are removed mainly by overlapping; this is the case that can be seen at the small-crater end of Table XLI, where the theoretical crater density will be close to n_0. When $t' < \tau_0$, erosion prevails and the calculated crater densities are less than n_0. Setting $t_0 = t'$ in Eq. (204), the probable values n' or n_e of differential crater density due to the combined removal by overlapping and hypothetical erosion have been calculated (9th or 8th columns for provisionals, 12th column for final solution in Table XLI). The last column in each subdivision of Table XLI contains the finally predicted cumulative frequency N_e of craters per 100 m².

The three components of the table do not overlap and their sum, obtained as shown in the 5th column of Table XLII, is then the predicted first approximation of the total cumulative crater density as derived from the influx rate of the projectiles, cratering theory, available knowledge of the mechanical properties of the lunar soil and bedrock, elimination through overlapping by larger craters (well defined), and erosion by smaller projectiles (provisional rate of erosion, emprically suggested by a discontinuity in the gradient of the crater frequency function). This can be compared with observed crater densities from three different sources as derived from the lunar probes (10–12th columns): Ranger VII and VIII (Shoemaker, 1966), Ranger VIII (Trask, 1966), and Surveyor I (Jaffe and NASA Team, 1966b).

A comparison of the first approximation (N, 5th column of Table XLII) and the observed (10–12th columns) crater densities seems to show convincingly that prediction, even with the provisional assessment of erosion, is in satisfactory accord with observation and that, like the distribution of large craters in the maria, the small-scale relief of the lunar surface can be well-accounted for theoretically in terms of the physical factors listed above. The divergencies among the different sources of crater counts are even greater than those between prediction and observation. Only within the 10- to 6-m diameter range does there seem to be a major discrepancy, the predicted numbers being some 5 times too high, but even this deviation is contradicted by the Surveyor data at 3 m, which show twice as many craters as those predicted, and 5 times the number derived from the Ranger photographs. The weak point of this prediction is the provisional and oversimplified treatment of erosion.

In Section X,D and E, a more sophisticated treatment of erosion is applied and the calculated results are given in the middle part (columns 6–9) of Table XLII. There is certainly better agreement now in the crater range (3–20 m) yielding the most discrepancies. However, the main features of the statistical balance of cratering on the Moon are not much altered by this more detailed theoretical study of erosion: the observational data (partly reflecting real differences on the lunar surface) are not in sufficient agreement to permit a check on the subtler details of the theory.

F. Mixing of Overlay

Each cratering event displaces a volume of $0.363x_0B_0^2$ [Eq. (1) with x_0 now standing for x_p], part of which is ejected while the remaining part falls back. This material becomes thoroughly mixed, to an average mixing depth h_0 over the crater area $\frac{1}{4}\pi B_0^2$, given by

$$h = h_0 = 0.462x_0 . \tag{206}$$

For a static overlay layer at depth h_0, the mixing efficiency per unit of time (year) equals $\sigma_B(x_0)$, the fractional area of the surface covered in unit time by craters reaching to and beyond central penetration depth x_0; this can be derived from the data of Tables XXXVI and XL although the latter table does not add much. Over a time interval of t years, the mixing factor Q_m (the effective number of times for a complete exchange of material, of this specific layer situated at depth h_0, with the overlying soil) is then

$$Q_m(h_0) = \sigma_B(x_0)t, \tag{207}$$

and the mixing time t_m, corresponding to $Q_m = 1$ or complete single mixing is

$$t_m = 1/\sigma_B . \tag{208}$$

However, the simple mixing process is complicated by the accretion of overlay, which not only adds new material to the surface but also provides an ever increasing protective layer. The average accretion of overlay on our "normal" region was estimated to equal, at present, 12 m per 4.5×10^9 yr, or 2.67×10^{-7} cm/yr; any marked layer at depth h_0 can be assumed to sink under the surface at this rate, so that its age in years is

$$t = t_0 = h_0/2.67 \times 10^{-7} = 3.75 \times 10^6 h_0 . \tag{209}$$

When $t_0 > t_m$, mixing is efficient; when $t_0 < t_m$, the time scales are such that the layer sinks at a faster rate than that at which it mixes, and thus becomes only partly mixed with the overlying strata, or not at all.

A question of identity arises for mixed strata. Physical identity is maintained only over short intervals of time, during which the rate of sinking is thus physically meaningful. The rate remains the same although the material content may change with efficient mixing.

The differential equivalent of Eq. (207) is $dQ/dt = \sigma$, and with the linear dependence of age on depth [Eq. (209)], the mixing factor can be integrated in terms of increments of either t or h_0, $Q_m = \int \sigma \, dt$ or

$$\Delta Q_m = \sigma_B(h_0) \, \Delta t = 3.7 \times 10^6 \sigma_B \, \Delta h_0 . \qquad (210)$$

The integral from $t = 0$ to $t = t_0$ yields the total mixing factor for the past history of the layer when its depth was less than h_0. The integral from $t = t_0$ to $t \to \infty$ defines the mixing factor for the future, Q_f; when this is small, mixing can be assumed to cease and the layer becomes stagnant. The probability of eventual subsequent mixing is

$$q_m = 1 - \exp(-Q_f); \qquad (211)$$

for small values it is close to Q_f, while for large values it approaches unity.

Table XLIII contains the calculated mixing probabilities that depend on the depth h_0 below the surface. The crater coverage σ_B is the sum for all four components of Table XXXVI, logarithmically interpolated when needed, for the chosen values of x_0.

The dividing line $t_m/t_0 = 1$ is at a depth of $h_0 = 11$ cm, while the one-half probability depth ($q_m = 0.5$) is near $h_0 = 8$ cm. At $h_0 < 4$ cm, the layers become well mixed, with mixing times running from a few million years to 160,000 years at a depth of from 2 to 0.5 cm. Below $h_0 > 25$ cm, the chance of ultimate mixing becomes very slight and the layers become stagnant, preserving the stratification once formed when they were near the surface. The stratification is washed out over a layer thickness of 8 cm (linear dispersion ± 4 cm); chronologically it reflects the average conditions (e.g., with respect to cosmic-ray intereactions) over a time interval of 30 million years. Fluctuations of a shorter period must be smoothed out and cannot be detected, unlike the high resolution in time of terrestrial sediments. With such a low resolution, the Quaternary and Pliocene would have been lost in the preceding Miocene, even the large subdivisions of the Tertiary and the Cretaceous–Paleocene transition would have been washed out.

X

EROSION LIFETIMES OF SURFACE FEATURES

A. Transport and Sputtering; Lifetime of Boulders

In Section IX,E, deletion of small craters (or other features of roughness) by later superimposed larger ones was evaluated on a firm statistical basis, while erosion by the continuous influx of small projectiles producing craters, smaller on a linear scale than a given roughness feature, was treated (as a first approximation) summarily by invoking an empirically adjusted unspecified smoothing process whose linear scale was assumed to be proportional to age. In this section we will consider theoretically the actual processes of this gradual erosion or "polishing," the final results, however, having been included in the middle part of Table XLII. Two main processes are at work: sputtering of large grains, boulders, and crater rims, and transport of the granular matrix.

Outstanding large grains or blocks, sufficiently larger than the impacting projectile, which do not behave as part of the granular matrix but retain their individuality (cf. Section IX,B), and unprotected crater rims, are sputtered by hard impacts. They are not much affected by the infalling accreting overlay (J_e) because of its low velocity (some grinding and collisional damage occurs; however, the mass affected is insignificant); only hypervelocity bombardment by the meteoritic components is relevant here and, because of the large mass influx rate (cf. Table XXXV), bombardment by micrometeorites (J_m) is the main agent and may be considered alone.

Transport, as distinct from plain mixing of the granular medium on a horizontal surface and discussed in the preceding section, works on slopes of craters or other elements of roughness. At a cratering impact, more grains are ejected farther downhill than uphill, which leads to a net downhill displacement or flow of the granular material. Also, out of a hole, fewer grains will be ejected by cratering events into the surrounding terrain than will be injected into the hole from the surroundings; this leads to a gradual filling of the hole (crater).

In transport, besides the micrometeorites (J_m), overlay influx (J_e) may be of some importance (cf. Table XXXV), somewhat enhanced by the higher velocities of ejection (u_s) due to greater strength (s_c) at greater penetration (x_0) than in the case of micrometeorites. In addition to momentum as determining the mass ejected, the transport efficiency can be assumed to be proportional to the flight distance L [Eq. (45)]. On this basis, the transport efficiency is measured by the product $E = kG_g w_0(d\mu/dt)L$, which thus differs from the plain momentum used in Table XXXV. The comparison between J_m and J_e can be made at the median (50%) cumulative mass (Table XXXVI). For J_m (this is at $R = 0.0205$ cm), $w_0 = 6 \times 10^5$ cm/sec, $k = 2$, $G_g = 0.41$, $B_0 = 6.7$ cm, $\bar{s}_c = 2820$ dyne/cm², and $d\mu/dt = 1.05 \times 10^{-8}$ gm/cm²-yr. For J_e, the typical parameters are $r = 1.0$ cm, $w_0 = 9.53 \times 10^3$ cm/sec, $k = 0.613$, $G_g = 0.56$, $B_0 = 12.1$ cm, $\bar{s}_c = 2.61 \times 10^4$ dyne/cm², and $d\mu/dt = 3.47 \times 10^{-7}$ gm/cm²-yr. The comparison, then, consists of two steps. An inner portion ($y_m' = 0.328$) of the J_m crater, which has the same shock velocity (u) as the entire ($y_e = 1$) harder J_e crater, yields (L calculated for $\lambda = 0.3$, $y = \frac{1}{2} y_m$ or $\frac{1}{2} y_e$) $L_e/L_m = 10.5/4.7 = 2.24$, $k_m/k_e = 3.26$, $w_m/w_e = 63$, $G_e/G_m = 1.37$, $\mu_e/\mu_m = 32.1$, and $E_e/E_m' = 0.476$. The total transport efficiency of the J_m crater (for $y_m = 1$)

is $E_m = 2.57 E_m'$, whence $E_e/E_m = 0.476/2.57 = 0.185$. The ratio turns out to be nearly the same as that arrived at in Table XXXV from a rougher estimate. To allow also for the other small components and going back to Table XXXV, the transport efficiency of the J_m component may be taken with an additional increase of 25% of its value. Thus, the predominance of micrometeoritic erosion greatly simplifies the calculation of transport, which cannot be estimated anyway to better than a close order of magnitude.

The experiments by Gold and Hapke (1966) as mentioned above could suggest that, simultaneous with dispersal of the granular substance through impacts, a build-up of surface roughness ("fairy castles") would take place, caused by adhesion. This can be only partly true for lunar overlay. The stickiness of the cement powder in the experiments was due to the very small grain size, of the order of 10^{-4} cm. It can be shown that, for a greatly simplified model consisting of two colliding equal spherical semielastic grains, the maximum relative velocity of encounter that can be balanced by the tensile elastic force as limited by cohesion, or the "velocity of inelastic capture," equals

$$v_a = (6/Y\delta)^{1/2}A_d/(\pi\lambda r^2), \tag{212}$$

where Y is Young's modulus, δ is the density, A_d the cohesive force between two grains, λ the linear elastic efficiency, and r is the radius. For silicate grains *in vacuo*, $Y = 6 \times 10^{11}$ dyne/cm^2, $\delta = 2.6$ gm/cm^3, $A_d = 1.0$ dyne (Smoluchowski, 1966; Ryan, 1966), and $\lambda = 0.5$ as for hard rock in a single collision, we have

$$v_a = 1.27 \times 10^{-6}r^{-2} \quad \text{cm/sec.} \tag{212a}$$

For $r = 10^{-4}$ cm as in the Gold–Hapke experiments, $v_a = 127$ cm/sec and the particles can be captured efficiently at moderate velocities of impact, while at $r = 10^{-2}$ cm, and $v_a = 0.0127$ cm/sec, they could hardly stick, especially when perturbed by other oncoming particles. Yet, according to Table XXXVI, Part D, the mass (y) of the small particles from 2×10^{-5} to 2×10^{-4} cm could only amount to 4.5% of the total mass of the overlay, while those larger than 10^{-2} cm account for 71%. There are not enough "sticky" particles in the overlay, and the build-up of "fairy castle" structures must be greatly inhibited, compared to the formation of regular impact-crater depressions, while the former are much more easily destroyed by "hard" impact than the latter.

We will first consider the leveling action of meteoritic bombardment on the granular elements of surface roughness. As a consequence of meteoritic (and other) impact, different parts of the overlay surface exchange material. (The influx from component J_e, or secondary ejecta from nearby craters, descends equally on all surface elements, leading to a continuous growth and simultaneous redistribution or filling, such as that considered in Section X,C.). For an element of surface placed at the same level as its surroundings (differences of level that matter are of the order of L, the flight distance of the ejecta), ejection and influx are obviously balanced. A surface element placed in a depression will receive more or less the same influx as that it would receive were it placed on level ground, but, on account of gravity, there will be some fallback at ejection, so that influx over the rim of the depression will exceed ejection and the depression will begin filling. On the contrary, an elevated area will eject over its rim the same amount as when placed on level ground, while receiving less from the surroundings; its height will decrease. We shall try to obtain a quantitative estimate for the time rate or "smoothing lifetime" for the elements of roughness, depending on their linear scale.

Only a crude approach is attempted. Strict evaluation of the integrals, subject to the conditions of the adopted model, is not justified, the model of cratering ejecta being itself only approximate and involving arbitrary quantitative relations [such as Eq. (27) for the ejection angle β and the use of a mean coefficient of elastic efficiency λ]. Here, as one of the simplifications, average quantities of ejection or influx over entire areas is used instead of the integrals. The mathematical error, perhaps some 10–20%, is probably much smaller than the uncertainty in the basic assumptions.

The granular surface is assumed to be horizontal on the average, except for the randomly distributed elements of roughness (chiefly craters). Meteorite impacts on inclined surfaces will lead to systematic flow downhill, a process to be considered separately.

For granular overlay, with the cratering parameters such as those adopted in Table XXXVI, Part A and Eq. (4), the marginal shock velocity is

$$u_{\mathrm{s}} = 2170^{1/2} = 46.6 \quad \mathrm{cm/sec}$$

and the equivalent volume of granular ejecta, proportional to $G_{\mathrm{g}} = 0.41$ (the fraction of granular encounters), which is increased by 25% to allow for other components of meteorite influx, and proportional to the ratio kw_0/u_{s}, according to Eq. (14), becomes $\chi = 0.669 \times 1.05 \times 10^{-8} \times 0.41 \times 1.25 \times 2 \times 6 \times 10^5/(46.6 \times 1.3)$, or

$$\chi = 7.13 \times 10^{-5} \quad \mathrm{cm/yr} \tag{213}$$

$(\mathrm{cm^3/cm^2}\text{-yr})$. In a successful granular impact, the ratio of mass ejected or disturbed (including fallback) to the micrometeorite projectile mass is then

$$M_{\mathrm{c}}/\mu = 1.72 \times 10^4. \tag{214}$$

The profile ratio B_0/x_0 is of the order of 60 for micrometeorite impact into the granular surface (Table XXXVI, Part A). The crater is extremely flat and the rim angle β_0 is nearer 90° (cf. Fig. 1); instead of 0.8 in Eq. (27), we set

$\sin \beta_0 = 1$. Within the same mathematical framework as that used for the determination of fallback in Section II,F, the average velocity of ejection in the direction β and at crater mass fraction y is

$$\bar{v} = \tfrac{2}{3}u_{\rm s}(\lambda/y) = 31.1\lambda/y \quad \text{cm/sec}, \tag{215}$$

the factor $\tfrac{2}{3}$ allowing for the assumed damping of ejection velocity with depth x [Fig. 1 and Eq. (24)]. The layer radiated from a horizontal level element of surface beyond a circle of radius L [Eq. (45)] around it and thus lost into the surroundings beyond L is then

$$\psi_{\rm L} = \chi[(1 + \zeta^2)^{1/2} - \zeta] \quad \text{cm/yr}, \tag{216}$$

where $\zeta = \alpha'L$ and

$$\alpha' = \tfrac{1}{2}g/v_0^2\lambda^2 = 0.084/\lambda^2. \tag{217}$$

The velocity $v_0 = 31.1$ is that corresponding to $y = 1$ and $\lambda = 1$, as in Eq. (215). The same amount $\psi_{\rm L}$ is, of course, gained by the element of surface through influx from beyond radius L, as follows from the condition of equilibrium, and can be proved directly by integration. In the framework of the prescribed conditions, Eq. (216) is mathematically exact.

Imagine now that the entire level circular portion of the surface of radius L radiates over its boundary. The case is more complicated than the fallback problem, because, in a single cratering event, the ejecta were supposed to fan out radially although at different angles, and the distance from crater rim was unique for each radiating spot, while in the presently considered process each spot emits ejecta in all directions along which the distances to the borderline are different. Instead of employing numerical integration, we estimate the average radiation from the entire surface, sent over the borderline L, to correspond to a point at $0.75L$ from the center of the area, and to equal the mean of two expressions [Eq. (216)], one for $L' = L/4$ and the other for $L'' = 7L/4$ (antipodal distance). Thus, for the entire level circular area of radius L, the average emission as well as influx over (or from over) its border becomes

$$\bar{\psi}_{\rm L} = \tfrac{1}{2}\chi[(1 + \zeta^2/16)^{1/2} + (1 + 49\zeta^2/16)^{1/2} - 2\zeta] \quad \text{cm/yr}. \tag{218}$$

For $\zeta > 5$, the bracketed expression can be approximated closely by $2[(8/7\zeta) - (4/\zeta^3)]$.

Consider now a cylindrical depression of radius L, flat at the bottom and of depth H. From Eq. (45), for the flight distance as compared with the vertical range of the flight trajectory, at $\beta = 45°$, a particle ejected from the center will just pass over the rim when $H = \tfrac{1}{2}L$. We make the schematic assumption that when $H \geqslant \tfrac{1}{2}L$, ejection is virtually blocked and Eq. (218) represents the net average accretion at the bottom. Further, when $H < \tfrac{1}{2}L$, we assume a linear decrease of the accretion balance from its maximum value [Eq. (218)] to zero at $H = 0$. Therefore, this defines the time scale for the filling of the depression to be

$$\tau_f = \tfrac{1}{2}(L/\bar{\psi}_L), \tag{219}$$

so that, when $H_1 < \tfrac{1}{2}L$ is the initial depth, the depth after time t will be

$$H_t = H_1 \exp(-t/\tau_f). \tag{220}$$

For $H > \tfrac{1}{2}L$, the accretion is constant,

$$dH/dt = -\bar{\psi}_L . \tag{221}$$

Equations (218)–(221) apply also to a cylindrical elevated circular plateau of radius L with a granular surface, which at a positive height $H > \tfrac{1}{2}L$ receives but a negligible influx from the lower placed surroundings and loses the net amount given by Eq. (218). Table XLIV contains the parameters and lifetimes of depressions (craters) or elevations (granular mounds).

These lifetimes are shorter than the hypothetical t' values [Table XLI(a)] for the small craters where $B_0 < 100$ cm and somewhat shorter than the overlapping lifetimes τ_0 for the crater diameter range from about 8 to 2000 cm, and, thus, must appreciably affect the crater statistics (Tables XLI and XLII, 1st versus 2nd approximation). Also, the condition $\bar{H} < \tfrac{1}{4}B_0$ may apply to practically all craters and mounds, so that Table XLIV may be considered to be of general applicability with respect to this particular process of erosion.

That part of meteorite flux that is not instrumental in granular cratering (i.e., a fraction $1 - G_g$ of the total) produces hard sputtering of single grains or exposed boulders. The soft component J_e is inefficient in this respect and only micrometeorites may be considered. With $s_c = 9 \times 10^8$, $\rho = 2.6$, $k = 2$, $w_0 = 6 \times 10^5$, and $u_s = 1.86 \times 10^4$ cm/sec, Eq. (14) yields, for the sputtering mass ratio, a value $1/400$ that of Eq. (214) and, therefore, negligible as a factor of mass transport. Also, with $\lambda^2 = 0.25$ and $y = 0.5$, the high-speed ejecta (and of very fine grain) are spread over a radius of more than 20 km and are not available for small-scale local smoothing. If Eq. (190) is accepted for the small craters of about 0.2 cm produced by the micrometeorites in rock, the effective strength of the material must be greatly increased, i.e., to about $s_c = 1.04 \times 10^{10}$ dyne/cm^2; this sets $u_s = 6.3 \times 10^4$ cm/sec and Eq. (14) then yields only

$$M_c/\mu = 12.7. \tag{222}$$

The layer carried away from an exposed horizontal grain or rock surface (density 2.6) by micrometeorite sputtering is then $\tfrac{1}{2}\chi_s$, when χ_s denotes the equivalent layer of overlay (density 1.3) created,

$$\chi_s = 1.05 \times 10^{-8} \times 12.7/1.3$$

or

$$\chi_s = 1.03 \times 10^{-7} \quad \text{cm/yr.} \tag{223}$$

This will be the ablation when $\bar{H} > \frac{1}{4}B_0$ for the block. A rocky surface on a level with the surroundings will be covered by overlay ejecta and thus protected from direct sputtering. For an intermediate height, the thickness of the protective layer will be such that micrometeorite bombardment will sweep it away as it comes in. As the influx is decreased in an assumed ratio of $\kappa = 4\bar{H}/B_0$ [basis of Eqs. (219) and (220)], the micrometeorites will spend a fraction $1 - \kappa$ of their momentum in sweeping away the thin protective sheet (actually it is an exponential function of κ, cf. Section X,B); the sputtering efficiency for the underlying rock will thus be κ. In addition, overlay showers on the block, burying its base at a rate of $1200/4.5 \times 10^9 = 2.67 \times 10^{-7}$ cm/yr. Thus, the outstanding height of such a block, with a flat top, decreases at a rate of

$$dH/dt = -2.67 \times 10^{-7} - 2.06 \times 10^{-7}(H/B_0),$$

whence

$$H = (H_1 + 1.29B_0)\exp(-2.06 \times 10^{-7}t/B_0) - 1.29B_0 . \tag{224}$$

The block is completely buried when $H = 0$ or

$$\exp(-2.06 \times 10^{-7}t/B_0) = [1 + 0.775(H_1/B_0)]^{-1}, \tag{224a}$$

when the initial average height $H_1 < \frac{1}{4}B_0$; the linear scale is in centimeters, the time in years.

As a typical example, set $B_0 = 50$ cm, $H_1 = 25$ cm $> \frac{1}{4}B_0$ in the beginning (cf. Figs. 5, 16, and 17). The surface is at first unprotected, being sputtered at $\frac{1}{2}\chi_s = 0.515 \times 10^{-7}$ cm/yr [Eq. (223)] and buried at 2.67×10^{-7} cm/yr, or a combined rate of 3.18×10^{-7} cm/yr. After an initial period of $12.5/3.18 \times 10^{-7} = 3.9 \times 10^7$ yr the outstanding height is reduced to 12.5 cm after which Eq. (224) applies, with $H_1 = 12.5$ and $B_0 = 50$ cm. The total lifetime of the block, until complete burial, becomes

$$3.9 \times 10^7 + 4.3 \times 10^7 = 82 \times 10^6 \quad \text{yr,}$$

at which time its thickness will be reduced to

$$2.67 \times 10^{-7} \times 8.2 \times 10^7 = 21.8 \text{ cm,}$$

having lost only 3.2 cm through sputtering. This typical case shows that blocks of this and other sizes are not ground to powder by meteorite impact before being buried in overlay; after a lifetime of 10–100 million years (according to size), they become incorporated in overlay, being no longer disturbed except in a rare large cratering event. Such hidden collections of blocks may then be the cause of thermal anomalies, even when they are not visible, having been covered entirely by overlay precipitation (cf. Section VI,C).

B. Downhill Migration of Dust

In Fig. 18, a micrometeorite strikes a granular surface SS, inclined under an angle α to the horizon HH. The meteorite MO impacts in an arbitrary unspeci-

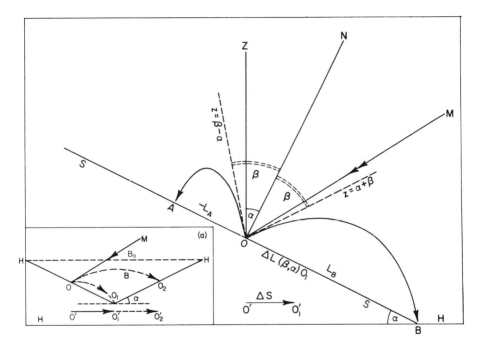

FIG. 18.   Downhill migration of dust. A micrometeorite MO strikes a slant surface SS. OA and OB are trajectories of particles ejected symmetrically with respect to the normal ON which makes an angle α with the vertical OZ. Insert (a): Cross section of conical crater HO_1H, with impact craterlet at O. Trajectories shorter than OO_1 lead to unrestricted downhill drift, those longer than OO_2 end uphill or outside the crater.

fied direction and causes a spray of ejecta from the point of impact O (which stands for an infinitesimal craterlet), symmetrical with respect to the normal ON, whatever the direction of OM. Two opposite, symmetrically directed (with respect to ON) jets (angle β) OA and OB are asymmetrical with respect to the vertical OZ, resulting in a greater downhill flight distance L_B than uphill L_A. The result is a net downhill displacement

$$OO_1 = \Delta L(\beta, \alpha) = \tfrac{1}{2}(L_B + L_A),$$

L_A being taken algebraically, i.e., negative when to the left of O. The displacement projected on the horizontal plane is evidently

$$O'O_1' = \Delta S = \Delta L \cos \alpha,$$

and this is the measure of migration of a mass fraction dy, ejected under an angle β such that

$$\sin \beta = y$$

[Eq. (27) with $\sin \beta_0 = 1$ as in the preceding section]. From elementary kinematical considerations we then find

$$\Delta S = (2v^2/g)(1 - y^2) \tan \alpha, \tag{225}$$

where v is the average ejection velocity as given in Eq. (215). Substituting this, integration over y yields the average displacement of the ejecta. However, there are some complications, which refere to the validity of Eq. (215) or Eq. (16), limited by the condition $y > y_Q$ [Eq. (17)]; for ejecta of micrometeorite impact, the limit (with $\lambda = 0.3$) corresponds to flight distances of the order of 5 km, far above the crater dimensions with which we are concerned. Such fast ejecta will go equally up- and downhill of smaller craters without a systematic drift, their effect being covered by the theory of the preceding section, while for downhill drift they are of no avail. Clearly, ejecta from the inner portions of the craterlet are irrelevant in the context. Only slow ejecta from the outer portions of the craterlet, whose flight distances are not large compared to crater diameter, will contribute to filling the crater by downhill drift. In Fig. 18a, a crater of diameter B_0 is schematically represented by a cone HO_1H, of constant slope α. Ejecta (OO_1) from a midpoint O (the micrometeor craterlet) on the slope will travel downward without difficulty when $\Delta S < O'O_1' = \frac{1}{4}B_0$ but will mount the opposite ledge $(\rightarrow OO_2)$ and even climb up or leave the crater when $\Delta S > O'O_2' = \frac{1}{2}B_0$. Rather than use proper-integration, the accuracy of which is not justified by the uncertainty in the basic data, we use averages as we have for many other cases in this treatise and set the lower limit of integration for Eq. (225) at a horizontal flight distance one-quarter that of the crater diameter,

$$v^2/g = 1/(4\alpha'y^2) \leqslant \tfrac{1}{4}B_0,$$

whence the lower limit of integration becomes

$$y_1 = (\alpha'B_0)^{-1/2}. \tag{226}$$

There is a lower limit to the validity of the treatment, i.e., $B_0 > 1/\alpha'$ (corresponding to the upper limit $y = 1$ which is not to be exceeded). Hence the average downhill displacement inside a crater of diameter B_0 and average slope

α, such that $\tan \alpha = 2x'/B_0$, with an additional mean factor of $2/\pi$ to allow for slant (nonmeridional) directions, becomes

$$(2/\pi) \int_{y_1}^{1} \Delta S \, dy,$$

or

$$\overline{\Delta S} = [2x'/(\pi\alpha'B_0)][(\alpha'B_0)^{1/2} + (\alpha'B_0)^{-1/2} - 2], \qquad (227)$$

with α' (cm^{-1}) defined by Eq. (217). The quantity $\alpha'B_0$ is thus dimensionless. The flow through 1 cm of crater circumference [πB, not necessarily πB_0, where $B = OO_2$ is an inner diameter (see Fig. 18a)] is then evidently

$$F_d = \chi \overline{\Delta S} [1 - e^{-\kappa}], \qquad (228)$$

where χ is the volume of granular material ejected per cm^2-sec [Eq. (213)] and κ the "kinetic depth" of overlay at the spot. Thus the expression in brackets denotes the fraction of projectile momentum spent in the (not infinite) granular layer, the remaining portion of the momentum being applied to the bedrock with much less sputtering efficiency [Eq. (223)], but of long range (10 km), and therefore of little consequence for the downhill migration problem.

Assuming the availability of a sufficient supply and thus maintaining a thick layer of overlay, $\kappa \to \infty$, the time scale for filling a conical depression of volume $V = (\pi x'B_0^2)/12$ by drift becomes

$$\tau_F = V/(\pi B_0 F_d)$$

or

$$\tau_F = \pi\alpha'B_0^2/\{24\chi[(\alpha'B_0)^{1/2} + (\alpha'B_0)^{-1/2} - 2]\}. \qquad (229)$$

With α' and χ as in the preceding section, this becomes

$$\tau_F = 1710B_0^2(0.966B_0^{1/2} + 1.035B_0^{-1/2} - 2)^{-1} \qquad (229a)$$

in years when B_0 is in centimeters. The formula is valid for $B_0 > 2$ cm, and the filling is supposed to proceed exponentially with time.

For craters completely imbedded in overlay, there is no shortage of supply to feed the flow downhill, rim and crater bed consisting equally of dust and rubble of great depth. From Table XXIX, the size limit for fulfillment of this condition ($H_p < 13$ m) is $B_0 < 160$ m for primaries and $B_0 < 36$ m for the "ray" secondaries. In this case, the flow sucks away the rim and the crater diameter increases, encroaching on the surrounding terrain, while the interior is filling. Shallow craters without elevated rims (as seen on the Surveyor and Ranger pictures) are produced in this way. Equations (229) or Eq. (229a) gives unconditionally the flow lifetime τ_F, for these overlay craters.

When supply is insufficient, and when, as for larger craters, there is bedrock underlying the crater profile, the flux defined by Eq. (228) adjusts itself to supply through a finite value of the overlay kinetic thickness κ; a thin layer of overlay then attains equilibrium with supply and downhill flow, especially in the outer portions of the crater where bare unprotected rock will be exposed and subjected to erosion by sputtering, at a rate of

$$dx/dt = \tfrac{1}{2}\chi_s\, e^{-\kappa} \quad \text{cm/yr,} \tag{230}$$

where χ_s is giveen by Eq. (223).

Supply to the outer regions of these larger craters can be assumed to consist of three main components:

(1) Low-velocity granular ejecta from the surroundings, however mostly nonsticking [cf. Eq. (212a)], rocicheting inward, and not apt to provide much of a supply near the crater rim; the rate of crater filling is measured by $1/\tau_f$ (Table XLIV) but only one-half of this should apply to the rim region.

(2) High-velocity sputtered material, of deposition rate χ_s [Eq. (223)], corresponding to a time scale (on the conical profile model) $\tau_s = \tfrac{1}{3}(x'/\chi_s)$ (no correction factor of $1 - G_g$ shall be used because the sputtering applies equally to granular and nongranular impacts); the fine-grained material (partly atomized) sticks to the spot without ricocheting and thus feeds the rim and the central regions equally.

(3) Low-velocity mostly coarse ejecta (J_e) of the accumulating overlay, with a time scale $\tau_e = \tfrac{1}{3}x'/J_e$ and $J_e = 2.67 \times 10^{-7}$ cm/yr as in the preceding section; these are even more mobile than the finer ejecta from the surrounding, and one-third of the rate can be assumed—somewhat arbitrarily—for the rim region. The condition of sufficient supply, and thus of the validity of Eq. (229) is then evidently

$$1/\tau_m = (1/2\tau_f) + (1/\tau_s) + (1/3\tau_e) > 1/\tau_F , \tag{231}$$

in which case the overlay thickness increases everywhere while the crater profile is gradually leveled out. When this condition is not fulfilled, partly or entirely unprotected rock is exposed, beginning from the rim inward, and the drift is adjusted to the supply through the proper value of κ in Eq. (228), i.e.,

$$1 - e^{-\kappa} = \tau_F/\tau_m . \tag{232}$$

The grinding of the incompletely protected rim proceeds, then, at a rate

$$dH/dt = -\tfrac{1}{2}\chi_s\, e^{-\kappa} = -5.2 \times 10^{-8}e^{-\kappa} \quad \text{cm/yr.} \tag{233}$$

The maximum rim erosion from this effect in 4.5×10^9 yr amounts thus to 234 cm, i.e., about $2\tfrac{1}{2}$ m.

Craters of less than 300-m diameter, which are eroded in less than 4.5×10^9 yr, would at present appear in various stages of erosion, according to age. With an initial depth to diameter ratio x'/B_0 of about 0.12 (cf. Fig. 4), in an average half-eroded crater, $x' = 0.06\, B_0$ can be assumed. With this the supply parameter, according to Eq. (231), becomes

$$1/\tau_{\rm m} = 1/(2\tau_{\rm f}) + (9.6 \times 10^{-6}/B_0).\tag{231a}$$

The calculated drift lifetimes $\tau_{\rm F}$ [Eq. (229a)] with the corresponding supply lifetimes $\tau_{\rm m}$ are given in Table XLV.

From this table we can see that the condition of Eq. (231), or $\tau_{\rm m} < \tau_{\rm F}$, is not fulfilled within the range of B_0 from about 6 cm to 28 m, where it is irrelevant because these small craters are completely built into overlay. Therefore, within the validity of our assumptions (with respect to which considerable uncertainty undoubtedly exists), the values of $\tau_{\rm F}$ [Eq. (229a)], or the drift rates $F_{\rm d}$ [Eq. (228) with $\kappa = \infty$], seem to be valid unconditionally.

This refers to the average flat crater rims. Step (or moderately steep) rims of larger craters with bedrock exposed will retain their unprotected rocky surfaces, while the inpouring ejecta roll or ricochet toward the interior. Alphonsus is an example (Figs. 7–9), although on a much larger scale and representing a more primitive stage. Another example is the boulder rim or stone wall of the Surveyor I crater on the horizon (Figs. 5 and 6).

C. Filling by Ricocheting Overlay Injection

The last, and most important, factor of erosion for the lunar surface features (craters) to be considered is the filling of depressions by incoming overlay. The ricocheting grains of overlay, as they lose kinetic energy in successive semielastic impacts, will have a preferential tendency to collect in "holes" from which they are unable to escape. Therefore the holes will receive more accretion than their surroundings and will gradually be filled at the latter's expense. This differential leveling action of accretion is superimposed on a continuously rising general level of overlay. With this, as well as with the two other types of erosion discussed earlier (Section X, A and B), the leveling of a depression takes place preferentially at the expense of its nearest surroundings; these surrounding areas are, so-to-speak, sucked in by the crater vortex and a secondary, wider but shallower, depression is formed around the original crater. Thus, the depression never disappears completely except when erased by a subsequent larger impact; it is only made increasingly shallower until it becomes unobservable.

The theory of these processes, although more or less straightforward when the initial conditions (coefficient of elasticity, λ, etc.) are defined, leads to complicated statistical integrations, amounting to an unjustified overdiscussion. In the following, a simplified artificial mechanical model is introduced, amply sufficient to estimate the trapping efficiency of a depression without pretending to describe the actual statistical complexity of trapping.

It also must be pointed out that the preferential trapping in depressions applies only to the nonsticking, ricocheting part of accretion. Micrometeorite material (J_m) retained by the Moon is not only quantitatively insignificant, compared with the overlay ejecta (J_e), but is fine-grained or even partly atomized and must stick at the spot where it settles, equally over a hole or an elevation; it covers the terrain with a uniform layer without a leveling action on its roughness profile. Similarly, the fine-grained component of overlay, say below $r = 5 \times 10^{-5}$ cm, which, according to Eq. (212a), would stick at an impact velocity as high as 5 m/sec, must be excluded as nonactive; according to Table XXXVI, Part D, this component accounts for 15% of the incoming mass. On the other hand, large projectiles do not fill but rather destroy a depression through overlap. For a given crater size B_0, one may set an upper limit for a "filling" projectile size to be that producing a crater twice the given size, i.e., one of diameter $2B_0$. (The limit is rough and is made to coincide with the lower limit assumed for overlapping; the conventional vagueness of it is of little practical consequence because of the slow variation of the cumulative mass of J_e with projectile radius.) Thus, for $B_0 = 21$ cm, Table XXXVI, Part D indicates $r = 4.8$ cm at $2B_0 = 42$ cm, and a cumulative mass fraction $y = 0.628$; subtracting 0.15 as for the sticking fine-grained fraction, the actively filling fraction of overlay for this size of crater becomes $y_f = 0.628 - 0.150 = 0.478$ (of the total stated to be 3.47×10^{-7} gm/cm²-yr).

Instead of the statistical complexity of particle size and velocity distributions (Table XXXVI, Part D), we choose a typical average size that best represents the entire particle spectrum. For a given crater diameter, the median particle size is that corresponding to half-mass or $\frac{1}{2} y_f$ as defined above. In Table XLVI some sample impact parameters for this median size are listed.

It will be found that, compared with the other two erosion processes, filling is important only for the larger craters. With this in view, we assume the typical projectile parameters to be $r_0 = r_m = 0.5$ cm, $\cos \gamma_0 = 0.926$, $\sin \gamma_0 = 0.378$, $\gamma_0 = 22°.2$, and $w_0 = 1.038 \times 10^4$ cm/sec; also $G_g = 0.52$ or, by convention, 0.5 at this size (Table XXXVI, Part D), which means alternating "soft" and "hard" ricochets with $\lambda = 0.3$ and 0.5, respectively, as represented in Table XXXVIII (1) and (2). As an additional schematization, we assume, at impact, the specular law of reflection for the angles, while the velocities are reduced by a factor of λ after each impact.

Let G_f denote the "gain factor" or the ratio of accretion trapped in a depression to that accreted by a level surface of equal area. Obviously, this is a function of the depth and profile of the depression. Provisionally, a single depression situated on an infinite level area is considered so that competition from other depressions may be disregarded.

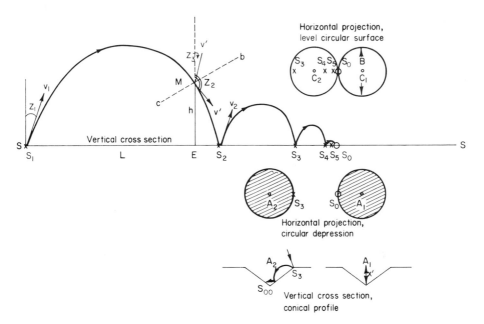

FIG. 19.   Trapping of incoming ricocheting overlay particles by a depression with a circular horizontal contour.

In Fig. 19, let S_1 be the point of first impact of the projectile upon level surface SS, and let S_2, S_3, S_4, and S_5 be the ricocheting projectile impacts that finally terminate at S_0, all the points, by convention, being assumed to align on a straight line (zig-zag paths will not essentially alter the gain factor; rectilinear path and specular reflection angle are convenient simplifications that should have little effect on the numerical results). A level circular area C_1 or C_2 of diameter B, in which the ricocheting path is assumed to pass through the middle, will contain the point S_0, and thus accrete the particle when the center of the area is displaced over a range of distance $C_1C_2 = B$ along the path SS, its "catch length" being B. A circular depression of equal diameter may trap the particle when impacting at S_3 (bottom part of Fig. 19, A_2; particle impinges at S_3, is reflected inward and is trapped at S_{00}), although unable to trap it at S_2 when the distance $S_2S_3 > B$; its catch length is then obviously $A_1A_2 = B + \Delta L_0$, where $\Delta L_0 = S_3S_0$ is the total tailpiece (which can be trapped) of the ricocheting particle path. Thus, obviously, we have from simple probability considerations,

$$G_f = A_1A_2/C_1C_2 = 1 + (\Delta L/B), \tag{234}$$

where $\Delta L = \Delta L_0$ in the particular case considered.

If ΔL_2, ΔL_1,... are the trapping lengths for impacts S_2, S_1 (which may not be zero in the general case), each of the preceding impacts adds to the probability. In such a case

$$\Delta L = \Delta L_0 + \Delta L_2 + \Delta L_1 + \cdots = \sum \Delta L_n. \qquad (235)$$

An overall rule of thumb, already applied in Section X,A, would set the extra trapping length equal to four times the average depth, which, for the conical cross section, equals one-half the maximum depth x'. Hence

$$\Delta L = \tfrac{1}{2}x' \times 4 = 2x',$$

or

$$G_f = 1 + (2x'/B). \qquad (236)$$

For an actually computed numerical case (mental experiment, see below) of $x'/B = 0.25$, $G_f = 1.529$ has been found while Eq. (236) yields 1.5, a surprisingly good confirmation of the rule of thumb.

For a noncentral path $S_1 S_0$ through a conical depression, the reflections cannot be kept in the same plane but, disregarding this finesse (in line with other simplifications), the ratio of depth to chord in a cross section remains equal to x'/B and Eq. (236), as well as its more sophisticated original equation [Eq. (234)], should remain valid for the entire depression area, and not only for its central section.

Of course, these expressions presume a depth-to-diameter ratio considerably smaller than unity, as is always the case with actual impact craters. In the case of a very deep or infinite hole, every particle entering it is trapped. The gain factor then is evidently

$$G_\infty = 1 + [(S_3 S_0)/B] + N, \qquad (237)$$

where N is the number of touchdowns preceding S_3 (when $S_2 S_3 > B$). (Here S_3 stands for the more general S_n, the nth touchdown.) With an average for the two types of ricocheting represented by cases (1) and (2) of Table XXXVIII, the gain factor is as shown in Table XLVII. Unlike the case of depressions having finite depth, the gain factor here is different for a chord and a diameter. The case, however, is only of academic interest so far as the Moon is concerned, although it probably helps in understanding the significance of the notion of gain factor.

Figure 20 explains in detail the conditions set up in the sample calculation of the gain factor for a triangular (conical) trap (cross section $\overline{ACB}$) having a depth equal to one-quarter its diameter, $\overline{OC}/\overline{AB} = x'/B = \tfrac{1}{4}$. Here, AB represents the ground level. The portion of the graph above this line pictures the transition from trapping to escape velocity (escape from the hole), for the lunar acceleration of gravity (162 cm/sec^2), in the form $\log(v_e^2/B)$ where v_e is the first rebound velocity *after* any impact at first entry (thus, in Fig. 19, v_2 is the velocity after touchdown S_2) just sufficient to make the rim of the depression, and $B = \overline{AB}$ is the diameter of the trap.

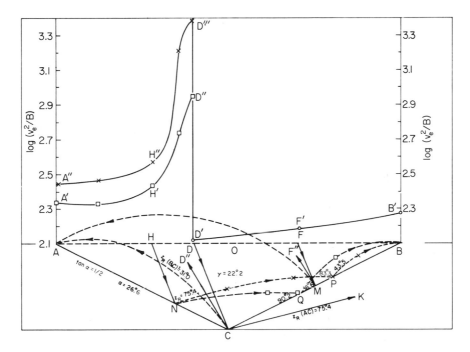

Fig. 20. Trapping mechanism by means of semielastic ricocheting in a depression of triangular (conical) vertical cross section. The impacting projectiles enter along HN, DC, and FM, at an angle $z - \gamma - 22°2$ from the vertical.

The succession of "soft" and "hard" impacts was assumed exactly as in Table XXXVIII, variant (1) ("soft" start) and variant (2) ("hard" start) being considered separately and an average of the two results then being taken. The elementary kinematical problem consists in the condition necessary for a projectile shot up from a point S_1 (Fig. 19, leg S_1S_2 as an example) under zenith angle z_1 with initial velocity v_1 to reach point M at distance $L = S_1E$ and altitude $h = ME$. The equation of the parabolic trajectory

$$h = L \cot z_1 - gL^2/(2v_1^2 \sin^2 z_1), \qquad (238)$$

to be represented also as a condition for the velocity

$$v_1^2 = gL/\{\sin 2z_1[1 - (h \tan z_1)/L]\} \qquad (238a)$$

when applied to the single or double reflected trajectories (CA, MA, NP: PB, NQ: QB, of Fig. 20) inside the depression, solves the problem of the velocity of escape from the trap. In the single or repeated (double) collisions, the degradation of velocity was assumed to proceed with strictly alternating elasticity factors $\lambda_1 = 0.3$ and $\lambda_2 = 0.5$ in continuation of the sequence of Table XXXVIII. If the point M (Fig. 19) were on the rim (points B or A; Fig. 20) of the depression and the velocity v_1 were a *minimum* compared to other combinations, then v_1 would be the escape velocity (v_e) required for a single impact. In Fig. 20, impacts on the back slope $\overline{CB}$ are indeed reflected in such a manner that the most efficient escape is achieved by the first reflection: $\overline{FM}$ reflected into $\overline{MA}$ with velocity F' (F''); $\overline{DC}$ reflected into $\overline{CA}$ with velocity D' (D''); etc. (The curvature of the incoming trajectories is neglected.) The escape velocity for these impacts is represented by the curve $\overline{D'F'B'}$ (Fig. 20, upper graph), based on three calculated points: for entry at D, at velocity (D'), $v_e^2/B = 131.0$; for entry at F, velocity (F'), $v_e^2/B = 152.6$; for entry at B, velocity (B'), $v_e^2/B = 183.4$. The curve $\overline{D'F'B'}$ shows the logarithms of these quantities.

For impacts on the front slope (somewhere all along $\overline{AC}$, Fig. 20), the first reflection (parallel to $\overline{CK}$) runs at zenith angle $z_R = z_1 = 75°4$ and either meets the opposite slope ($\overline{CB}$), or, even when not meeting, requires a double touchdown for escape at minimum velocity. If $\overline{cb}$ (Fig. 19, first leg again) is the opposite slope $\overline{CB}$ (of Fig. 20), the impacting zenith angle z_2 is determined from

$$\cot z_2 = dh/dL = \cot z_1 - [gL/(v_1{}^2 \sin^2 z_1)], \tag{239}$$

the impact velocity from

$$(v')^2 = v_1{}^2 - 2gh, \tag{240}$$

and the reflected zenith angle (Fig. 19) from

$$z_3 = 180° - 2\alpha - z_2, \tag{241}$$

where α is the inclination of the slope $\overline{cb}$ or $\overline{CB}$ (26°6 in the present case).

The left-hand side of the velocity diagram in Fig. 20 (above $\overline{AD}$) represents the escape velocities [$\log(v_e^2/B)$] for impacts on the front slope $\overline{AC}$. Two different cases occur, corresponding to two possible values of the elasticity factor at second impact λ^+.

Thus, a projectile entering at H along $\overline{HN}$ (Fig. 20) will leave the depression at point B with mimimum velocity H'' by double trajectory $\overline{NP: PB}$ when $\lambda^+ = 0.3$ at point P of second impact, and with minimum velocity H' along $\overline{NQ: QB}$ when $\lambda^+ = 0.5$ at point Q. This bifurcation of escape velocity at impact on the front slope $\overline{AC}$ is represented in the upper graph of Fig. 20 by curve $\overline{A''H''D'''}$ for $\lambda^+ = 0.3$ and by curve $\overline{A'H'D''}$ for $\lambda^+ = 0.5$

For a given touchdown $S_n(S_1, S_2, S_3, ...,$ Fig. 19), the catch length is then equal to the abscissa interval, in the upper portion of Fig. 20, over which the actual rebound velocity v_n falls below v_e. Thus, for $v_e^2/B = 100$, $\log(v_e^2/B) = 2.00$, this condition is fulfilled over the entire length $\overline{AB}$, or the catch length $\varDelta L = B$. For $v_e^2/B = 200$, $\log(v_e^2/B) = 2.3$, the condition is fulfilled only over the left-hand portion $\overline{AD}$, or for impacts on slope $\overline{AC}$ while those impacting on slope $\overline{CB}$ all escape; $\varDelta L = 0.408B$ obtains in this case.

Calculations of the gain factor at $x'/B = \frac{1}{4}$, along these principles, for the same chosen set of diameters as in Table XLVII, yielded values of G_f from 1.40 to to 1.67, fluctuating nonsystematically over the entire range of B from 2 to 8×10^4 cm, i.e., the range for which $B < S_1 S_2$ (Fig. 19) (the first ricochet length, Table XXXVIII), and an expected systematic decrease only for larger depression diameters. The fluctuations were due to a "resonance" or "interference" effect between the diameter and the set of ricochet intervals. Otherwise the absolute value of B was irrelevant and the individual values of G_f found for each B were considered as fair random samples of the gain factor. An average of

$$G_f = 1.529 \pm 0.089$$

was obtained. Of this, the unit part is accounted for by the length B itself, a fraction of the decimal portion (0.387) is the average of the tailpiece $\varDelta L_0/B$ ($S_3 S_0$ in Fig. 19), and a fraction (0.142) is contributed by higher order touchdown (chiefly S_2, Fig. 19). By analogy with the rule of thumb equation (236), we may set (when $x'/B < 0.5$)

$$G_f = 1 + (2.116x'/B), \tag{242}$$

the coefficient being based on the outcome of our "numerical experiment." This gain factor is not dependent on absolute dimension when $B < 800$ m and, thus, applies, in practice, to craters of all sizes that can be eroded in 4.5×10^9 years or less and that are the object of our interest.

Gain in accretion inside a crater must be compensated for by a loss in its neighborhood. The trapped tailpiece of the ricochet, $\varDelta L = S_3 S_0$ (Fig. 19), would have passed on level ground to a distance $\varDelta B$ ranging from 0 to $\varDelta L$ beyond the rim of the crater and thus be subtracted from a ring around the crater, bounded by the radii $\frac{1}{2}B$ and $\frac{1}{2}B + \varDelta L$, the distribution function being uniform over this range (of the linear displacement of the source S_1, Fig. 19). In a first approximation, $\varDelta L \sim Gf - 1$ and, therefore, when G_f decreases, the range $\varDelta L$ decreases also and the withdrawal is effected chiefly by narrowing the ring of withdrawal, while its depth is changed very little.

From the same numerical experiment, i.e., filling a crater with overlay at a depth $x'/B = \frac{1}{4}$ (for a progression of the diameters as in Table XLVII), the distribution of the deficits (withdrawals) in the surroundings was obtained as represented in Table XLVIII (each particle trapped in the depression corresponding to one missing from its prospective landing point outside the depression). With $\xi = \Delta L/B$ denoting the relative radial extension of the catch area, each individual (unity) event contributing ΔG_f to the gain factor inside the depression (B) is, by convention, to be spread uniformly over ΔL and contributes over this length to a uniform deficit $\Delta G_f/\xi$. The cumulative sum of these deficits ω_e, normalized to unity, is given in Table XLVIII for each ξ value without smoothing, i.e., exactly as calculated for single chosen B values.

The deficit integrated over the interval $\xi = 0$ to $\xi = \infty$ (radius $0.5 + \xi$ to ∞ in B-units) must equal the gain $G_f - 1$ over an area of $\frac{1}{4}\pi$ (in B^2 units). From this condition, accretion χ/χ_0 (in units of average accretion χ_0) around the depression (crater) is given by

$$\chi/\chi_0 = 1 - [\tfrac{1}{4}\pi(G_f - 1)(d\omega_e/d\xi)/\pi(1 + 2\xi)],$$

or

$$\chi/\chi_0 = 1 - [\tfrac{1}{4}(G_f - 1)(d\omega_e/d\xi)/(1 + 2\xi)]. \tag{243}$$

Here $1 + 2\xi$ is the relative radius, or the relative diameter of the zone. The gradient $d\omega_e/d\xi$ was determined graphically from the smoothed data of Table XLVIII, and the resulting relative accretion function (slightly smoothed) is given in Table XLIX, for the original case of $x'/B = \frac{1}{4}$ and for a number of other depression (crater) profiles based on Eqs. (242) and (243) and a homology relation following from them when $\xi \sim G_f - 1$,

$$(\chi_0 - \chi_b)/(\chi_0 - \chi_a) = (1 + 2\xi_a)/(1 + 2\xi_b), \tag{244}$$

where

$$\xi_b/\xi_a = (x'/B)_b/(x'/B)_a . \tag{244a}$$

Figure 21 represents the distribution of accretion rates inside and around a crater, according to Table XLIX.

The data of Table XLIX can be used to calculate the evolutionary changes in the crater profile as it is filling, x' decreasing while a conical cross section is assumed to be maintained. If $dH = \chi_0 \, dt$ is the increment of total accretion, the local increment is $dh = \chi \, dt$ and that at the crater edge (initially level outward, without raised rim) is $dh_0 = 0.69 \, dH = 0.69\chi_0 \, dt$, "expression (a)." Filling of the crater depth by an amount $-dx'$ relative to the edge (simultaneously raised by dh_0) requires the addition of an average accreted layer over the crater area equal to $\frac{1}{3} \, dx' + dh_0$, whence

$$\chi_0 \, G_f \, dt = \tfrac{1}{3}dx + dh_0$$

or, eliminating the time dt as well as χ_0 from expression (a), we obtain

$$0.69 \, dH/dx' = dh_0/dx' = 0.23/(G_f - 0.69). \tag{245}$$

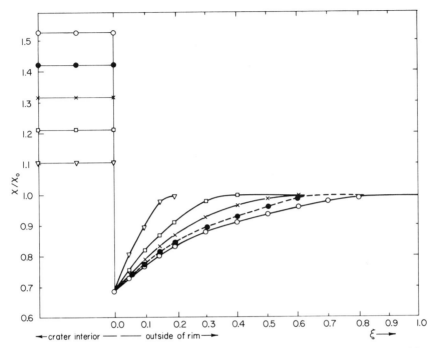

Fig. 21. Relative accretion rates of overlay, χ/χ_0 (ordinates), inside and outside a rimless crater. ξ (abscissa) is the distance reckoned outward from the crater edge, in units of crater diameter. Profile ratios: circles, $x_1'/B = 0.25$; dots, $x_1'/B = 0.20$; crosses, $x_1'/B = 0.15$; squares, $x_1'/B = 0.10$; triangles, $x_1'/B = 0.05$.

Also, for any point at distance ξ from the crater edge, with its proper accretion rate χ, the relative increment of accretion becomes

$$d(h - h_0)/dx' = \tfrac{1}{3}(\chi/\chi_0 - 0.69)/(G_f - 0.69). \qquad (245a)$$

Starting with $x'/B = 0.25$ or the first case of Table XLIX, the evolution of the crater profile, as calculated by approximate numerical integration of Eqs. (245a) and (245), is represented in Table L.

The consecutive stages of evolution by filling are represented in Fig. 22, according to the data of Table L. The gradual degradation of the crater at the expense of its nearest surroundings leads to the formation of a depression around the crater border, gently sloping toward it and without sharp outlines, reminding one of some "washed-out" crater structures on Ranger photographs (shallow depressions and "dimple" craters, Fig. 23). At stage V (Table L), the crater and its surroundings have melted into one such shallow structure of increased diameter and indefinite outline. This is the limiting stage for which the inte-

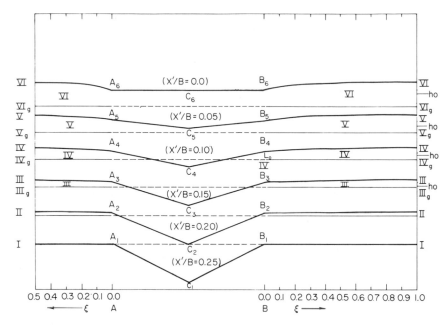

Fig. 22. Evolution of rimless crater profile and its surroundings by trapping and filling. Initial profile ratio $x_1'/B = 0.25$. Qualitatively valid also for erosion.

gration of Eq. (245a) is self-consistent and can be trusted. As for stage VI, it is the result of a linear extension of Eq. (245a) and is of but qualitative or symbolic significance.

Although derived for the process of filling by incoming overlay, the sequence of evolution of a crater profile as shown in Fig. 22 would also apply to the two other processes of degradation (filling by spray, and downhill migration) because they, too, are working at the expense of the nearest surroundings of a crater; only the appropriate time scales of the processes will be different. The presence, initially, of a raised "soft" rim, consisting of the same overlay rubble, will not essentially alter the time scales of degradation, although geometrically there will be some difference while the rim is eroded simultaneously with the filling of the crater.

As to a hard rocky rim, such could be expected at present for craters larger than 150 m, a two-staged process for their degradation will be considered separately.

By the nature of the filling process, the absolute rate of filling is roughly proportional to the relative depth x'/B, implying the rate of degradation to be an approximately exponential function of H, the total accretion (3rd column

FIG. 23.   Ranger IX photograph, 24 March 1965, showing shallow eroded craters in Alphonsus. Frame 1.6 × 1.4 miles. Smallest craters are 9 m in diameter. (Photograph by courtesy of NASA.)

of Table L). Multiplying these values by a factor of $2/\pi = 0.637$, as for an average chord of a circular crater, in order to obtain the accretion in units of B_0,

$$x'/B_0 = (x_1'/B_0) \exp(-H/He),$$ (246)

where x_1', as before, denotes the initial depth. The average linear measure of degradation H_e, corresponding to the different intervals of the table, is found as follows:

Interval, x'/B	0.25–0.05	0.20–0.05	0.15–0.05	0.10–0.05
$\bar{H}_e/B_0$	0.0437	0.0410	0.0375	0.0328

The variation of this parameter partly reflects the approximative character of Eq. (245a), in which changes in the surroundings are not taken into account, and only partly seems to be due to a real acceleration of the process at shallower profiles. This detail is only of academic interest, and we can assume safely an overall value of $H_e/B_0 = 0.0375$. The exponential relaxation time for filling is then

$$\tau_e = H_e/\chi_0$$

or, with the nonsticking influx of overlay being given by

$$\chi_0 = 2.66 \times 10^{-7}(y_2 - 0.15) \quad \text{gm/cm}^2\text{-yr,} \tag{247}$$

where y_2 is the cumulative mass fraction in the 2nd line of Table XXXVI, Part D for double crater size,

$$\tau_e = 1.41 \times 10^5 B_0/(y_2 - 0.15) \tag{248}$$

in years for B_0 in centimeters. These values are calculated and used jointly with the two other processes of erosion in the following section.

D. Erosion Lifetime of Soft-Rimmed Craters

As distinct from τ_f, τ_F, and τ_e, or the exponential time scales of erosion and filling, the erosion lifetime used in Eq. (204) (and which is to be set against elimentation by overlapping of larger craters) is the time interval during which the crater profile becomes so shallow that it becomes practically unrecognizable in the crater counts. For this we had already set, by convention, a limit of degradation $x'/B_0 = 0.02$. Hence the erosion lifetime of a crater starting with a profile ratio of $x'B_0$ can be set equal to

$$t_e = \tau_E \ln[(x'/B_0)/0.02], \tag{249}$$

where

$$1/\tau_E = (1/\tau_e) + (1/\tau_F) + (1/\tau_f), \tag{249a}$$

and τ_E is the total crater degradation time scale.

In Section VII,C, when discussing the accretion of overlay, a bend in the frequency of lunar craters at $B_1 = 285$ m suggested a lifetime of 4.5×10^9 yr at this size and a linear dependence of the lifetime on crater diameter for smaller craters,

$$t' = 1.58 \times 10^5 B_0 \quad \text{yr} \tag{250}$$

when B_0 is in centimeters. The formula was meant to apply only to larger craters, $285 > B_0 > 20$ m. This provisional lifetime (still without allowance for removal by overlapping) is given in the last line of Table LI. At the largest size (138 m), where the linear effect of filling (τ_e) prevails, the two figures are close enough for this sort of data, while for smaller sizes the *a priori* calculated values of t_e become rapidly shorter than the rough linear approximation t' because of the nonlinear effects of flow (τ_F) and spray (τ_t). The t_e values carry, of course, a greater weight than t', a rough approximation.

E. Erosion Lifetime of Hard-Rimmed Craters

A raised rocky rim can at present only be an attribute of moderately large craters, measuring hundreds of meters or more. Smaller craters will be completely built in overlay, with the projectiles not reaching down to the bedrock (as those of Table XXXVI). A rocky wall will isolate the crater from its surroundings so far as spray and flow are concerned, while the mechanism of filling by ricocheting overlay will be impeded less. We assume that, while the wall lasts, the crater bowl is filled by all the incoming overlay (2.66×10^{-7} cm/yr) without excluding any part of it (the "sticking" fraction is taken care of by the flow and spray mechanism inside the crater walls), plus straightforward micrometeorite accretion (8×10^{-9} cm/yr), which makes an average accretion of

$$\chi_i = 2.74 \times 10^{-7} \quad \text{cm/yr,}$$

while outside the crater, in view of the outward slope (Fig. 1), the accretion will increase outward, being negligible on the wall top. This would lead to a leveling out of the outward terrain and burying of the rocky wall, accelerated by direct sputtering as considered in Section X,A. With some protective layer being present, we assume one-half of the maximum sputtering rate of 5.3×10^{-8} cm/yr. This gives the rate at which the wall is buried into a level terrain as

$$\chi_e = \chi_i + 2.6 \times 10^{-8} = 3.00 \times 10^{-7} \quad \text{cm/yr.} \tag{251}$$

A rocky rim of height h_r above the terrain will thus be buried in a time interval of

$$t_1 = h_r/\chi_e , \tag{252}$$

which represents the duration of the first stage of erosion.

According to Eq. (1), with $x' + h_r$ standing for x_p , the average depth of a typical crater reckoned from the wall top is $0.463(x' + h_r)$. During the first stage, this decreases by $\chi_i t_1$, whence the depth at the end of the first stage becomes

$$x_1 = x' + h_r - \chi_i t_1/0.463. \tag{253}$$

The second, rimless stage begins at this point and is to be treated according to the rules of Section X,D. The relaxation times τ are assumed to depend solely on crater diameter as before, while the erosion lifetime in the second stage t_e , is calculated for a degradation of the profile from x_1'/B_0 to 0.02 [Eq. (249)].

The results are collected in Table LII. The total lifetime then equals the sum of the two time intervals

$$t_t = t_e + t_I. \tag{254}$$

The provisionally estimated linear lifetime t' [Eq. (250)] differs (by chance) very little from the values of t_e or t_I of Case B, Table LII. Contrary to what was found for the small rimless craters, the erosion lifetimes of the craters having a hard rim are found here to be longer than the t' values: by 20–30% in Case A (primaries) and by a factor of about 2.5 in Case B (ray secondaries). The turning point in the frequency of craters would then be expected to take place at $B_0 = 214$ m when the primary craters (Case A) begin to be completely eroded in 4.5×10^9 yr, and a second turning point is predicted at about $B_0 = 112$ m when the deeper profiles of the secondaries (Case B of the table) are erased during this interval of time. Of course, the transition in the frequency function of crater-area densities is expected to take place gradually because of the spread in the physical and geometrical parameters of cratering. The empirically suggested start of complete erosion at $B_1 = 285$ m is thus not in contradiction but in satisfactory agreement with the prediction, which cannot pretend to suggest anything more than a close order of magnitude.

Of interest is the case of $t_I > 4.5 \times 10^9$ yr, which defines the survival of an original stone-walled rim since the "beginning." In Table LII, Case A (asteroidal), the lower limit of unconditional survival of a hard rim is $B_0 > 700$ m, while in Case B (secondaries) the deeper profile would allow the rocky rim to survive when $B_0 > 270$ m.

A striking example, almost a test case, is presented by the stone-walled crater on the horizon of Surveyor I pictures (Fig. 6). The boulder wall is apparently the crest of a buried rocky rim of a crater that has come near the end of the first stage of erosion. Although there is much freedom in the interpretation, some limitations can be discerned. With $B_0 = 450$ m, its age must be less than t_I which is 7.5×10^9 yr when a secondary (Case B) and 3.0×10^9 yr when a primary crater (Case A). If a secondary, its actual age cannot exceed 4.5×10^9 yr, during which time its outward rim height, chiefly buried and partly eroded, must have decreased by $\chi_e \times 4.5 \times 10^9$ cm [Eq. (251)] or 13.5 m. An original height of $h_r = 0.05B_0 = 22.5$ m would leave thus 9.0 m of a stony rim towering above the surrounding plain, and more if the age were shorter; the actual height of the stone wall at its conspicuous part is 1.7 m, and the all-round average is less, perhaps 1.0 m. Case B is difficult to reconcile with the data and appears to be improbable. It remains to assume that the crater is a primary one, of an age less than about three billion years. There is, of course, an uncertainty in the initial profile ratio, but assuming the typical Case A of Table LII, the initial rim height may have been $h_r = 0.02B_0$ or 9.0 m, of which, however, the top may have consisted of overlay. The finer ingredients of overlay are rapidly removed by micrometeorite impacts, leaving the coarser fraction, about 60% of its mass (Table XXXVI, Part A; $1 - G_g = 0.60$ average), to be sputtered as hard rock. The effective height of the rocky rim must therefore be decreased by 0.4 of the overlay layer. If t is the age of the crater as a fraction of 4.5×10^9 yr, the overlay thickness at the time of impact can be set at $14(1 - t)$, the effective initial altitude of the hard wall at $9.0 - 0.4 \times 14(1 - t)$m, which is to be buried and eroded at a rate of $3.00 \times 10^{-7} \times 4.5 \times 10^9$ cm or 13.5 m per chosen unit of time. [The initial rate of overlay formation, 14 m before the cratering event, is purposely taken larger than its later or present rate, $2.66 \times 10^{-7} \times 4.5 \times 10^9$ cm or 12 m per unit time (aeon).] If the wall height has been decreased by burial and erosion to an average altitude of 1.0 m, this leads to an equation for the determination of age:

$$9.0 - 5.6(1 - t) - 1.0 = 13.5t,$$

which yields

$$t = 0.30 = (1.35 \pm 0.5) \times 10^9 \quad \text{yr},$$

or roughly 1.5 billion years for the age of the stone-walled crater of Surveyor I.

F. Overlay Accretion: Second Approximation

In Table XXIX, the overlay volume was calculated from observed crater volume statistics with provisional allowance being made for the disappearance of smaller craters through erosion, and for two limiting cases of a protective layer: A, for zero overlay thickness; and B, for 12 m present thickness. The linear equation [Eq. (250)] for the lifetime was used. Now we are in possession of erosion lifetimes, calculated *a priori* by more sophisticated methods that can be applied to a revision of the expected accumulated volume of overlay. Only those craters whose lifetimes are shorter than 4.5×10^9 yr are affected by the lifetime condition. Part B of Table XXIX has been recalculated accordingly, using the new lifetime data of Tables LI and LII, separately for primaries and secondaries. Overlapping is a minor effect for these large craters, affecting their numbers by but a fraction of a percent and is disregarded here. The crater areal densities are then simply proportional to the lifetime t_e. Table LIII contains the results of the revision.

The result does not differ essentially from the first one and firmly indicates an overlay layer of about 14 m at present. The total sum in the last line of the table, 1535 cm replacing the former result of 1307 cm, is increased chiefly at the expense of small secondaries, which actively affected the bedrock for only a very short time after the formation of the maria. This detail is highly conjectural, and we may leave the subject at that, being satisfied that the new refined treatment of crater lifetimes has little affected our original estimate of overlay thickness.

XI

SUMMARY

(1)   Impact cratering and erosion are the prevailing factors that have been shaping the lunar surface for the past four billion years.

(2)   Impact melting has produced the lava flows of the maria, at an early stage, 4.5 billion years ago. The maria were also the seat of primeval volcanism, testimony for which is given by some of the less conspicuous surface details such as the domes and dykes.

(3)   No traces of contemporary orogeny or volcanism on the Moon are indicated. The Alphonsus event of 1958 was not a gaseous eruption but a case of fluorescence of the solid crater peak.

(4)   Cratering formulas are proposed as derived from first principles, with little empirical adaptation. The main arguments are momentum of the projectile and strength of the target, rather than energy, which has often been used in limited interpolations of experimental results. The range of application of the formulas is almost unlimited, for velocities from tens of centimeters to tens of kilometers per second. Without using empirical coefficients of proportionality, the formulas represent the cratering dimensions—penetration and volume— to better than $\pm 20\%$. Special formulas are derived and empirically tested for low-velocity impact of rigid projectiles into granular targets, with direct application to the lunar surface layer. The cratering formulas, including throwout and fallback equations, are used to derive the cohesive strength of lunar rocks, in its bearing on the origin and history of the lunar surface.

(5)   Formulas for the encounter probabilities, lifetimes, and statistical accelerations of particles in planetary encounters are given, with emphasis on small relative velocities as for near-circular nearly coplanar orbits. The damping effect of an orbiting ring of particles upon its individual members, and its bearing on accretion of larger bodies is considered.

(6)   The problems of the origin of the Moon are analyzed with the help of the theories of cratering and planetary encounters. The mathematical theory of tidal evolution can describe the past history of the Earth-Moon system with some confidence only as far back as to a "zero hour," corresponding to the Moon's approach to a distance near Roche's limit, somewhat less than three Earth radii, at which time the Moon began to recede. From geologic evidence, the date of this phase could not have been later than 3.5 billion years ago, the age of the oldest dated terrestrial rocks; most probably, it coincided with the age of the Earth, 4.5 billion years, because all the initial events connected with the origin of the Moon must have evolved on a short time scale of 10^5–10^7 years. Tidal friction at the time of closest approach, working on a time scale of 10^3 years

or less (too short for significant cooling by radiation), must have melted the outer mantle of the Earth, erasing all previous geologic records.

(7)   The history of the Earth–Moon system prior to this zero hour is open to conjecture, because neither the identity of the interacting bodies (of which most have disappeared) nor their masses and initial orbits can be ascertained. If, however, the theories are to conform to the meagre observational evidence, requiring (a) that the craters on the continentes were formed on the receding Moon by projectiles orbiting the Earth at about 5 Earth radii (as testified to by their ellipticities, and by the lack of an excess of crater numbers on the preceding hemisphere of the Moon), and (b) that the surface of the Moon at that time (which although solid was soft) was hot but only insignificantly or moderately melted by the impacts, then two models appear more probable than the others. The first is that of Model 5, Table IX, implying an origin from debris orbiting inside Roche's limit, either analogous to the rings of Saturn, or thrown off through instability of the rotating Earth; with sufficient mass load in the rings, cohesive clumping of the debris enables them slowly to work their way out tidally, beyond Roche's limit, to collect first into some six-odd intermediate moonlets, and ultimately into one lunar body at a distance of about 5 Earth radii. The second is Alfvén's adaptation of Gerstenkorn's model of tidal capture, in which the incoming Moon, originally captured into a retrograde orbit and put into synchronous rotation, passes slightly outside Roche's limit at closest approach where it sheds its outer and lighter mantle and retains the denser core. Thus, while receding, it again collects most of the lost material. The formation of the maria on the earthward side of the Moon, through a belated impact of a moonlet (broken up tidally before impact on the Moon), previously formed from the material ejected inward, is also plausibly accounted for by Alfvén's model. As to the time of the event, it could never have happened as recently as 700–1000 million years ago, for reasons stated under point (6).

(8)   The formation of a mare is explained by impact melting of a hot crust in a cratering collision, on a linear scale sufficiently large for the melt to fall back into the crater; on a smaller scale, the liquid is sprayed over the crater walls in all directions and cannot form one coherent fluid body of lava.

(9)   Crater profiles, orographic differences in level, and the secondary craters produced by the ejecta of ray craters can be interpreted consistenly if it is assumed (a) that the postmare craters were produced in a relatively cool rocky target of the strength of granite or basalt, by interplanetary projectiles (asteroidal bodies and comet nuclei) at velocities of 20–40 km/sec, (b) that the premare craters were produced by slow projectiles, about 3 km/sec, impacting on a hot and relatively soft surface, about one-tenth of the strength of granite, and (c) that the orographic differences of level on the Moon were formed during the same period of primeval bombardment when the crust was hot and soft.

(10)   The statistics of craters (larger than 1–2 km) in lunar maria are consistent with astronomical observation, cratering theory, and theory of planetary encounters when target rock is assumed to be of the strength of granite or basalt. There is no basis whatever for attributing the origin of the overwhelming majority of the craters to causes other than impacts.

(11)   Details in the frequency function of the diameters of smaller craters suggest a minimum survival limit of about 300-m diameter against erosion during 4.5×10^9 yr. This roughly agrees with theoretical calculations of the rate of erosion on the Moon.

(12)   The statistics of small craters in Alphonsus are consistent with their impact origin from a mixed population of premare projectiles, among which slow secondary ejecta prevailed; the scarcity of craterlets on the peak and wall of Alphonsus is explained by the hardness of these targets (bare rock or rock under a thin protective cover), while the floor of Alphonsus carrying 15–20 times more craterlets per unit area is consistent with a loose target about the strength of terrestrial desert alluvium. The collapse or caving-in hypothesis of the craterlets is unacceptable, both because of their prevailing circular shape, and because of the relative uniformity of their distribution, the crater density down to the same diameter being similar in distant regions of the maria as well as in Alphonsus. The floor of Alphonsus, formed probably without coherent melting, may have spread out onto a level surface in a kind of "ash flow." Its peak (as well as the peaks of many other craters) can be interpreted to be a surviving remnant (compacted at impact) of the rear portion of the projectile that produced the crater.

(13)   The top layer of lunar soil consists of a heterogeneous mixture of particles having a broad distribution of sizes. Effective values of different physical quantities may depend on particles of different size. Thus the thermal conductivity of the upper 10 cm, consisting of three components, i.e., the bulk conductivity through a grain, the radiative conductivity between grains, and the contact conductivity (depending on contact area, increasing with pressure and depth), can be accounted for by a constant effective grain diameter of 0.033 cm, while the strong radar reflectivity and thermal inertia of the hot spots require the presence of a prominent component of sizeable boulders, imbedded in the rubble as well as strewn over its surface. The normal radar reflectivity, pointing to a bulk dielectric constant of 2.6–3.0, is compatible with an average density of 1.3, or 50% porosity for a basaltic composition. Cohesion of grains *in vacuo* is sufficient to balance the lunar gravity of grains smaller than about 0.13 cm; these are responsible for the "fairy-castle" structure of the top layer and determine the optical properties of the Moon, especially the dominance of phase angle and the strong backscatter at zero phase, in the visible portion of the spectrum. On a scale of centimeters and meters, the top soil is polished by micrometeorites into a gently undulating surface with specular reflectivity.

(14)   The dependence of radiative conductivity on temperature leads to a day–night asymmetry and a positive thermal gradient in the top soil even at zero flux. If this is taken into account, radio observations of the thermal gradient in the lunar soil lead to a net flux from the Moon's interior of 3.4×10^{-7} cal/cm²-sec, compared to 4.3×10^{-7}, which is the Earth's value decreased in proportion to the lunar radius (thus corresponding to equal content of radioactive sources if thermal equilibrium is assumed).

(15)   The average temperature on the lunar equator is −65 C on the surface, −49 C at one meter, and −23 C at 7-m depth. The pressure-dependent increase of conductivity with depth prevents the top layer from playing any significant insulating role, so that the thermal state of the Moon's crust is not much affected by it: at equal depth, the crust is only about 50 C warmer than it would have been without the insulating top layer.

(16)   Impact erosion leads to the leveling out of lunar surface features without relevance to an "angle of repose."

(17)   The amount of lunar surface material sputtered to space by solar wind, about 45 gm/cm² in 4.5 billion years, is nearly equal to the gain from the slow micrometeorites of zodiacal dust. The meteoritic material is admixed to an average accumulated overlay layer of about 13 m or 1700 gm/cm² (over the maria and outside the range of ejecta of large craters). The other, fast meteoric components lead to a loss of about 23 gm/cm² of lunar material.

(18)   In the build-up of overlay from the underlying rock, three penetrating components of meteorite flux are relevant: the Apollo-meteorite component (pseudoasteroidal), the comet nuclei, and the asteroids deflected from Mars crossings. Frequency formulas for the three fluxes, depending on particle radii, are given and the corresponding crater densities (numbers per unit area) calculated. The observed excess in the densities of small craters is consistently interpreted to be due to a fourth component, namely, to secondary ejecta from violent cratering events (ray craters). With little dependence on this interpretation, the overlay thickness and its statistical distribution over maria and continentes is calculated from the volume excavated by the actually observed craters, the numbers of those smaller than 300 m being corrected for survival from erosion.

(19)   From cratering theory and an empirical relation for the dependence of the strength of brittle materials on particle size (diameter to the $-\frac{1}{4}$th power), the exponent of the differential frequency (per unit of volume or mass) of particle radii in cratering ejecta is found to be $n = 3.875$, which is in good agreement with particle counts in lunar overlay obtained from spacecraft landings.

(20)   The mechanical properties of lunar soil are similar to those of terrestrial sand. The bearing strength at equal depth, and the kinetic efficiency at impact are nearly one-half of those of typical terrestrial beach gravel. The bearing

strength (frontal resistance) is about 5×10^4 dyne/cm^2 at the surface, 6×10^5 at 5 cm, and 2.5×10^6 at 10 cm penetration. The cohesive lateral resistance (crushing strength) is about 1/18th of the bearing strength at equal depth.

(21)   Electrostatic transport is theoretically limited to particles of submicron size. The absence of blurring of detail (less than 0.5 km for demarcation lines in Alphonsus) indicates that such particles and "electrostatic hopping" do not play a significant role on the lunar surface.

(22)   The ballistic fluxes impinging on the lunar surface consist of five interplanetary components (J_M, the micrometeorites; J_0, the dustball meteors; J_1, the Apollo-meteorite group; J_2, comet nuclei; J_3, Mars asteroids deflected to Earth crossing) and of secondary ejecta from primary cratering events (component J_e). The quantitative characteristics of the interplanetary components are deduced from observation as corrected for selectivity, while J_e is assessed from the excavated cratering volume as corrected for interplanetary impacts and erosion.

(23)   Cratering parameters for the ballistic fluxes are calculated for overlay. For component J_e, quantitatively assessed ricocheting is viewed as amplifying crater generation in overlay. Only a fraction Gg(0.4–1.0, depending on particle size) of the impacts are of the granular target type while the rest are into larger grains or boulders and are of the hard target type.

(24)   Components J_M and J_0 produce craters in overlay that are too shallow to be observed; the action of these components is limited to erosion, 82% of which can be accounted for by them. Component J_e accounts for 97% of the ballistic mass, but for only 15% of the impact momentum and erosive capacity. Components J_2 and J_3 (comet nuclei and the Mars asteroids) are negligible for small cratering in the subkilometer to meter range, where J_e and the meteorite groups J_1 are of sole importance, but they—chiefly J_2—gain in importance and dominate in large cratering events (above 5 km).

(25)   With a conventional limit of observability of 0.02 for the crater depth-to-diameter ratio, the *a priori* calculated crater generation rates, as set against deletion through overlapping and degradation through filling and erosion, lead to theoretical crater areal densities in the diameter range from 3 cm to 150 m and beyond, which are in satisfactory agreement with observation from space probes.

(26)   Mixing of overlay proceeds more slowly than its accumulation. The mixing thickness is about 8 cm, corresponding to a difference or "blurring" in the age of the strata of about 30 million years; this represents the "stratigraphic resolving power" of overlay. There is practically no interchange of material between layers separated by more than 25 cm or 100 million years.

(27)   The ballistic astronautical hazard on the lunar surface is negligible, being by orders of mangitude smaller than the hazards we are willing to accept in everyday life on Earth.

(28)   Crater and boulder degradation rates from filling by overlay, and from several types of erosion (spray from micrometeorite impact, downhill migration of dust, sputtering of crater rims and boulders, and their burial by overlay) have been quantitatively assessed theoretically from first principles and from the observed properties of interplanetary populations. The results are in satisfactory accord with the observed areal densities of small craters on the Moon.

(29)   Ablation of exposed rock on the lunar surface is estimated to be about 5×10^{-8} cm/yr, while the average rate of burial into overlay (which is, however, a widely fluctuating quantity, according to cratering events in the vicinity) is about 2.7×10^{-7} cm/yr, so that it is buried before becoming eroded. Rocks lying on the surface are secondary ejecta which have come to rest after several ricochets. Craterlets in overlay, left behind by the ricocheting impacts of secondary ejecta, are relatively deep (Fig. 5, the 3-m crater), contrary to those made by primary interplanetary impacts, which are too shallow to be observable.

(30)   As the result of filling and erosion, which takes from the surroundings the filling material, the crater profile becomes shallower while the effective diameter increases.

(31)   Some examples of theoretical degradation lifetimes of craters, or the time of reduction to a profile ratio of $x'/B_0 = 0.02$ are as follows.

Crater diameter, B_0 (cm)	3.3	45	650	1.5×10^3
Origin and description	Secondary, soft rim	Secondary, soft rim	Secondary, soft rim	Interplanetary primary, soft rim
Lifetime, years	5.4×10^4	1.5×10^6	3.0×10^7	2.3×10^7
Crater diameter, B_0 (cm)	4×10^3	1.4×10^4	1.15×10^4	2.1×10^4
Origin and description	Interplanetary primary, soft rim	Interplanetary primary, soft rim	Secondary, hard rim	Interplanetary primary, hard rim
Lifetime, years	2.7×10^8	1.4×10^9	4.5×10^9	4.5×10^9

XII

ADDENDUM

Publication of this chapter has been delayed. Meanwhile spectacular research, culminating in the two Apollo landings, has been going on. It may be asked whether, as a consequence, additions or changes in the text could be contemplated.

To the author's satisfaction, solid isotopes of the lunar rock samples indicate an age of the solidified maria surface close to the 4.5 billion years assumed throughout this work. The cosmic-ray exposure age of the top 0–25 cm is, of course, much less, ranging from 15 to 100 million years. In general, nothing need be changed in the calculations of erosion except for the addition of one more theoretical factor, as pointed out by Kopal (1968). It consists in the sifting action of moonquakes, during which fine-grained material tends to slip under the shaking boulders and gradually lift them up. According to Kopal, the efficiency of the process is locally limited to elevations—mounds and slopes—from which the incoming rain of overlay is swept away as described in this monograph. On level ground, the boulders must be eroded at the top and buried by overlay at the bottom faster than their lift-up by moonquakes, otherwise the entire lunar surface would be covered by boulders rising from the bottom of the overlay, which certainly is not the case. Besides, boulders lying on top of the lunar soil need not be ascribed necessarily to the lifting action of Moon tremors; in leap-frog ricocheting, the boulder is not buried inside its first impact crater, but, as a rule, lands gently on top of the soil at the end of its trajectory, leaving behind a series of craterlets. When this happens on a down slope, the loss of kinetic energy at impact may be restored in flight so that a long ricocheting path may result—such as that shown on an Orbiter 5 photograph near the crater Vitello (Kopal, 1968, Fig. 5); the path is not that of a continuously *rolling* stone, but that of one that was bouncing and thus leaving a chain of discrete crater impressions behind. On the other hand, the stone-wall remains of a filled crater such as that in Fig. 6, while cleared of rubble at the top by the regular erosion processes we have already described, apparently testify to the action of Moon tremors that have also cleared the interstices between the boulders, which otherwise are well shielded from micrometeorite erosion.

The tremors themselves must be caused chiefly by meteorite impacts, to which the solidified lava sheet of a lunar mare is highly susceptible, as has been shown by the seismic experiment of Apollo 12.

Although it seems that the results of the lunar landings have had little effect on our pre-Apollo conclusions, this by no means belittles the tremendous importance of the mere fact that man has set his feet on the Moon. In the light of future developments, the fact that the first men to land on the Moon were not professional scientists may be of more far-reaching consequence than having a "real" scientist, say a geologist, take part in the venture. The participation of "professionals" will, undoubtedly, be desirable in future landings, but, in the present case, the absence of a scientist with all his preconceived ideas and selectivity may even have been a bonus in ensuring that the samples of lunar material brought back to Earth were representative and nonselective. The presence at the first lunar landings of a scientist with such built-in prejudices would have been of very doubtful advantage.

XIII

APPENDIX: TABLES

TABLE I

RELATIVE CRATER DIMENSIONS FOR VERTICAL IMPACT
INTO SOLID ROCK; $\gamma = 0°$

w_0 (km/sec)	3	6	10	15	20	25	30	35	40	50	75
Stone impact into stone, $\delta/\rho = 1$											
k	2.03	2.14	2.36	2.78	3.31	3.79	4.18	4.44	4.44	4.44	4.44
p	2.03	2.12	2.20	2.26	2.30	2.33	2.37	2.39	2.41	2.45	2.51
D	4.82	6.86	9.12	12.0	14.9	17.8	20.3	22.4	23.9	26.5	32.1
Iron impact into stone											
k	2.05	2.22	2.56	3.12	3.75	4.21	4.52	4.70	4.70	4.70	4.70
p	3.25	3.69	3.82	3.93	4.00	4.06	4.11	4.15	4.19	4.26	4.37
D	6.63	9.17	12.5	16.7	20.9	24.6	28.2	30.4	32.4	35.9	43.4

TABLE II

COMPARISON WITH EXPERIMENT: ALUMINUM → ALUMINUM, $\rho = \delta = 2.7$

μ (gm)	1.265	0.378	0.158	0.047	0.376	0.376	
w_0 (km/sec)	5.22	6.87	8.63	9.05	7.80	6.58	
s_t (10^9 dyne/cm^2)	1.01	1.01	1.01	1.01	5.07	2.84	
k (calc)	2.11	2.19	2.28	2.30	2.23	2.17	
x_p + lip (obs, cm)	2.36	1.86	1.58	1.04	1.50	1.41	
$1.16x_p$ (calc, cm)	2.40	1.65	1.24	0.84	1.56	1.57	Aver
x_p (ratio calc/obs)	1.02	0.89	0.79	0.81	1.04	1.11	0.94
B_0 (obs, cm)	4.44	3.32	2.76	1.82	2.71	2.68	
B_0 (ratio calc/obs)	1.06	1.09	1.12	1.16	0.98	1.04	1.08
B_0 (calc, cm)	4.71	3.62	3.10	2.12	2.67	2.78	

TABLE III: Mechanical Strength and Cratering in Natural Beach Gravel

Site	I	II	III	IV	V
Average grain size (mm)	5	3.5	4	1.5	1.8
State of moisture	dry	dry	dry	wet	moist
Assumed density (gm/cm³)	1.7	1.7	1.7	2.0	1.7
Friction (f_s)	0.63	0.63	0.63	0.40	0.63

(a) Static experiments with craters measured ($x' = 0.15x_0$ average)

No.	Site	x_0 (cm)	B_0 (cm)	σ (cm²)	F	s_p(max) (dyne/cm²)	s_p (dyne/cm²)	s_c (dyne/cm²)	S_p (dyne/cm⁴)	S_c (dyne/cm⁴)	$\bar{s}/s_c$
1	I	5.5	11.5	3.63	0.117	1.15^6	4.33^5	1.96^4	35800	1620	22.1
2	I	15.0	12.5	3.63	0.052	1.97^7	6.66^6	1.02^5	86500	1330	65.3
3	I	1.75	23.5	198	0.094	3.69^5	2.20^5	1.75^4	72800	5800	12.6
4	I	8.5	22.5	29.3	0.105	2.42^6	8.51^5	3.82^4	32600	1460	22.3
5	I	5.2	10.5	3.25	0.176	6.27^5	2.38^5	8.9^3	21600	810	26.8
6	I	6.5	9.9	3.25	0.161	9.02^5	3.28^5	1.28^4	20400	800	25.6
					Mean 0.118					Log mean	25.5
										Probable deviation ratio	1.36

2a   The rod of Experiment 2 was excavated *in situ* and the same load $s_p = 1.97 \times 10^7$ was applied. The additional penetration was $x_0' = 5.0$ cm (thus reaching a total of $15.0 + 5.0 = 20.0$ cm below the undisturbed surface).

(b) Other static experiments

No.	Site	x_0 (cm)	σ (cm²)	s_p(max) (dyne/cm²)	S_p (dyne/cm⁴)
7	I	0.9	3.63	2.82^5	100000
8	I	1.25	3.63	4.31^5	121000
9	I	2.82	3.63	6.16^5	61900
10	I	4.9	3.63	1.15^6	44200
11	I	17.8	3.63	1.96^7	61600
12	I	0.28	29.3	1.65^5	79300
13	I	1.7	3.25	2.46^5	50400
14	I	2.0	3.25	3.50^5	58300

No.	Site	x_0 (cm)	σ (cm²)	s_p(max) (dyne/cm²)	S_p (dyne/cm⁴)
15	II	2.5	145	4.76^5	57800
16	II	4.0	46	1.50^6	83300
17	III	1.4	140	4.80^5	121000
18	IV	0.90	140	6.5^4	23200
19	IV	2.25	140	4.8^5	68000
			Log mean of Exts. (1)–(19)		55500
			Probable deviation ratio		1.46

TABLE III (*continued*)

(c) Dynamic (impact) experiments

No.	Site	x_0 (cm)	B_0 (cm)	m (gm/cm²)	σ (cm²)	w_0 (cm/sec)	(w_1/w_0)	k	s/s_a	S_p (dyne/cm⁴)	S_c (dyne/cm⁴)	S_p/S_c
20	I	5.8	54	34.5	525	577	0.759	0.253	1.89	31500	3250	9.7
21	I	2.6	21	18.4	108	221	0.788	0.211	1.82	19900	2520	7.9
22	I	3.9	26	18.4	108	479	0.788	0.250	1.98	32000	2980	10.7
23	II	2.3	8.8	8.05	15.8	372	0.809	0.260	1.79	26200	4230	6.2
24	II	2.4	10.0	8.05	15.8	525	0.809	0.267	1.96	47000	5400	8.7
25	II	2.3	13.0	8.05	15.8	578ᵃ	0.809	0.271	2.41	60300	3380	17.9
26	III	2.9	14.7	7.55	22.1	1266	0.770	0.317	2.03	132000	9750	13.5
27	IV	0.9	20.1	12.93	140	220	0.660	0.257	1.43	59100	8420	7.0
28	IV	1.2	20.1	12.93	140	372	0.660	0.268	1.43	112000	14100	7.9
29	IV	2.2	23.5	12.93	140	525	0.660	0.300	1.49	70600	8530	5.7
30	V	3.2	13.3	10.5	9.0	1266	0.880	0.237	4.31	220000	5210	42.2
31	V	2.5	12.5	10.5	9.0	525	0.880	0.205	4.60	73400	1530	48.0
32	V	1.8	9.5	10.5	9.0	372	0.880	0.169	3.87	77200	3080	25.1
33	V	2.6	12.1	4.87	19.4	1266	0.677	0.397	1.47	63800	9960	6.4
34	V	2.1	10.9	4.87	19.4	525	0.677	0.366	1.48	20800	3340	6.2
35	V	1.7	10.4	4.87	19.4	372	0.677	0.339	1.53	17700	2610	6.8
36	V	3.7	16.1	7.55	22.1	1266	0.770	0.344	2.00	60700	3870	15.7
37	V	2.65	12.0	7.55	22.1	525	0.770	0.304	1.78	29300	3580	8.1
38	V	2.22	11.4	7.55	22.1	372	0.770	0.281	1.81	23500	2780	8.5
Logarithmic mean, Experiments 20–38										48600	4380	10.9
Probable deviation ratio										1.65	1.54	1.61

ᵃ Experiment 25 at oblique incidence.

TABLE IV

Typical Gravel Cratering Parameters

Experiments 23, 24, 36, 37 (weighted)

	B_0 (cm)	$s^{1/4}$	s (dyne/cm^2)	$w_0^{\frac{1}{2}}$	w_0 (cm/sec)	x_0 (cm)	x_0/B_0	Lip height h (cm)	h/B_0	x_p/B_0
Average	11.7	12.0	2.07×10^4	25.7	660	2.65	0.226	0.66	0.056	0.286

Throwout estimated characteristics, in units of B_0 experiment 26 only; $B_0 = 14.7$ cm

	Outer crater wall	Massive throwout	Considerable throwout	Farthest noticeable throwout	s	k	w_0
B/B_0	1.355	1.96	3.9	6.8	4.87×10^4	0.317	1266

TABLE V

"Teapot" Nuclear Crater Ejecta (Fallback + Throwout) Distribution

Ballistic distance

B/B_0 $\leqslant$	1.000	1.183	1.304	1.414	1.581	1.732	∞

Calculated with $a = 1.92$, $b = 0.80$, $\sin \beta_0 = 0.800$ for ballistic distance

F_B	0.346	0.400	0.436	0.468	0.517	0.561	1.000

Observed (Fig. 1 volume $\times$ 0.883)

F_B	0.274	0.389	0.490	0.570	0.665	0.731	(1.000)

Calculated distance reduced by atmospheric drag

B'/B_0	1.000	1.175	1.289	1.386	1.522	1.607
$F_\mathrm{B(obs)}$	0.274	0.384	0.477	0.550	0.631	0.676
$F_\mathrm{B(calc)}$	0.346	0.400	0.436	0.468	0.517	0.561

TABLE VI

Variation of Mass Ratio of Two Accreting Nuclei

μ_1	1.000	0.512	0.216	0.125	0.064	0.027	0.008	10^{-3}	10^{-6}	10^{-9}
$\mu_1 \mid \mu_2$	81.5	49.8	27.5	19.0	12.7	8.0	4.6	2.37	1.10	1.01

TABLE VII

HYPOTHETICAL HISTORY OF ACCRETION OF THE MOON FROM
SIX MOONLETS WITH HIGH INCLINATIONS

Time (yr)		0	70	140	210	280	350	420	490	560
		_	_	_	Distance a, Earth radii					
Moonlet	i									
I	42°	2.86	3.37	3.66						
II	39°		2.86	3.37						
III	36°			2.86	3.37	3.66				
IV	33°				2.86	3.37				
V	30°					2.86	3.37	3.66		
VI	27°						2.86	3.37		
I + II				3.52	3.96	4.26	4.48	4.66	4.81	4.95
III + IV						3.52	3.96	4.26	4.48	4.66
V + VI								3.52	3.96	4.26

TABLE VIII

IDEALIZED HISTORY OF ACCRETION OF THE MOON FROM
SIX MOONLETS ORBITING IN THE EARTH'S EQUATORIAL PLANE

Time (yr)	0	70	140	210	280	350	420	490
			Geocentric distance a, Earth radii					
Moonlet								
I	2.86	3.37						
II		2.86						
III			2.86	3.37				
IV				2.86				
V					2.86	3.37		
VI						2.86		
I + II		3.12	3.76	4.11	4.37			
III + IV				3.12	3.76			
V + VI						3.12	3.76	4.11
I + II + III + IV					4.06	4.54	4.85	5.11
U (km/sec)						0.75	0.48	0.40
σ_0 (Earth radii)						0.72	1.11	1.29

TABLE IX
Synopsis of Origin and Heating

Hypothesis	1	2	3	4	5
Accretion source	Inter-planetary	Inter-planetary	Inter-planetary	Inter-planetary	Fission or ring
First accretion place	Inter-planetary	In Earth-bound orbit	Inter-planetary	Inter-planetary	Inside Roche's limit
Mode of capture	Nontidal	Formed in orbit outside Roche's limit	Tidal	Tidal	Formed in place
Final accretion place	Inter-planetary	In Earth-bound orbit	Earth-bound	Earth-bound	Earth-bound
Distance (Earth radii)	10	10	5	5	5
Inclination	Any	Any	$10°$	$34°$	$0°$
Time scale (yr) Minimum (av)	50,000	50,000	350	350	350
Temperature, T_s (K)	404	850	1680	1300	1260
Final cratering impact velocity (km/sec)	2.5	4.9	3.1	2.4	2.4
Maximum melted fraction θ	0.0016	0.046	0.838	0.355	0.301

TABLE X
Sample Calculations of Thermal Conditions in a Body of Lunar Size[a]

Age (yr)	Central temp. (K)	Average temp. (K)	Radioactive heating, temp. rise (C/10^6 yr)
	A		
0	300	300	3.24
4.5×10^9	5300	3750	0.70
	E		
0	300	300	0.74
4.5×10^9	1400	1100	0.18
	G		
0	1600	1600	0.35
4.5×10^9	1670	1260	0.10

[a] See Allan and Jacobs (1956).

TABLE XI

CUMULATIVE FREQUENCY OF CRATERS IN LUNAR MARIA IN THE TWO HEMISPHERES

Hemisphere	Area (km²)	B > 1.6 km		B > 3.2 km	
		Number	Number per 10⁵ km²	Number	Number per 10⁵ km²
Eastern	1.52×10^6	583	38.3 ± 1.1	273	18.0 ± 0.8
Western	1.29×10^6	555	43.0 ± 1.3	276	21.3 ± 0.9
Ratio E/W			0.89 ± 0.04		0.84 ± 0.06

TABLE XII

DISTRIBUTION OF CRATER CLASSES BY SIZE IN BALDWIN'S LIST

Class	1	2	3	4	5	All	Class 1 (%)
Diameter (miles)				Number			
>40	35	24	20	3	30	112	31
20–40	66	18	7	0	28	120	55
10–20	44	4	0	0	4	52	85
<10	59	0	0	0	0	59	100

TABLE XIII

DISTRIBUTION OF BALDWIN'S MEASURED CRATERS ON MARIA AND CONTINENTES

Diameter (miles)	>40		20–40		10–20		<10	
	Contin.	Maria	Contin.	Maria	Contin.	Maria	Contin.	Maria
Ray Craters (Class 1)								
Number	2	3	3	6	5	2	0	3
Postmare (%)	100	100	100	100	100	100		100
Class 1 (excluding ray craters)								
Number	29	1	47	10	13	24	11	45
Postmare (%)	3	100	22	100	100	100	100	100
Classes 2, 3, 4 (with one ray crater)								
Number	47	0	25	1	1	3	0	0
Postmare (%)	0		0					
Class 5								
Number	28	2	13	15	1	3	0	0

TABLE XIV

CALCULATED CRATER DEPTH TO DIAMETER RATIOS (H/B_0)
FOR LUNAR GRAVITY[a]

x_p (km)	0.25	0.5	1.5	5.0	15.0	25.0	50.0
Model A[b]							
s (10^8 dyne/cm²)	1.08	1.12	1.28	1.84	3.44	5.08	9.04
B_0 (km)	1.78	3.52	10.2	31.2	80.0	121	209
$H/B_0 = 1.8x'/B_0$	0.210	0.177	0.113	0.0490	0.0317	0.0330	0.0349
Model B[c]							
$H/B_0 = 1.8x'/B_0$	0.222	0.196	0.144	0.0826	0.0586	0.0614	
Model C[d]							
s (10^8 dyne/cm²)			9.3	9.8	11.4	13.0	17.0
B_0 (km)			6.75	22.2	64.2	103.5	194
$H/B_0 = 1.8x'/B_0$			0.304	0.196	0.0759	0.0441	0.0231
Model D[e]							
s_c (10^8 dyne/cm²)			3.0	2.0	1.5	1.2	1.0
s (10^8 dyne/cm²)			3.2	2.8	3.9	5.2	9.0
B_0 (km)			8.14	28.1	77.5	120	209
$H/B_0 = 1.8x'/B_0$			0.212	0.0703	0.0236	0.0181	0.0194
Model E[f]							
$H/B_0 = 1.8x'/B_0$			0.228	0.0887	0.0321	0.0264	0.0287
Model F[g]							
s_c (10^8 dyne/cm²)			5.0	3.3	2.5	2.0	1.7
s (10^8 dyne/cm²)			5.2	4.1	4.9	6.0	9.7
B_0 (km)			7.22	25.5	73.1	116	206
$H/B_0 = 1.6x'/B_0$			0.263	0.127	0.0445	0.0288	0.0278
Model G[h]							
s_c (10^8 dyne/cm²)			7.5	5.0	3.8	3.0	2.5
s (10^8 dyne/cm²)			7.7	5.8	6.2	7.0	10.5
B_0 (km)			6.54	23.4	69.0	111	204
$H/B_0 = 1.6x'/B_0$			0.318	0.182	0.0735	0.0421	0.0313

[a] Friction $f_s = 0.78$, average angle of incidence $\gamma = 45°$; premare conditions: $w_0 = 3$ km/sec, $\delta = 1.3$, $\rho = 2.6$ gm/cm³, and $k = 2$.

[b] Parameters: $s_c = 1.04 \times 10^8$ dyne/cm² $=$ const, $\lambda^2 = 0.35$, $s_p = 2 \times 10^8$, $p = 1.093$.

[c] Parameters: B_0, s_c, s, and p have same values as in Model A; $\lambda^2 = 0.50$.

[d] Parameters: $s_c = 9 \times 10^8$, $\lambda^2 = 0.12$, $s_p = 2 \times 10^9$, and $p = 1.014$.

[e] Parameters: s_c decreases with depth, $s_p = 2 \times 10^8$, $p = 1.093$, $\lambda^2 = 0.25$.

[f] Parameters: B_0, s_c, s, and p have same values as in Model D; $\lambda^2 = 0.30$.

[g] Parameters: s_p and p are the same as in Model D; s_c decrease with depth, $\lambda^2 = 0.28$.

[h] Parameters: s_p and p are the same as in Model D; s_c decreases with depth but is still larger for any given depth, $\lambda^2 = 0.28$.

TABLE XV

CALCULATED CRATER DEPTH TO DIAMETER RATIOS (H/B_0)
FOR LUNAR GRAVITY[a]

x_p (km)	0.125	0.25	0.5	1.25	2.5	5.0	12.5	25.0
s (10^8 dyne/cm²)	9.02	9.04	9.08	9.2	9.4	9.8	11.0	13.0

Model P (asteroidal bodies)[b]

D	14.9	14.9	14.9	14.8	14.7	14.6	14.2	13.6
B_0 (km)	1.14	2.28	4.57	11.3	22.6	44.6	109	208
$H/B_0 = 1.6x'/B_0$	0.170	0.169	0.161	0.143	0.120	0.0881	0.0381	0.0182

Model Q (cometary nuclei)[c]

D	26.6	26.6	26.5	26.4	26.3	26.0	25.3
B_0 (km)	2.21	4.43	8.83	22.0	43.8	86.7	211
$H/B_0 = 1.6x'/B_0$	0.0889	0.0881	0.0828	0.0740	0.0604	0.0369	0.0126

[a] Friction $f_s = 0.78$, average angle of incidence $\gamma = 45°$, $\sin \beta_0 = 0.80$; postmare conditions: $s_p = 2 \times 10^9$, $s_c = 9 \times 10^8$ dyne/cm².
[b] Parameters: $\delta = \rho = 2.6$, $w_0 = 20$ km/sec, $k = 3.31$, $p = 1.63$, and $\lambda^2 = 0.22$.
[c] Parameters: $\delta = 2.0$, $\rho = 2.6$, $k = 4.44$, $w_0 = 40$ km/sec, $p = 1.50$, and $\lambda^2 = 0.28$.

TABLE XVI

LOWER LIMIT OF COMPRESSIVE STRENGTH[a] FROM
OROGRAPHIC FEATURES FOR MOON AND EARTH

Moon limb												
Position angle (degrees)	95	103	114	125	129	146	161	169	266	276	285	325
(a) Δh (km)		3.0	2.4		2.6	2.2	2.7		3.4			3.1
(a) s_p ($\geqslant$)		1.3	1.0		1.1	0.9	1.1		1.4			1.3
(b) Δh_e (km)	2.6	2.3	3.9	3.0	2.6	4.1	4.3	3.2	6.9	4.1	3.0	
(b) s_p ($\geqslant$)	0.9	0.8	1.3	1.0	0.9	1.4	1.5	1.1	2.3	1.4	1.0	

	Moon contour map				
Feature	NW of Eratosthenes (Appenines–Mare Imbrium)	N of Copernicus (Mare Imbrium)	SE of Archimedes (Mare Imbrium)	N of Mare Humorum	Sinus Iridum
Length of slope (km)	80	80	50	100	90
Δh_e (km)	6.1	5.2	5.2	6.9	6.1
s_p ($\geqslant$)	2.1	1.7	1.7	2.3	2.1

(*continued*)

TABLE XVI (*continued*)

Feature	Earth					
	Andes 30°S and the Pacific	W Mexican coast 17°N and the Pacific	Karakorum and the Indian plain	Philippines Deep 10°N	Tuscarora Deep off Japan 43°N	Penguin Deep Kermadec Trough 30°S
Δh (km)	8.8	7.7	5.2	7.5	5.7	5.3
s_p ($\geqslant$)	23	20	14	20	15	14

[a] s_p (10^8 dyne/cm^2).

TABLE XVII

CUMULATIVE NUMBER OF CRATERING IMPACTS ON WESTERN MARE IMBRIUM (465,000 km^2)[a]

Crater diameter (km) ($\geqslant$)	1.19	2.48	5.40	12.7	34.3	70.6
Observed number uncorrected	732	207	35(34)	10(9)	3(2)	1(0)
Observed number, corrected for secondaries	560	180	35(34)	10(9)	3(2)	1(0)
Calculated number (4.5×10^9 yr)	1050	202	35	5.0	0.44	0.10

[a] From Öpik (1960).

TABLE XVIII

CUMULATIVE NUMBER OF PRIMARY CRATERS ON LUNAR MARIA
(3.63×10^6 km^2)

Crater diameter (km) ($\geqslant$)	1.61	3.22	6.44	12.9	25.7	51.5	103.0
Observed number	1493	732	289	91	29	7	2
Calculated number (4.5×10^9 yr)	3350	870	190	38	7.3	1.5	0.35

[a] Data from Shoemaker *et al.* (1963) and Baldwin (1964b).

TABLE XIX

CRATER COUNTS ON ALPHONSUS FLOOR AND MARE NUBIUM[a]

| | Alphonsus (strips E–W) | | | | Alphonsus (all) | Predicted postmare primary craters |
	Southern-most	South of middle	North of middle	Northern-most		
Area (km²)	1230	1940	1420	1400	5990	
Number	47	61	82	86	276	
Number (per 10⁴ km²)	382 ± 39	314 ± 28	577 ± 45	614 ± 47	460 ± 20	42

| | Mare Nubium (strips) | | | | |
	Southern (triangular)	Middle	Northern (bounded by reticle marks)	Mare Nubium (all)	
Area (km²)	1300	2470	2700	6510	
Number	9	12	26	47	
Number (per 10⁴ km²)	69 ± 16	49 ± 10	94 ± 13	72 ± 7	42

[a] Effective diameter limit 0.91 km.

TABLE XX

CRATER COUNTS TO LIMITING DIAMETER (0.270 km) IN CENTRAL REGION (45 × 42 km) OF ALPHONSUS[a]

	West (left) of peak	Center	Imbrian ridge	East of peak
	W–E, on line with peak[b]			
Number	28	38	1 (peak)	9	34
	W–E, just north of peak[b]			
Number	21	52	46	18	40
	W–E, near northern edge of frame[b]			
Number	27	43	68	16	25

	Average density[c]
Peak	400 ± 300
Imbrian ridge	5950 ± 660
Other areas	16000 ± 550
Interplanetary postmare impacts on peak, predicted (without erosion)	600t

[a] Data obtained by the author from Fig. 8.
[b] Number of counts per 24.0 km².
[c] Number of craters per 10⁴ km².
t Allowance is made for projected area and shadow of the peak.

TABLE XXI

DENSITY OF FRESH DUNITE POWDER AT LUNAR GRAVITY

Depth (cm)	0.16	1.0	5	10	100	1000	10^4	(∞)
Density (gm/cm³)	0.40	0.43	0.50	0.55	0.71	1.0	1.3	(3.3)

TABLE XXII

EFFECTIVE MEAN THERMAL CHARACTERISTICS OF THE LUNAR SOIL
(THERMAL INFRARED, $10\,\mu$)

Source	τ_t (sec)	γ_t	k_t	L_t (cm)	d_g (cm)
Eclipse[a]	1.8×10^4	1000	3.8×10^{-6}	0.5	0.035
Lunar Cycle	2.5×10^6	750	6.8×10^{-6}	8.0	0.031

[a] From Wesselink (1948).

TABLE XXIII

MEAN TEMPERATURE AND DEPTH, LUNAR EQUATOR

Depth[a] (cm)	0	25	50	100	200	400	700
Temperature (C)	—65	—59	—54	—49	—43	—33	—23

[a] The depth is calculated from the wavelength according to Eq. (146a).

TABLE XXIV

THICKNESS OF CENTRAL OVERLAY, Δz, FROM FALLBACK IN POSTMARE CRATERS

Diameter B_0 (km)	2.2	4.5	22	44	210
Asteroidal impact Δz (m)	8	34	780	2500	23000
Cometary impact Δz (m)	1.8	6	230	840	10800

TABLE XXV

OVERLAY (X_1) FROM LARGE CRATERS (> 2.5 km) IN WESTERN MARE IMBRIUM, AVERAGED OVER S = 465,000 km²

B_0 [crater diameter (km)]	Number		ΔX_1 (cm)	X_1 [cumulative (cm)]	Average separation (km)
	n (interval)	N (cumulative)			
2.48		207		0	47
	97		14.1		
2.92		110		14	65
	36		8.3		
3.35		74		22	79
	21		8.0		
4.10		53		30	94
	19		15.1		
5.40		34		46	117
	9		14.6		
6.70		25		60	136
	7		20.4		
7.99		18		80	161
	3		15.4		
9.72		15		96	176
	6		62.2		
12.7		9		158	227
	0		0		
16.6		9		158	227
	1		50		
21.4		8		209	241
	4		432		
27.0		4		647	340
	2		423		
34.3		2		1064	480
	1		425		
43.0		1		1489	680
	1		888		
55.5		0		2377	
	0		0		
∞				2377	

TABLE XXVI

OBSERVED AND CALCULATED OVERLAY (X_1) FROM LARGE CRATERS (>2.5 km) IN WESTERN
MARE IMBRIUM, AVERAGED OVER $S = 465,000$ km²

Crater diameter (km)	Apollo group	X_1 calc. (cm)			X_1 obs. (cm)
		Comet nuclei	Mars asteroids	All	
2.48–5.40	37	27	1.0	65	46
2.48–21.4	124	140	6.8	271	209
2.48–43.0	184	266	18	468	1489
21.4–43.0	60	126	11	197	1280

TABLE XXVII

DISTRIBUTION OF OVERLAY THICKNESS IN LUNAR MARIA
BASED ON VOLUME AND RANGE OF EJECTA

Thickness (m)	13	14.5	18.5	23	29	39	60	180	330	550	880	1470
Area (%)	61.7	22.6	7.1	1.7	1.4	0.6	1.4	0.3	1.3	0.8	0.5	0.6

TABLE XXVIII

CUMULATIVE CRATER NUMBERS[a] PER 2.22×10^5 km²

$\log B_0$ (m)	1.247	1.636	1.947	2.270	2.512	2.716	2.935	3.071	3.282	3.636
$\log N$ (obs.)	7.332	6.590	6.000	5.414	4.758	4.071	3.402	2.873	2.437	1.798
$\log R_a$ (cm)	1.775	2.162	2.473	2.796	3.038	3.242	3.461	3.597	3.808	4.162
$\log R_c$ (cm)	1.521	1.910	2.221	2.544	2.786	2.990	3.209	3.345	3.556	3.910
$\log(N_a + N_c)$	7.220	6.212	5.483	4.609	4.010	3.512	2.984	2.659	2.164	1.338
N_a/N_c	4.48	2.86	2.01	1.39	1.05	0.83	0.64	0.55	0.43	0.29

[a] N observed, Mare Cognitum according to Shoemaker (1966). [Shoemaker gives his crater densities "per 10^6 km² and 10^9 yr," introducing the hypothetical element of an age of 4.5×10^9 yr and a uniform incidence into his straightforward counts. If the total age is more or less correct, the effective age is shorter on account of erosion, at least for the smaller craters (cf. Table XXIX). His figures are actually counted numbers per $10^6/4.5$ km².] N_a and N_c are calculated values for primaries on hard ground; B_0, lower limit of crater diameter.

TABLE XXIX

CALCULATION OF MARIA OVERLAY $(X)^a$
FROM EJECTA OF CRATERS SMALLER THAN 2.5 KM

B_0 , av. (m)	2190	1500	1010	759	590	462	366	283	214
n_0 obs.	107	472	1770	2920	6400	1.41^4	3.14^4	6.5^4	1.37^5
(per 2.22×10^5 km²)									
E_f	1	1	1	1	1	1	1	1.04	1.33
n_c corrected	107	472	1770	2920	6400	1.41^4	3.14^4	6.8^4	1.82^5
(per 10^6 km², 10^9 yr)									

Part (A)^b

n_p primary (calc.)	69	310	508	806	1480	2520	4470	1.02^4	2.02^4
$n_s = n_c - n_p$, secondary	38	162	1262	2114	4920	1.16^4	2.69^4	5.8^4	1.62^5
ΔX_p (cm per 4.5×10^9 yr)	13	19	10	6	6	5	4	4	4
ΔX_s (cm per 4.5×10^9 yr)	38	51	125	89	97	110	127	128	153
$X = \Sigma \Delta X$	51	121	256	351	454	569	700	832	989
Rubble penetr. H_p primaries (m)	169	116	78	59	46	36	28	22	16.6
Rubble penetr. H_s secondaries (m)	790	540	370	270	210	170	133	102	78

Part (B)^c

Av. age t_f (in units of 4.5×10^9)	0.5	0.5	0.5	0.5	0.5	0.5	0.5	0.48	0.38
Av. overlay X_a (m)	6.0	6.0	6.0	6.0	6.0	6.0	6.0	6.2	7.4
n_p primary (calc.)	70	319	540	907	1750	3280	6250	1.85^4	4.76^4
$n_s = n_c - n_p$, secondary	37	153	1230	2010	4650	1.08^4	2.49^4	4.95^4	1.34^5
η_p (efficiency, primaries)	1.00	1.00	0.99	0.99	0.98	0.97	0.95	0.93	0.87
η_s (efficiency, secondaries)	1.00	1.00	1.00	1.00	1.00	1.00	1.00	1.00	0.99

New overlay

ΔX_p (cm) primaries	13	20	10	7	7	6	6	7	8
ΔX_s (cm) secondaries	37	48	122	85	92	102	117	109	125
$X = \Sigma \Delta X$ (cm)	50	118	250	342	441	549	672	788	921

Total present rate of ejecta^d

ΔX_p (cm)	13	20	10	7	7	6	6	8	9
ΔX_s (cm)	37	48	122	85	92	102	117	109	127
X (cm)	50	118	250	342	441	549	672	789	927

(*continued*)

TABLE XXIX (continued)

B_0 av. (m)	164	129	100	79	62	49	37	28	20
n_0 obs. (per 2.22×10^5 km²)	1.47^5	2.32^5	3.62^5	5.7^5	9.0^5	1.41^6	2.98^6	5.4^6	9.4^6
E_t	1.73	2.22	2.85	3.63	4.60	6.56	7.64	10.3	13.9
n_c corrected (per 10^6 km², 10^9 yr)	2.54^5	5.15^5	1.03^6	2.07^6	4.14^6	9.26^6	2.28^7	5.56^7	1.31^8

Part (A)[b]

n_p primary (calc.)	3.88^4	7.7^4	1.48^5	2.28^5	3.99^5	7.0^5	1.91^6	4.15^6	9.0^6
$n_s = n_c - n_p$ secondary	2.15^5	4.38^5	8.82^5	1.84^6	3.74^6	8.56^6	2.07^7	5.14^7	1.22^8
ΔX_p (cm per 4.5×10^9 yr)	3	3	4	2	2	2	2	2	1
ΔX_s (cm per 4.5×10^9 yr)	91	91	85	87	86	97	101	109	94
$X = \Sigma \Delta X$	1083	1177	1266	1355	1443	1542	1645	1756	1851
Rubble penetr. H_p primaries (m)	12.7	10.0	7.8	6.1	4.8	3.8	2.9	2.2	1.5
Rubble penetr. H_s secondaries (m)	60	47	35	29	22	18	13.4	10.1	7.3

Part (B)[c]

Av. age t_t (in units of 4.5×10^9)	0.29	0.23	0.20	0.14	0.11	0.08	0.07	0.05	0.04
Av. overlay X_a (m)	8.5	9.2	9.6	10.3	10.7	11.0	11.2	11.4	11.5
n_p primary (cal.)	1.49^5	3.77^5	8.14^5	1.25^6	2.19^6	3.85^6	1.05^7	2.28^7	5.0^7
$n_s = n_c - n_p$ secondary	1.05^5	1.38^5	2.16^5	8.2^5	1.95^6	5.41^6	1.23^7	3.28^7	8.1^7
η_p (efficiency primaries)	0.78	0.62	0.48	0.38	0.30	0.24	0.18	0.14	0.09
η_s (efficiency secondaries)	0.99	0.98	0.97	0.96	0.93	0.89	0.80	0.63	0.46

New overlay

ΔX_p (cm) primaries	9	9	7	4	3	2	2	1	1
ΔX_s (cm) secondaries	44	29	20	37	42	55	48	44	29
$X = \Sigma \Delta X$ (cm)	974	1012	1039	1080	1125	1182	1232	1277	1307

Total Present Rate of Ejecta[d]

ΔX_p (cm)	12	15	15	11	10	8	10	9	7
ΔX_s (cm)	45	30	21	39	45	61	60	70	62
X (cm)	984	1029	1065	1115	1170	1239	1309	1388	1457

[a] Density 1.3 gm/cm³.
[b] Upper limit of overlay: all impacts on hard rock.
[c] Assumed: present overlay thickness of 12 m and uniform accretion with time.
[d] Given in cm/4.5×10^9 yr (from overlay plus bedrock).

TABLE XXX

CRATER STATISTICS[a] AND OVERLAY DISTRIBUTION ON THE LUNAR CONTINENTES

Crater diameter (km) (average of group) B_0	Number per 10^5 km² n_i	Crater area coverage		Crater ejecta fringe coverage			Distribution		
		Single (km²) S_0	Fractional	Single (km²) S_r	Fractional τ	Cumul. fractional coverage $\Sigma\tau$	(%)	Overlay (m) X	Averaged ΔX (m)
412	0.106	135,000	0.143	140,000	0.148	0.148	13.8	43,000[b]	5990
206	0.458	33,200	0.152	36,500	0.167	0.315	13.2	20,800[b]	2750
103	1.99	8300	0.165	10,000	0.199	0.514	13.2	9480[b]	1250
51.5	8.67	2070	0.179	2970	0.258	0.772	11.0	4000[b]	440
25.8	37.7	520	0.196	1060	0.400	1.172	17.9	1410	252
12.9	164	130	0.213	411	0.674	1.846	15.0	455	68
6.4	712	32	0.223	211	1.500	3.346	12.4	108	13
3.2	3100	8	0.247	137	4.240	7.586	3.45	20.8	1
1.6		2.0		105			0.05	3.4	0
All							100.0		10,764

[a] From Baldwin (1964b).

[b] These layer thicknesses are so great that compaction under their own weight must have taken place as well as partial deep-layer melting. These layers may have slumped to two-thirds or less of the indicated thickness and have reached a density of 2.0 or more.

TABLE XXXI

Cratering Parameters for Postmare Conditions with Size-Dependent Lateral Strength[a]

x_p (km)	0.125	0.25	0.5	1.25	2.5	5.0	12.5	25.0
s_c	3.9	3.3	2.7	2.2	1.8	1.5	1.2	1.0
s	3.92	3.34	2.78	2.4	2.2	2.3	3.2	5.0
B_0 (km)	1.40	2.93	6.15	16.2	32.4	64.2	148	264
B_0 (Table XV)	1.14	2.28	4.57	11.3	22.6	44.6	109	208
Ratio	1.23	1.28	1.35	1.43	1.43	1.44	1.36	1.27

[a] The values of s, s_c are given in 10^8 dyne/cm². See Table XV.

TABLE XXXII

Mechanical Strength and Cratering in Top Lunar Soil

(a) Static bearing strength experiments with Surveyor III Scoop[a]

Test No.	s_p (10^5 dyne/cm²)	x_0 (cm)	$a^2 = 2$ cm² (assumed)		$a^2 = 1$ cm² (assumed)		$a^2 = 3$ cm² (assumed)	
			S_p (10^4 dyne/cm⁴)	s_p (10^5 dyne/cm²)[b]	S_p (10^4 dyne/cm⁴)	s_p (10^5 dyne/cm²)[c]	S_p (10^4 dyne/cm⁴)	s_p (10^5 dyne/cm²)[d]
1	3.92	2.5	4.75	2.65	5.41	2.68	4.24	2.60
2	2.16	2.5	2.62	2.65	2.98	2.68	2.34	2.60
3	1.76	1.9	3.13	1.80	3.82	1.71	2.66	1.86
4[e]	2.16	0.6	9.15					
5	2.16	2.2	3.15	2.20	3.70	2.16	2.76	2.20
6	0.72	(0.0)[f]		0.64		0.37		0.84
7	2.16	2.9	2.07	3.34	2.30	3.48	1.89	3.21
8	2.32	1.9	4.13	1.80	5.03	1.71	3.51	1.86
Log. mean			3.21		3.70		2.81	
Prob. dev. ratio			1.23	1.23	1.24	1.28	1.22	1.23

(b) Dynamic (impact) experiments[g]

Test	x_0 (cm)	w_0 (cm/sec)	B_0 (cm)	s_p (max) 10^5 dyne/cm²	s/s_a	S_p (dyne/cm⁴)	S_c (dyne/cm⁴)	S_p/S_c
Surveyor I (Footpad 2)	5.8	364	42	7.87	5.14	22100	1190	18.6
Surveyor I (Footpad 3)	5.1	364	37	6.35	3.39	22700	1830	12.4
Surveyor III (Footpad 1 2nd touchdown)	2.5	(53)	32.6	2.01	21.1	24400	930	26.2

(continued)

TABLE XXXII (*continued*)

Test	x_0 (cm)	w_0 (cm/ sec)	B_0 (cm)	s_p (max) 10^5 dyne/ cm^2	s/s_a	S_p (dyne/ cm^4)	S_c (dyne/ cm^4)	S_p/S_c
Surveyor III (Footpad 2 3rd touchdown)	2.5	(60)	26.5	3.30	25.2	40000	2380	15.8
Surveyor III (Footpad 3 3rd touchdown)	5.0			5.18		19200		
Log. mean						24800	1490	17.6
Prob. dev. ratio						1.19	1.39	1.25

[a] From NASA (1967).
[c] Values calculated with $S_p = 3.70 \times 10^4$.
[b] Values calculated with $S_p = 3.21 \times 10^4$.
[d] Values calculated with $S_p = 2.81 \times 10^4$.
[e] Bearing test 4 made on an excavated trench bottom 5–7.5 cm below the surface. Not rised in mean.
[f] Compressed clod.
[g] Parameters: $\sigma = 324$ cm^2 ($m = 16$ gm/cm$^2\pm$, $R = 10.15$ cm, $\rho = 1.3$ gm/cm^3 (assumed), $w_1/w_0 = 0.704\pm$); $F = 0.118$ and $a^2 = 2$ cm^2 (assumed).

TABLE XXXIII

SURVEYOR I, FOOTPAD 2: RECONSTRUCTION OF MOTION DURING PENETRATION

(a) Strain gage data

t (sec)	0.0000	0.0166	0.0705	0.1140
F_d (dyne)	0	1.74×10^8	2.12×10^8	2.55×10^8
s_p (dyne/cm^2)	0	5.37×10^5	6.52×10^5	7.87×10^5
x (cm)	0	4.72	5.24	5.80
Av. velocity $\bar{w}$ (cm/sec)	284	9.6	12.9	

(b) Interpolated velocities[a]

t (10^{-3} sec)	0.5	1.5	2.5	3.5	4.5	5.5	6.5	7.5	8.5	9.5
w (cm/sec)	256	256	268	284	299	319	338	348	352	351
t (10^{-3} sec)	10.5	11.5	12.5	13.5	14.5	15.5	16.0	—70.5	70.5	—114
w (cm/sec)	345	331	304	275	190	108	10		13	

[a] Parameters: $w_0 = 364$ cm/sec[b], $w_1 = 256$ cm/sec, and $dw_1/dt = 6850$ cm/sec^2 ($\rho = 1.3$ gm/cm^3, $K_a = 0.75$).
[b] The NASA team (Jaffe and NASA Team (1966b, p. 69) gives $3.6 \pm$ m/sec.

TABLE XXXIV

AVERAGE MECHANICAL PROPERTIES OF LUNAR SOIL AND NATURAL TERRESTRIAL GRAVEL[a]

	f_s (Coeff. of Friction)	Bearing (compressive) strength			Lateral (crushing) strength			λ^2
		S_p Av.[b]	S_p Probable range[b]	s_{p0} Surface strength[c]	S_c Av.[b]	S_c Probable range[b]	s_0 Surface strength[c]	Kinetic efficiency
			Dynamic			Dynamic		
Lunar soil		2.5 ± 0.2	2.1–3.0	5.0	0.14 ± 0.02	0.10–0.20	0.28	0.08
Terr. gravel		4.9 ± 0.6	3.0–8.0	9.7	0.44 ± 0.05	0.28–0.68	0.87	0.16
			Static			Static		
Lunar soil	(1)	3.2 ± 0.3	2.6–3.9	6.4				
Terr. gravel	0.63	5.6 ± 0.6	3.8–8.1	11.1	0.22 ± 0.03	0.16–0.30	0.44	

[a] To be used in particular with Eqs. (37), (37a), and $a^2 = 2$ cm.

[b] Values for S_p and S_c are given in units of 10^4 dyne/cm^4.

[c] Values for s_{p0} and s_0 are given in units of 10^2 dyne/cm^2.

TABLE XXXV

PERCENTAGE RELATIVE SWEEPING OR STIRRING MOMENTUM

Source	J_M Micro- meteorites	J_0 Dustball meteors	J_1 Meteorites Apollo group	J_2 Comet nuclei	J_3 Mars asteroids	J_e Secondary ejecta	Total
Stirring power (%)	79.5	2.8	2.3	0.2	0.0	15.2	100.0

TABLE XXXVI (A–D)

INCIDENT FLUXES AND CRATERING PARAMETERS IN OVERLAY AT
PRESENT ($\rho = 1.3$); "NORMAL" OVERLAY SURFACE,
REMOTE FROM LARGE CRATERS

TABLE XXXVI A

MICROMETEORS OF THE ZODIACAL CLOUD (J_M)

R (cm)	0.035	0.0205	0.0121	0.0060	0.0010
Cumul. mass fraction	0	0.50	0.75	0.90	0.99
Cumul. number	0	1.88^{-5}	6.54^{-5}	2.45^{-4}	5.22^{-3}
Cumul. crater coverage σ_B	0	4.72^{-4}	8.39^{-4}	1.23^{-3}	1.81^{-3}
B_0 (cm)	11.8	6.7	4.1	2.1	0.3
x_0 (cm)	0.207	0.121	0.072	0.035	0.006
x' (cm)	0.125	0.092	0.063	0.033	0.006
B_0/x' (profile ratio)	95	73	65	63	60
Fraction of granular target G_g	0.44	0.41	0.38	0.35	0.28

TABLE XXXVI B

VISUAL DUSTBALLS (J_0)

R (cm)	0.061	0.109	0.194	0.416	0.743	1.32	2.84	5.08	9.08
Cumul. mass fraction	1.00	0.50	0.25	0.10	0.05	0.025	0.010	0.005	0.0025
Cumul. number	3.72^{-8}	3.28^{-9}	2.86^{-10}	1.18^{-11}	1.04^{-12}	9.12^{-14}	3.75^{-15}	3.30^{-16}	2.89^{-17}
$\bar{s}_c$	2830	2870	2970	3520	5100	10030	36100	135000	344000
B_0 (cm)	27.8	49.5	87.3	180	294	449	716	935	1350
x_0 (cm)	0.177	0.317	0.565	1.20	2.12	3.67	7.60	13.0	22.4
x' (cm)	0.044	0.022	0.01	0	0	0	0	0	0
B_0/x'	630	2500	9000	∞	∞	∞	∞	∞	∞
Fraction of granular target G_g	0.49	0.52	0.57	0.63	0.67	0.72	0.79	0.85	0.92
Cumul. crater coverage σ_B	2.21^{-5}	6.64^{-6}	2.02^{-6}	3.17^{-7}	7.28^{-8}	1.55^{-8}	1.51^{-9}	2.67^{-10}	4.50^{-11}

TABLE XXXVI C
APOLLO GROUP METEORITES (J_1)

R (cm)	200	100	50	20	10	5	2	1	0.5	0.2
Cumul. mass fraction $R < R_0$	1.000	0.813	0.659	0.501	0.408	0.330	0.251	0.205	0.165	0.126
Cumul. number $R < \infty$	5.15^{-20}	3.34^{-19}	2.18^{-18}	2.57^{-17}	1.68^{-16}	1.09^{-15}	1.29^{-14}	8.40^{-14}	5.44^{-13}	6.49^{-12}
$\bar{s}_c$	3.64^{7}	3.12^{7}	2.04^{7}	4.20^{6}	1.16^{6}	3.23^{5}	5.95^{4}	1.81^{4}	6830	3390
B_0 (cm)	13800	7140	3940	2270	1530	1030	612	404	254	117
Cumul. crater coverage	0	2.18^{-11}	6.26^{-11}	2.28^{-10}	6.06^{-10}	1.66^{-9}	6.51^{-9}	1.69^{-8}	4.26^{-8}	1.29^{-7}
σ_B (yr^{-1}) x_0 (cm)	880	443	224	94.4	49.6	26.0	11.0	5.71	2.96	1.21
x' (cm)	830	432	219	87.7	41.5	17.0	1.9	0.2	0.0	0.0
Cumul. crater number	5.15^{-20}	3.34^{-19}	2.18^{-18}	2.57^{-17}	1.65^{-16}	1.01^{-15}	1.08^{-14}	6.42^{-14}	3.82^{-13}	4.07^{-12}
B_0/x'	16.6	16.5	18.0	25.9	36.9	60.7	322	2000	∞	∞
G_σ	1.00	1.00	1.00	1.00	0.96	0.88	0.78	0.72	0.66	0.58

TABLE XXXVI D

EJECTA FROM PENETRATING CRATERING EVENTS (J_e)

(a) Primary Impacts

r (cm)^a	200	100	50	20	10	5	2	1	0.5	0.2
Cumul. mass fraction y	1.000	0.916	0.841	0.750	0.687	0.631	0.562	0.515	0.473	0.422
Cumul. number $r < r_0$	0	1.10^{-15}	9.14^{-15}	1.29^{-13}	9.50^{-13}	6.96^{-12}	9.49^{-11}	7.13^{-10}	5.24^{-9}	7.30^{-8}
$\cos \gamma$	0.600	0.681	0.740	0.800	0.835	0.863	0.893	0.911	0.926	0.941
$w_\bullet$ (cm/sec)	4905	5350	5830	6550	7250	7770	8740	9530	10380	11620
L_{max} (km)	2.88	3.55	4.20	5.10	5.98	6.50	7.59	8.41	9.30	10.6
x_0 (cm)	69.1	58.5	47.9	34.4	25.6	18.2	10.90	7.02	4.36	2.19
k	0.189	0.219	0.256	0.302	0.398	0.472	0.567	0.613	0.646	0.656
V_d/V	0.612	0.611	0.619	0.653	0.712	0.762	0.831	0.873	0.907	0.929
B_c (cm)	1460	655	339	143	80.0	44.6	21.0	12.1	7.00	3.26
x' (cm)	62.7	53.6	43.7	30.6	22.3	15.0	7.89	4.52	2.68	1.11
B_0/x'	23.3	12.2	7.76	4.62	3.59	2.97	2.66	2.68	2.61	2.92
G_g	1.00	1.00	0.92	0.81	0.75	0.69	0.62	0.56	0.52	0.46
Gran. impact cumul. number primary (cm^{-2} yr^{-1})	0	1.10^{-15}	8.82^{-15}	1.13^{-13}	7.54^{-13}	5.08^{-12}	6.27^{-11}	4.28^{-10}	2.88^{-9}	3.66^{-8}

r (cm)^a	0.1	0.05	0.02	0.01	0.005	0.002	0.001	0.0005	0.0002	0.0001	5×10^{-5}	2×10^{-5}
Cumul. mass fraction y	0.386	0.355	0.316	0.290	0.266	0.237	0.217	0.200	0.178	0.163	0.150	0.133
Cumul. number $r < r_0$	5.35^{-7}	3.92^{-6}	5.48^{-5}	4.02^{-4}	2.94^{-3}	4.11^{-2}	0.304	2.21	30.8	228	1660	23100
$\cos \gamma$	0.951	0.959	0.968	0.973	0.977	0.982	0.985	0.987	0.990	0.992	0.993	0.994
w_0 (cm/sec)	12,700	13,800	15,500	16,900	18,500	20,700	22,600	24,600	27,600	30,200	32,700	36,900
L_{max} (km)	11.7	12.8	14.6	16.0	17.5	19.8	21.7	23.6	26.4	28.9	31.6	35.8
G_g	0.42	0.39	0.35	0.32	0.29	0.26	0.24	0.22	0.19	0.18	0.16	0.15

(continued)

TABLE XXXVI D (continued)

(b) Primary Plus Ricocheting Impacts[b]

B_0 (cm)	1460	655	339	143	80.0	44.6	21.0	12.1	7.00	3.26
Cumul. Crater number ($cm^{-2}\text{-}yr^{-1}$)	0	1.23^{-15}	1.09^{-14}	1.48^{-13}	1.02^{-12}	7.10^{-12}	9.04^{-11}	6.32^{-10}	4.34^{-9}	5.65^{-8}
Cumul. Crater coverage (yr^{-1})	0	1.18^{-9}	3.33^{-9}	1.00^{-8}	2.00^{-8}	4.17^{-8}	1.20^{-7}	2.58^{-7}	5.72^{-7}	1.76^{-6}
Aver. x_0 (cm)	61.2	51.7	42.3	30.4	22.6	16.1	9.6	6.2	3.9	1.9
Aver. x' (cm)	52.3	44.7	36.4	25.5	18.6	12.5	6.57	3.77	2.24	0.92
Aver. B_0/x'	27.9	14.6	9.3	5.6	4.3	3.6	3.2	3.2	3.1	3.5

[a] Here r is used for the radius of a secondary fragment, while R would denote the radius of an impacting interplanetary body.

[b] Numbers reduced to same crater diameter limit B_0 as in the 10th row of part (a). As compared to the numbers of primary granular impacts [14th row of Part (a)], they are increased by ricocheting of larger projectiles than those of the tabular r-limit.

TABLE XXXVII[a]

Expected Number of Punctures on the Moon from Meteorite Hazard

Magnesium sheet thickness (mm)	0.4	1.2	1.4	4.2	14	42
Steel sheet thickness (mm)	0.14	0.42	0.5	1.5	5	15
Number of punctures per 100 m² and 10 yr	2000	200	1.5	0.015	1.5×10^{-4}	1.5×10^{-6}

[a] From Öpik (1961a).

TABLE XXXVIII

SAMPLE CALCULATED RICOCHETING CRATER CHAINS, COMPONENT J_e

(1) r = 0.5 cm; G_g = 0.52; intermittent granular and hard target (G_g = 0.5) calculated; granular start

No. of impact	w_0 (cm/sec)	x_0 (cm)	B_0 (cm)	k	x' (cm)	B_0/x'	L (cm)	Area (0.785 cm²)	Cratering volume (0.363 cm³) total	ejected	Radial momentum (cm/sec-gm)
S1 Gran.	10,380	4.36	7.00	0.646	2.68	2.61	4.20^4	49.0	214	130	6710
S2 Hard	3110						1.05^4				
S3 Gran.	1555	2.24	4.28	0.481	1.13	3.79	944	18.3	41	21	750
S4 Hard	466						236				
S5 Gran.	233	0.46	2.73	0.201	0.36	7.7	21.2	7.5	3	3	50
S6 Hard	70						5.3				
S7 Gran.	35	0.014	1.13	0.108	0.014	83	0.5	1.3	0	0	0
S8 Stop								0	0	0	0
Chain total							5.35^4	76.1	258	154	7510
Amplification ratio							1.27	1.55	1.17	1.16	1.12

(2) r = 0.5 cm; G_g = 0.52; intermittent granular and hard target (G_g = 0.5) calculated; hard start

No. of impact	w_0 (cm/sec)	x_0 (cm)	B_0 (cm)	k	x' (cm)	B_0/x'	L (cm)	Area (0.785 cm²)	Cratering volume (0.363 cm³) total	ejected	Radial momentum (cm/sec-gm)
S1 Hard	10,380						1.16^5				
S2 Gran.	5190	3.55	5.92	0.629	1.87	3.16	1.05^4	35.0	124	65	3260

(continued)

TABLE XXXVIII (continued)

(2) (continued)

S3 Hard	1555						2620				
S4 Gran.	778	1.53	3.60	0.376	0.82	4.39	236	13.0	20	11	290
S5 Hard	233						59				
S6 Gran.	117	0.145	2.14	0.135	0.127	16.7	5.3	4.6	1	1	20
S7 Hard	35						1.3				
S8 Gran.	17.5	0.004	(0.77)	0.106	0.004	192	0.1	(0.6)	0	0	0
S9 Stop								0	0	0	0
Chain total							1.29^{5}	53.2	145	77	3570
Amplification ratio							3.07	1.09	0.68	0.59	0.53

(3) $r = 0.5$ cm; $G_g = 0.52$;   all granular target ($G_g = 1$) calculated

1	10,380	4.36	7.00	0.646	2.68	2.61	4.20^{4}	49.0	214	130	6710
2	3110	2.98	5.12	0.567	1.54	3.32	3780	26.2	78	40	1760
3	933	1.71	3.76	0.403	0.90	4.18	340	14.1	24	13	370
4	280	0.58	2.90	0.232	0.41	7.0	30.6	8.4	5	3	70
5	84	0.075	1.86	0.127	0.068	17.4	2.7	3.5	0	0	10
6	25	0.0070	(0.93)	0.107	0.0068	137	0.2	(0.9)	0	00	0
7 Stop							000	000	0	0	0
Chain total							4.61^{4}	102.1	321	186	8920
Amplification ratio							1.10	2.08	1.50	1.43	1.33

TABLE XXXVIII (*continued*)

(4) $r = 50$ cm; $G_g = 0.92$; all granular target ($G_g = 1$) calculated

1	5830	47.9	0.256	43.7	7.8	1.88^4	1.15^5	5.50^6	5.03^6	1490
2	1749	23.3	0.192	18.6	15.7	1690	0.86^5	2.01^6	1.60^6	340
3	525	11.3	0.153	5.7	42	152	0.59^5	0.67^6	0.34^6	80
4	158	5.0	0.129	0.54	354	14	0.36^5	0.18^6	0.02^6	20
5 Stop	(47)						0	0	0	0
Chain total						2.07^4	2.96^5	8.36^6	6.99^6	1930
Amplification ratio						1.10	2.57	1.52	1.39	1.30

(5) $r = 200$ cm; $G_g = 1.00$; all granular target ($G_g = 1$) calculated

1	4905	115.2	0.189	62.7	23	1.29^4	2.13^6	2.46^8	1.34^8	930
2	1472	56.3	0.149	26.0	40	1160	1.10^6	0.62^8	0.29^8	220
3	442	15.0	0.124	4.7	179	105	0.71^6	0.11^8	0.03^8	50
4 Stop	(133)						0	0	0	0
Chain total						1.41^4	3.94^6	3.19^8	1.66^8	1200
Amplification ratio						1.09	1.85	1.30	1.24	1.29

TABLE XXXIX

COMPARISON OF RICOCHETING AMPLIFICATION RATIOS[a]

	Crater area	Volume excavated	Volume ejected	Radial momentum	All (av.)
Average[b]	1.32	0.925	0.875	0.825	
Case (3)[c], $A_0 =$	2.08	1.50	1.43	1.33	
Correction factor to case (3)	0.635	0.616	0.612	0.620	0.621

[a] $r = 0.5$ cm.
[b] [Case (1) + Case (2)]/2; ($G_g = 0.50$). See text.
[c] For all soft target, $G_g = 1.00$. See text.

TABLE XL

CUMULATIVE AREA COVERAGE[a] BY LARGE CRATERS[b]

B_0 (m)	σ_{cB}, primaries	σ_{cB}, secondaries ($x_0 = 0.35B_0$)	σ_{cB}, total	B_0 (m)	σ_{cB}, primaries	σ_{cB}, secondaries ($x_0 = 0.35B_0$)	σ_{cB}, total
325	0	0	0	55.0	1.09^{-11}	2.34^{-11}	3.43^{-11}
246	1.16^{-12}	3.11^{-12}	4.27^{-12}	43.3	1.09^{-11}	3.36^{-11}	4.45^{-11}
186	2.87^{-12}	7.92^{-12}	1.08^{-11}	32.1	1.09^{-11}	4.68^{-11}	5.77^{-11}
145	6.01^{-12}	1.01^{-11}	1.61^{-11}	23.8	1.09^{-11}	6.70^{-11}	7.79^{-11}
113.5	1.09^{-11}	1.19^{-11}	2.28^{-11}	17.7	1.09^{-11}	9.25^{-11}	1.03^{-10}
88.5	1.09^{-11}	1.36^{-11}	2.45^{-11}	<17.7	1.09^{-11}	9.25^{-11}	1.03^{-10}
69.7	1.09^{-11}	1.76^{-11}	2.85^{-11}				

[a] σ_{cB} per year.
[b] For component J_c (supplementary to Table XXXVI), which contains component J_1 only down to $B_0 > 139$ m, the rest of J_1 being represented by Table XXXVI, Part C.

TABLE XLI

BALANCE OF CRATER CREATION BY IMPACT AND DELETION BY
OVERLAPPING AND EROSION TO PROFILE RATIO $B_0/x' > 50$

(a) Component J_e

B_0 (cm)	B_0/x'	ν	τ_0	F_1	n_0	N_0	t'	n'	N'	t_e	n_e	N_e
1460	27.9	1.92^{-10}	5.2^9	1.23^{-15}	2.4	0	1.55^8	0.19	0	3.54^7	0.044	0
655	14.6	1.02^{-9}	9.8^8	9.70^{-15}	4.9	2.4	7.45^7	0.67	0.19	2.22^7	0.22	0.044
339	9.3	3.76^{-9}	2.66^8	1.37^{-13}	16.6	7.3	3.48^7	4.2	0.86	1.05^7	1.38	0.26
143	5.6	1.81^{-8}	5.5^7	8.77^{-13}	28	23.9	1.89^7	12.5	5.1	4.62^6	3.8	1.64
80.0	4.3	5.52^{-8}	1.81^7	6.08^{-12}	62	52	9.45^6	38	17.6	2.23^6	12.4	5.4
44.6	3.6	1.74^{-7}	5.75^6	8.33^{-11}	230	114	4.85^6	190	56	9.07^5	64	17.8
21.0	3.2	7.55^{-7}	1.32^6	5.42^{-10}	480	344	2.52^6	450	246	3.94^5	172	82
12.1	3.2	1.71^{-6}	5.4^5	3.71^{-9}	690	820	1.45^6	690	700	2.03^5	460	254
7.00	3.1	1.70^{-5}	5.9^4	5.22^{-8}	1200	1510	7.55^5	1200	1390	8.70^4	1180	710
3.26	3.5	1.11^{-4}	9.0^3			2700			2590			1890

(b) Component J_1

B_0 (cm)	B_0/x'	ν	τ_0	F_1	n_0	N_0	t'	n'	N'	t_e	n_e	N_e
13800	16.6	2.7^{-12}	3.7^{11}	2.82^{-19}	1.27^{-3}	0	1.57^9	4.4^{-4}	0	9.37^8	2.6^{-4}	0
7140	16.5	1.61^{-11}	6.2^{10}	1.85^{-18}	7.9^{-3}	1.27^{-3}	8.38^8	1.55^{-3}	4.4^{-4}	4.11^8	7.6^{-4}	2.6^{-4}
3940	18.0	4.47^{-11}	2.24^{10}	2.35^{-17}	0.092	9.2^{-3}	4.73^8	1.11^{-2}	1.99^{-3}	1.52^8	3.6^{-3}	1.02^{-3}
2270	25.9	9.1^{-11}	1.10^{10}	1.36^{-16}	0.44	0.101	2.95^8	0.040	1.31^{-2}	4.43^7	6.0^{-3}	4.6^{-3}
1530	36.9	1.77^{-10}	5.6^9	7.58^{-16}	2.0	0.54	1.98^8	0.148	0.053	1.8^7	0.014	1.06^{-2}
1030	60.7	3.89^{-10}	2.57^9			2.54			0.201			0.025

(continued)

TABLE XLI (*continued*)

(c) Component S

B_0 (cm)	ν	τ_0	F_0	n_0	N_0	t'	n'	N'	F_i	t_e	n_e	N_e
14500	1.38^{-12}	7.3^{11}			0			0				0
11350	4.01^{-12}	2.5^{11}	1.38^{-20}	6.2^{-5}	6.2^{-5}	2.03^{9}	3.2^{-5}	3.2^{-5}	1.23^{-20}	4.5^{9}	5.5^{-5}	5.5^{-5}
8850	1.24^{-11}	8.1^{10}	2.16^{-20}	9.7^{-5}	1.6^{-4}	1.58^{9}	3.4^{-5}	6.6^{-5}	2.04^{-20}	4.21^{9}	8.6^{-5}	1.41^{-4}
6970	1.79^{-11}	5.7^{10}	8.2^{-20}	3.7^{-4}	5.3^{-4}	1.24^{9}	1.03^{-4}	1.69^{-4}	5.02^{-20}	3.32^{9}	1.67^{-4}	3.08^{-4}
5500	3.03^{-11}	3.3^{10}	1.95^{-19}	8.5^{-4}	1.38^{-3}	9.78^{8}	1.88^{-4}	3.57^{-4}	2.96^{-19}	9.88^{8}	2.92^{-4}	6.0^{-4}
4330	4.18^{-11}	2.4^{10}	5.41^{-19}	2.2^{-3}	3.6^{-3}	6.85^{8}	3.7^{-4}	7.3^{-4}	6.97^{-19}	7.09^{8}	4.9^{-4}	1.09^{-3}
3210	5.66^{-11}	1.77^{10}	1.23^{-18}	4.8^{-3}	8.4^{-3}	5.88^{8}	7.2^{-4}	1.45^{-3}	2.36^{-18}	4.62^{8}	1.09^{-3}	2.18^{-3}
2380	8.57^{-11}	1.17^{10}	3.28^{-18}	1.28^{-2}	2.12^{-2}	4.36^{8}	1.46^{-3}	2.91^{-3}	8.22^{-18}	2.58^{8}	2.12^{-3}	4.3^{-3}
1770	2.31^{-10}	4.3^{9}	8.1^{-18}	0.026	0.047	3.24^{8}	2.6^{-3}	4.5^{-3}	3.70^{-17}	1.08^{8}	4.00^{-3}	8.3^{-3}

[a] Outside the ejecta from craters larger than 590 m.

TABLE XLII

COMPARISON OF PREDICTED (N', N_e) AND OBSERVED (N) CUMULATIVE AREA DENSITIES OF SMALL CRATERS (B_0, CRATER DIAMETER, CM) HAVING A PROFILE RATIO $B_0/x' < 50$[a]

B_0 (cm)	N' predicted 1st approxim. component				N_e predicted 2nd approxim. component				N observed		
	S	J_1	J_e	N' all	S	J_1	J_e	N_e all	(Shoemaker) (1966)	(Trask) (1966)	(Jaffe) (1966)
7140	1.6^{-4}	4.4^{-4}		6.0^{-4}	2.9^{-4}	2.6^{-4}		5.5^{-4}	6.7^{-4}	10.6^{-4}	
3940	9.1^{-4}	19.9^{-4}		2.9^{-3}	1.50^{-3}	1.02^{-3}		2.5^{-3}	2.1^{-3}	3.5^{-3}	
2270	3.3^{-3}	13.1^{-3}		1.64^{-2}	5.1^{-3}	4.6^{-3}		9.7^{-3}	6.0^{-3}	10.4^{-3}	
1530	5.0^{-3}	5.3^{-2}		5.8^{-2}	8.0^{-3}	1.06^{-2}		1.86^{-2}	1.28^{-2}	2.30^{-2}	
1030	5.0^{-3}	0.201	0.090	0.296	8.0^{-3}	0.025	0.020	0.053	0.027	0.050	
655	5.0^{-3}	0.201	0.190	0.396	8.0^{-3}	0.025	0.044	0.077	0.067	0.126	
339	5.0^{-3}	0.201	0.86	1.07	8.0^{-3}	0.025	0.22	0.25	0.22	0.47	2.0
143	5.0^{-3}	0.201	5.1	5.3	8.0^{-3}	0.025	1.64	1.67	1.13	2.64	9.5
80.0	5.0^{-3}	0.201	17.6	17.8	8.0^{-3}	0.025	5.4	5.4			27.0
44.6	5.0^{-3}	0.201	56	56	8.0^{-3}	0.025	17.8	17.8			77
21.0	5.0^{-3}	0.201	246	246	8.0^{-3}	0.025	82	82			300
12.1	5.0^{-3}	0.201	700	700	8.0^{-3}	0.025	254	254			810
7.00	5.0^{-3}	0.201	1390	1390	8.0^{-3}	0.025	710	710			2170
3.26	5.0^{-3}	0.201	2590	2590	8.0^{-3}	0.025	1890	1890			8590[b]

[a] Thus sufficiently deep to be observed.

[b] The counted value at $B_0 = 3.26$ cm was 3850 (Jaffe and NASA Team, 1966b), but, somewhat arbitrarily, allowance was made in this case for the incompleteness in the counts of these small craters and the number was increased accordingly.

TABLE XLIII
Cumulative Mixing Factor (Q_t) and Eventual Mixing Probability (q_m) of Overlay at Present Depth $h_0{}^a$

$x_0{}^b$	224	94.4	69.1	58.5	47.9	34.4	25.6	18.2	10.9
$h_0{}^a$	104	43.6	32.0	27.0	22.2	15.9	11.8	8.40	5.04
$\sigma_B{}^c$	6.26^{-11}	2.28^{-10}	3.64^{-10}	1.65^{-9}	3.96^{-9}	1.10^{-8}	2.16^{-8}	4.44^{-8}	1.26^{-7}
$t_m{}^d$	1.60^{10}	4.39^{9}	2.75^{9}	6.06^{8}	2.53^{8}	9.09^{7}	4.63^{7}	2.25^{7}	7.94^{6}
$t_0{}^e$	3.90^{8}	1.63^{8}	1.20^{8}	1.01^{8}	8.32^{7}	5.96^{7}	4.32^{7}	3.15^{7}	1.89^{7}
Q_t	0.052	0.079	0.091	0.107	0.112	0.128	0.152	0.55	1.49
q_m	0.051	0.076	0.087	0.101	0.106	0.120	0.141	0.42	0.77

x_0		7.02	4.36	2.19	1.08	0.54	0.27	0.135	0.0675
h_0		3.25	2.02	1.01	0.50	0.250	0.125	0.0625	0.0312
σ_B		2.69^{-7}	5.98^{-7}	1.87^{-6}	6.26^{-6}	2.05^{-5}	6.64^{-5}	6.06^{-4}	1.45^{-3}
t_m		3.72^{6}	1.67^{6}	5.35^{5}	1.60^{5}	48,800	15,100	1650	690
t_0		1.22^{7}	7.58^{6}	3.78^{6}	1.88^{6}	9.4^{5}	4.7^{5}	2.35^{5}	1.17^{5}
Q_t		2.72	4.54	8.5	Large →				
q_m		0.93	0.99	1.00	1	1	1	1	1

^{a} Values of h_0 are given in centimeters.
^{b} Values of x_0, the crater penetration depth, are given in centimeters.
^{c} σ_B is the relative crater creation area per year.
^{d} t_m is the time scale of mixing in years.
^{e} t_0 is the accretion age in years.

TABLE XLIV
Time Scale (τ_t) of Craters or Granular Mounds^a of $H < \frac{1}{2}L$ with Respect to Stirring of Overlay by Small Projectiles

$\lambda^2 = 0.09$; $\alpha' = 0.933$; $B_0 = 2.14\zeta$; $\tau_t = \frac{1}{4}B_0/\bar\psi_L$

ζ	1	2	3	4	5	10
$\bar\psi_L/\chi$	0.523	0.379	0.315	0.242	0.202	0.110
$\bar\psi_L$ (10^{-5} cm/yr)	3.73	2.70	2.25	1.73	1.44	0.785
B_0 (cm)	2.14	4.28	6.42	8.56	10.7	21.4
$d\log\tau_t/d\log B_0$	1.47	1.54	1.71	1.77	1.87	1.94
τ_t (yr)	14300	39600	71300	12400	186000	682000

ζ	20	50	100	200	500	1000^b
$\bar\psi_L/\chi$	0.057	0.0228	0.0114	0.0057	0.00228	0.00114
$\bar\psi_L$ (10^{-5} cm/yr)	0.407	0.163	0.081	0.0407	0.0163	0.0081
B_0 (cm)	42.8	107	214	428	1070	2140
$d\log\tau_t/d\log B_0$	2.00	2.00	2.00	2.00	2.00	
τ_t (yr)	2.62^{6}	1.64^{7}	6.60^{7}	2.62^{8}	1.64^{9}	6.60^{9}

^{a} Crater or mound diameter $B_0 = 2L$. $\lambda^2 = 0.09$, $\alpha' = 0.933$, $B_0 = 2.14\zeta$, $\tau_t = \frac{1}{4}B_0/\bar\psi_L$.
^{b} For larger value, $\tau_t \sim B_0{}^2$.

TABLE XLV
Drift Relaxation Time τ_F of Craters[a]

B_0 (cm)	2.14	21.4	214	2140	4280	2.14^4	4.28^4 [b]
τ_F (yr)	64900	2.92^5	6.41^6	1.84^8	5.13^8	5.62^9	1.59^{10}
$d \log \tau_F / d \log B_0$		0.65	1.34	1.46	1.47	1.49	1.50
τ_m (yr)	25300	8.48^5	1.91^7	2.19^8	4.42^8	2.23^9	4.46^9

[a] Valid either when $\tau_F > \tau_m$, or when crater diameter $B_0 < 39$ meters (produced by "ray secondaries") or $B_0 < 160$ m (produced by interplanetary primaries). The conditions of validity are always fulfilled.

[b] For larger values, $\tau_F \sim B_0^{1.5}$ and $\tau_m \sim B_0$.

TABLE XLVI
Median Particle Radius r_m, Impact Velocity w_0 and Impact Angle cos γ of Overlay Accretion, Filling Crater of Diameter B_0

B_0 (cm)	r_m (cm)	$B_0/2r_m$	w_0 (cm/sec)	cos γ
$\geqslant 655$	2.4	>135	$\leqslant 8400$	0.890
80	0.37	216	11000	0.930
21	0.1	210	12700	0.951
3.26	0.017	191	16000	0.969

TABLE XLVII
Trapping Gain Factor for Infinitely Deep Circular Hole of Average Chord B, or Diameter $B_0 = \frac{1}{2}\pi B$

B (cm)	2	10	50	200	1000	5000	10^4	2×10^4	4×10^4	8×10^4	1.6×10^5
G_∞	7.50	6.63	5.83	5.23	4.28	3.41	3.21	2.63	2.06	1.92	1.57

TABLE XLVIII
Cumulative Deficit or Negative Gain Factor ω_e[a]

ξ [b]	0	0.121	0.168	0.241	0.292	0.330	0.47	0.54	0.61
ω_e [a]	0	0.257	0.346	0.474	0.558	0.608	0.762	0.834	0.891

ξ [b]	0.70	0.91	1.01	1.74	2.08	2.75	4.66	5.85	11.1
ω_e [a]	0.932	0.954	0.959	0.979	0.985	0.992	0.997	0.999	1.000

[a] ω_e (normalized to unity) is taken around a depression with $x'/B = \frac{1}{4}$.

[b] $\xi = \Delta L/B$ is the relative distance from the edge, so that the corresponding distance from the center of the depression is $B(\frac{1}{2} + \xi)$.

TABLE XLIX

RELATIVE ACCRETION χ/χ_0 OUTSIDE A CRATER AT
DISTANCE $\Delta L = \xi B_0$ FROM THE RIM; INSIDE THE AVERAGE VALUE IS G_t [a]

ξ	0.0	0.05	0.10	0.15	0.20	0.3	0.4	0.5	0.6	0.7	0.8	1.0
					$x'/B = 0.25$;	$G_t = 1.529$						
χ/χ_0	0.690	0.730	0.769	0.805	0.835	0.882	0.912	0.938	0.962	0.983	0.995	0.998
					$x'/B = 0.20$;	$G_t = 1.423$						
χ/χ_0	0.690	0.735	0.779	0.818	0.850	0.896	0.931	0.964	0.989	0.996	0.998	
					$x'/B = 0.15$;	$G_t = 1.317$						
χ/χ_0	0.690	0.744	0.794	0.838	0.871	0.922	0.969	0.987	0.998			
					$x'/B = 0.10$;	$G_t = 1.212$						
χ/χ_0	0.690	0.758	0.825	0.872	0.911	0.984	0.997					
					$x'/B = 0.05$;	$G_t = 1.106$						
χ/χ_0	0.690	0.809	0.897	0.981	0.997							

[a] The depth to diameter ratio of the circular depression is x'/B.

TABLE L

EVOLUTION OF CRATER PROFILE BY FILLING[a]

			ξ									h_0/B
			0.1	0.2	0.3	0.4	0.5	0.6	0.7	0.8	1.0	
Stage	x_1'/B	H/B					$10^5(h - h_0)/B$					
I	0.25	0	0	0	0	0	0	0	0	0	0	0
II	0.20	0.0212	178	322	422	492	553	606	635	650	657	0.0146
III	0.15	0.0457	414	739	959	1129	1254	1351	1390	1410	1417	0.0315
IV	0.10	0.0747	762	1322	1722	1979	2136	2237	2289	2309	2316	0.0515
V	0.05	0.1103	1370	2162	2662	3035	3233	3340	3392	3412	3419	0.0761
VI	0.00	0.1562	2537	3576	4085	4458	4656	4768	4815	4835	4842	0.1078

[a] For $\bar{B} = (2/\pi)B_0$.

TABLE LI
FILLING TIME SCALE τ_e, TOTAL DEGRADATION TIME SCALE τ_E, AND DEGRADATION LIFETIME t_e FOR CRATERS IN OVERLAY WITHOUT HARD RIM

(a) Component J_e with Ricochets

B_0 (cm)	3.26	7.00	12.1	21.0	44.6	80.0	143	339	655	1460
x_1'/B_0 (av.)	0.286	0.323	0.313	0.313	0.278	0.232	0.179	0.108	0.068	0.036
$y_2 - 0.15$	0.32	0.38	0.42	0.48	0.55	0.61	0.67	0.77	0.84	0.85
τ_e (yr)	1.44^6	2.60^6	4.06^6	6.17^6	1.14^7	1.85^7	3.01^7	6.21^7	1.10^8	2.42^8
τ_F (yr)	8.70^4	1.41^5	2.01^5	2.92^5	7.77^5	1.70^6	3.73^6	1.26^7	3.29^7	1.05^8
τ_t (yr)	2.66^4	8.25^4	2.34^5	6.57^5	2.84^6	9.17^6	2.92^7	1.66^8	6.20^8	3.08^9
τ_E (yr)	2.01^4	5.10^4	1.05^5	1.96^5	5.78^5	1.33^6	2.99^6	9.86^6	2.43^7	7.15^7
t_e (yr)	5.35^4	1.42^5	2.89^5	5.39^5	1.53^6	3.26^6	6.55^6	1.67^7	2.96^7	4.22^7

(b) Component J_1 (Apollo Meteorites)[a]

B_0 (cm)	1530	2270	3940	7140	13,800
x_1'/B_0 (av.)	0.027	0.039	0.056	0.060	0.060
τ_e (yr)	2.54^8	3.77^8	6.54^8	1.19^9	2.29^9
τ_F (yr)	1.13^8	1.97^8	4.54^8	1.10^9	2.92^9
τ_t (yr)	3.38^9	7.43^9	2.23^{10}	7.36^{10}	2.75^{11}
τ_E (yr)	7.63^7	1.28^8	2.62^8	5.67^8	1.28^9
t_e (yr)	2.29^7	8.59^7	2.70^8	6.24^8	1.41^9
t' (yr)	2.42^8	3.59^8	6.23^8	1.13^9	2.18^9

[a] $y_2 - 0.15 = 0.85$.

TABLE LII
EROSION AND FILLING DEGRADATION LIFETIMES t_t OF CRATERS WITH A HARD RIM[a]

B_0 (cm)	4.62^4	3.66^4	2.83^4	2.14^4	1.64^4	1.29^4	1.00^4
τ_e	7.67^9	6.07^9	4.70^9	3.56^9	2.72^9	2.14^9	1.66^9
τ_F	1.78^{10}	1.26^{10}	8.53^9	5.61^9	3.76^9	2.62^9	1.79^9
τ_E	5.40^9	4.10^9	3.03^9	2.18^9	1.58^9	1.18^9	8.62^8

A. Typical primary craters (asteroidal)[b]

t_e	7.56^9	5.74^9	4.24^9	3.05^9	2.21^9	1.69^9	1.51^9
t_I	3.08^9	2.44^9	1.89^9	1.43^9	1.09^9	0.86^9	0.68^9
t_t	10.64^9	8.18^9	6.13^9	4.48^9	3.30^9	2.55^9	2.19^9

B. Typical secondary craters (ray crater ejecta)[c]

t_e	12.7^9	9.63^9	7.12^9	5.13^9	3.72^9	2.78^9	2.54^9
t_I	7.7^9	6.11^9	4.73^9	3.58^9	2.73^9	2.15^9	1.67^9
t_t	20.4^9	15.74^9	11.85^9	8.71^9	6.45^9	4.93^9	4.21^9

[a] Erosion relaxation times are independent of profile $\tau_e = 1.66 \times 10^5 B_0$ for filling by overlay; $\tau_F = 1790 B^{1.5}$ yr for downhill transport; $\tau_t = 1440 B_0^2$ yr for spray over the rim (too long a time to be considered) $\tau_E = (1/\tau_0 + 1/\tau_F)^{-1}$ the total relaxation time.
[b] For $x' = 0.100 B_0$, $h_r = 0.020 B_0$, $t_I = 6.67 \times 10^4 B_0$, $x_1' = 0.081 B_0$, and $t_e = 1.40 \tau_E$.
[c] For $x' = 0.25 B_0$, $h_r = 0.05 B_0$, $t_I = 1.67 \times 10^5 B_0$, $x_1' = 0.201 B_0$, $t_e = 2.35 \tau_E$.

TABLE LIII
RECALCULATION OF OVERLAY ACCRETION, CASE B OF TABLE XXIX[a]

B_0 (m)	$\geqslant 214$	164	129	100	79	62	49	37	28	20
n_p (primaries)[b]		1.49^5	3.77^5	8.14^5	1.25^6	2.19^6	3.85^6	1.05^7	2.28^7	5.00^7
t_e (primaries)[c]		1.68^9	1.32^9	9.46^8	7.11^8	5.11^8	3.67^8	2.39^8	1.35^8	5.61^7
E_f (primaries)[d]		2.68	3.41	4.77	6.33	8.81	12.3	18.8	33.4	80.8
$n_\mathrm{p}/E_\mathrm{f}$ (primaries)[e]		5.6^4	1.09^5	1.71^5	1.98^5	2.48^5	3.13^5	5.59^5	6.83^5	6.19^5
n_0 (total observed)		1.47^5	2.32^5	3.62^5	5.70^5	9.00^5	1.41^6	2.98^6	5.40^6	9.40^6
$n_0 - n_\mathrm{p}/E_\mathrm{f}$ (secondaries)[f]		9.1^4	1.23^5	1.91^5	3.72^5	6.52^5	1.10^6	2.41^6	4.72^6	8.88^6
t_e (secondaries)[g]		$>4.5^9$	$>4.5^9$	4.21^9	3.32^9	9.88^8	7.09^8	4.62^8	2.58^8	1.08^8
E_f (secondaries)		1	1	1.07	1.35	4.55	6.34	9.75	17.4	41.6
n_s (secondaries)[h]		9.1^4	1.23^5	2.04^5	5.02^5	2.96^6	6.97^6	2.36^7	8.22^7	3.70^8
					Accretion					
ΔX_p (unchanged, cm)		9	9	7	4	3	2	2	1	1
ΔX_s (cm)		38	26	19	23	64	71	92	110	133
$X = \sum \Delta X$ (cm)	921	968	1003	1029	1056	1123	1196	1290	1401	1535

[a] Calculated assuming the present overlay thickness to be 12 m and a uniform rate of accretion (data which are not quoted are identical with those of Table XXIX) (the crater numbers are for an area of 2.22×10^5 km²).

[b] True number, calculated, for 4.5×10^9 yr.

[c] All "soft," in years.

[d] Depletion ratio.

[e] Predicted observable number of primaries.

[f] Concluded observable number of secondaries.

[g] For hard-rimmed craters when $B_0 \geqslant 79$ m; soft rimmed when $B_0 \leqslant 62$ m, 1.93 times the value for primaries.

[h] True number.

Table XXXVI is given in four parts (A–D). The explanatory material (including the parameter assumed in each part) is presented here. The tabulated data (for each part) then follow.

Part A. Micrometeors of the Zodiacal Cloud (J_M)

The projectile J_M is explosively destroyed. Parameters: $w_0 = 6.0$ km/sec; $\delta = 3.5$ gm/cm³; upper limit radius $R_0 = 0.035$ cm; $n = 2.7$; cumulative number flux, $R < 0.035$ cm, $dN/dt = 1.27 \times 10^{-5}[(R_0/R)^{1.7} - 1]$ cm⁻² horizontal surface and yr⁻¹; cumulative mass flux, $d\mu/dt = 1.05 \times 10^{-8}[1 - (R/R_0)^{1.3}]$

gm/cm²-yr. Cratering parameters: penetration negligible; $\bar{s}_p = 5.0 \times 10^4$, $\bar{s}_c = 2.82 \times 10^3$ dyne/cm²; $k = 2.0$, $\gamma = 45°$, $p = 3.52$, $D = 169$, $D/p = 48.0$, $B_0 = 338R$, $x_p = 7.04R$, $x_0 = 5.92R < 0.2$ cm, $x' = x_0(1 - F_B)$; cumulative crater coverage (cm²/cm²-yr) $\sigma_B = 0.00231[1 - (R/R_0)^{0.425}]$ per year.

Part B. Visual Dustballs (J_0)

The projectile J_0 is explosively destroyed. Parameters: $w_0 = 18$ km/sec; $\delta = 0.65$ gm/cm³; lower limit radius $R_0 = 0.061$ cm; $n = 5.2$; cumulative number flux, $R > 0.061$ cm, $dN/dt = 2.93 \times 10^{-13}R^{-4.2}$ cm⁻² and yr⁻¹, cumulative mass flux, $R > R_0$, $d\mu/dt = 8.02 \times 10^{-11} (R_0/R)^{1.2}$ gm/cm²-yr. Cratering parameters: penetration of majority small; $\bar{s}_c = 2820(1 + 1.47R^2)$ dyne/cm²; $k = 3.1$; $\gamma = 45°$; $p = 1.82(1 + 1.47R^2)^{-1/30}$; $D = 228(1 + 1.47R^2)^{-0.233} = B_0/2R$; $x_0 = 2.91(1 + 1.47R^2)^{-1/30}R$; $x' = x_0(1 - F_B)$; cumulative crater area coverage $\sigma_B = 0.785 \int B_1 B_2(Gg\,\Delta N)$.

Part C. Apollo Group Meteorites (J_1)

The projectile J_1 is explosively destroyed. Parameters: $w_0 = 20$ km/sec; $\delta = 3.5$ gm/cm³; radius nonpenetrating upper limit, $R_0 = 200$ cm (when penetrating, not limited); $n = 3.7$; cumulative number flux, $R > 0$, $dN/dt = 8.4 \times 10^{-14}R^{-2.7}$ cm⁻² and yr⁻¹ cumulative mass $d\mu/dt = 5.5 \times 10^{-11}(R/R_0)^{0.3}$ gm/cm²-yr. Cratering parameters: $\bar{s}_p = 2.48 \times 10^4(2 + \frac{1}{3}x_0^2)$ dyne/cm² when $x_0 < 173$ cm and $\bar{s}_p = 7.4 \times 10^8[1 - (115/x_0)]$ when $x_0 > 173$ cm; $\bar{s}_c = \bar{s}_p/17.6$, $B_0 = 2DR$, $x_p = 2pR$, $x_0 = 1.6pR$, $x' = x_0(1 - F_B)$; $k = 3.3$, $\gamma = 45°$.

Part D. Ejecta from Penetrating Cratering Events (J_e)

The projective J_e is not destroyed; $\delta = 2.6$.

(a) Primary Impacts

Contributing impacts (Apollo Meteorite type) are assumed with R from 400 to 1600 cm (larger impacts are outside reach of "normal" surface sample); typical "feeding" impact $R = 800$ cm, $B_0 = 30R = 24,000$ cm, yielding largest ejecta blocks $r_{max} = \frac{1}{4}R = r_0 = B_0/120 = 200$ cm, $n = 3.875$. Cumulative number influx of ejecta, $r < r_0$, $dN/dt = 1.73 \times 10^{-16}[(r_0/r)^{2.875} - 1]$ cm⁻²-yr⁻¹ cumulative mass influx of ejecta, $d\mu/dt = 3.47 \times 10^{-7}y$ gm/cm²-yr, with $y = (r/r_0)^{0.125}$. Maximum velocity of ejection, $w_{max}^2 = u_s^2\lambda^2/y^2$, and average $w_0^2 = \frac{1}{2}w^2_{max}$; for parent crater, $\bar{s}_c = 5.7 \times 10^8$ dyne/cm², $\rho = 2.6$, $u_s^2 = 2.19 \times 10^8$, $\lambda^2 = 0.22$; $\sin\gamma = 0.8y$; maximum distance of flight from parent crater, $L_{max} = (w^2_{max}/g)2\sin\gamma\cos\gamma$.

Impact into overlay, $w_1/w_0 = 0.842$, $\bar{s}_p = 2.48 \times 10^4(2 + x^2)$, $\bar{s}_p$ and $\bar{s}_c$ as in Part C of this table, $P = 0.562/r$ cm⁻¹, $Q = 8.05r^2\cos^2\gamma$ (cm/sec)²,

$x_0 = (\xi_0 \cos \gamma)/P$, $x_p = x_0 + \frac{1}{2}r \cos \gamma$, $\sigma = \pi r^2$ cm^2, $\mu = 10.9r^3$ gm, $m = 3.47r$ gm/cm^2, $a^2P^2 = 0.632/(r \cos \gamma)^2$, $x_p B_0^2 = V_p$ (pressure component) $+ V_d$ (dynamic component) $= V$ (total volume), $V_p = 5.72\sigma x_0 \sec \gamma$, $V_d = 10.13k\mu w_0(\bar{s}_p)^{-1/2}$, $x' = x_0(1 - F_B)$, all to be used with the equations of Section II, E. The blocks under oblique incidence ricochet and usually settle on the undisturbed surface not far from the crater.

(b) *Primary Plus Ricocheting Impacts*

Contains crater statistics for the sum total of ricocheting chains.

Cumulative number of craters to indicated limit (B_0), $dN_r/dt = 1.57 \sum (Gg \, \varDelta N)(1.5 - 0.5Gg)$ cm^{-2}-yr^{-1} (primaries + ricocheting chain), where $\varDelta N$ is the differential number of primary impacts.

Table XLI is given in three sections (a–c). Explanatory material (parameters, etc.) is presented here; the tablulated data follow.

Parameters (consult Table XXXVI for impact parameters, Tables LII and LIII for erosion): F_i(cm^{-2}yr^{-1}), differential influx; ν, deletion expectation per year (from all five sources); B_0/x', crater profile ratio at impact; $\tau_0 = 1/\nu$(yr); n_0 is the differential number of craters per 100 m^2 which would survive if uneroded for 4.5×10^9 yr; $t' = 1.58 \times 10^5 B_0$(yr) is a rough first-approximation erosion lifetime as used in Table XXIX, and t_e the lifetime according to the final solution (Tables LII, LIII). The differential crater densities per 100 m^2, predicted from $n_i = 10^6 F_i[1 - \exp(-t\nu)]/\nu$, are n_0, n', n_e, corresponding to 4.5×10^9, t', and t_e, respectively. $N(N_0, N', N_e)$ are the predicted cumulative crater densities per 100 m^2. B_0 is the crater diameter, x', crater depth.

(a) Component J_e : secondary ejecta from nearby penetrating cratering events (i.e., those penetrating the overlay); primary and ricochets combined.

(b) Component J_1 : Apollo meteorites.

(c) Component S: major secondary ejecta from ray craters, corresponds either to n_s in the 4 th row of Table XXIX, B $(F_0, n_0$, and 1st approximation $n')$ or in the 4th row from bottom of Table LIII $(F$, 2nd approximation, $n_e)$; identical with component J_c after exclusion of overlapping component J_1. $B_0/x' \sim 4$ to 6 initially.

LIST OF SYMBOLS

(a) *Cratering Symbols*

a, b	fallback parameters; also semiaxes of ellipse
$a^2 = 2$ cm^2	coefficient of surface strength of granular target
B_0	rim-to-rim diameter of crater

$\frac{1}{2}B$	throwout distance from center of crater
d	spherical equivalent diameter of projectile
$D = B_0/d$	relative crater diameter
F	coefficient of lateral transmission of penetration (pressure) work against cohesive resistance
f_b	differential fallback fraction
F_b	integrated fallback fraction
f_s	coefficient of friction
f_g	vaporized fraction in central funnel
g	acceleration of gravity
H	rim-to-bottom crater depth
K_a	drag coefficient of inertial (hydrodynamic) resistance
k	coefficient of radial momentum transfer
L	flight distance of ejecta
M	total cratering mass affected
M_c	cratering mass crushed
m	mass load of projectile per unit cross section
$p = x_p/d$	the relative penetration
P, Q	penetration parameters for granular target
Q	symbol of central funnel
q	shock heating per unit mass
R	equivalent radius of projectile front surface (cross section)
s, s_p	lateral (crushing) and frontal (compressive) strength of target, respectively
s_c	cohesive component of lateral strength
s_t	tensile strength
S_c, S_p	strength parameters for granular target
u	shock front velocity
u_s	ultimate (destructive) shock velocity of target
V, V_d, V_p	total cratering volume affected, and its dynamic and pressure components, respectively
V_c	cratering volume crushed
v, v_0	ejection velocity from depth and from surface, respectively
w_0	initial impact velocity of projectile
$w_1 = \psi w_0$	velocity of entry into granular target as decreased by shock
w_m	minimum velocity for destruction of projectile
x_0	depth of penetration of front of projectile below original target surface
x_p	maximum depth of crater bottom affected
x'	apparent depth of crater
x_c	average depth of crushing
y	fraction of cratering mass inside shock front
y_Q	fraction of cratering mass in central funnel
z	zenith angle of incidence or ejection
β, β_0	angle and maximum angle of ejection relative to normal, respectively
γ	angle of incidence of projectile relative to normal
δ	density of projectile
ϵ	ellipticity of crater
λ	coefficient of kinetic efficiency (elasticity) at ejection
λ_x, λ_0	same, at depth x and in central funnel, respectively
λ_r	same, in ricocheting
μ	mass of projectile

ρ	density of target (rock, soil)
σ	cross section of projectile
σ_s	strength of target at secondary ray crater

(b) *Planetary Encounters and Cosmogony Symbols*

$A = a/r$	relative semimajor axis of orbit
A_p	total accretion rate (on a satellite)
a	orbital semimajor axis
a_m ; a_1 , a_2	distance of Moon from Earth; Earth radii
c_1 , c_2	average specific heat of solid and liquid, respectively
D_t	a distance inside Roche's limit
D_r	Roche's limit of distance for tidal disruption
e, e_0 , e'	orbital eccentricity
f_m	melted fraction of cratering material
f_∞	fraction of particles ejected out of the system in gravitational encounters
G	gravitational constant
H_0	premelting heat per unit mass
H_t	heat of fusion
i_e	inclination of the orbits of two colliding bodies to the resultant orbit
i, i_p , i_0	orbital inclinations
J_m	rate of mass accretion per unit area
K_p	coefficient of P_0 , or of the average encounter probability
k_t , k_0	thermal conductivity, and same of compact rock
k_s	Stefan's radiation constant
L_r	radial damping length for collision of planetesimals
m, m_0 , m_c	mass load per unit cross section, of preplanetary ring, of planetesimal, respectively
m_c'	linear mass load of planetesimal
N	number density
N_t	number of fragments
N_r	number of fragments in one coherent ring
P_e , P_m	mathematical expectation of encounter per orbital revolution, and its upper limit, respectively
P_0	average P_e for a random U vector
Q_0	rate of black-body radiation
R_a	radius of sphere of gravitational action
R_0	radius of planet (Earth)
R_p and R_c	radius of satellite (Moon) or planet (Earth), and of planetesimal, respectively
R_t	upper limit of radius of fragments surviving tidal breakup
r	distance from main body (heliocentric, geocentric)
T_s , T_e , T_0	temperature: subsurface, surface radiative, and initial, respectively
T_a	angular deflection parameter
T_m	temperature of fusion
t_A	time scale of increase of semimajor axis of accelerated particles
t_a	time of outward tidal drift of the Moon
t_F	average time of free fall of particle upon satellite
t_1	period of precession of orbital plane
t_z	damping lifetime of planetesimal in ring at finite inclination
t_r	radial damping time at $i \sim 0°$

t_8	synodic orbital period
$t(\sigma)$	lifetime for encounter at parameter σ
$t(\pi)$	period of the motion of perigee
$t(\omega)$	period of the argument of periastron (perigee, perihelion)
u_0	equatorial velocity of rotation
U, U_m	Jacobian velocity of encounter, in units of circular velocity, and its average, respectively
U_r	radial component of U
v	encounter velocity in metric units
v_c	orbiting circular velocity
v_h	heliocentric velocity
v_∞	escape velocity
w_f	minimum impact velocity to cause fusion
δ, δ_0, δ_p	density: of planetesimal, of Earth, and of Moon, respectively
Δh	thickness of lava crust
$\zeta(U)$	fraction of particles surviving (eluding) physical collisions
η_m	mass fraction retained by preplanetary ring
η_t	probability of encounter for time interval t
θ, θ_0, θ_{max}	melted fraction
μ	mass of "planet" or "satellite" revolving around central body of mass $1 - \mu$
μ_1, μ_2	masses of two competing accreting nuclei
ρ, ρ_0	space density of particulate matter in preplanetary ring
σ_0	target radius for physical collision
σ_e	target radius or encounter parameter (in units of r)
σ_a	target radius for angular deflection of $90°$
ω	angular distance of periastron from node, or argument of periastron
ω_0	orbital angular velocity
ω_f	angular velocity of rotation

(c) Symbols in Mixed Context

A_a, A_v, A_e, A_0	ricocheting amplification factors
A_d	cohesive force between grains
A_r	reflectivity at normal incidence
d_g	diameter of grain
dN/dt	number flux
E_1, E_e, E_m'	transport efficiency
E_f	statistical erosion (survival) factor of craters
F_i	meteorite incidence; creation rate of craters
F_d	downhill flow (of dust, of overlay)
G_g	fraction of granular target at impact into overlay
G_f, G_∞	gain factors in filling of depression by overlay
H	depth of depression; also total layer of accreted overlay
H_p, H_8	impact penetration into overlay
h_r	height of rim
J	radial (cratering) momentum flux
J_M, J_0, J_1, J_2, J_3, J_e	the six components of the lunar surface cratering (radial) momentum flux in overlay, as well as the symbols of the components themselves: micrometeorites, dustball meteors, Apollo meteorites, comet nuclei, Mars asteroids, secondary ejecta, respectively

J_c supplementary to preceding, overlay—penetrating flux relating to intermediate and large craters: $S = J_c - J_1$

J_r total radial momentum in a cratering event

L_d electrostatic screening length in plasma

L_e effective depth from which radiation is emitted

L_m radius of spread of cratering ejecta

L_t effective depth of thermal wave in soil

n power index in differential frequency function of particle radii (frequency index of radii)

N_a , N_c cumulative number of impacts: by Apollo group, by comet nuclei, respectively

n_i , n_0 , n_c , n_p , n_s number of impacts; crater areal density

Q_a amplitude of heat content per cm² of surface

Q_m , Q_t mixing factor of overlay, for past and future, respectively

R (with proper subindices)   impinging projectile radius

r , r_0 ejecta particle radii as distinct from R

S area

S_0 , S_t area of crater, area covered by ejecta, respectively

t_t relative age of craters

t_e , t' degradation lifetime of craters, and provisional value, respectively

t_I erosion lifetime of rim

t_t total degradation lifetime of rimmed craters

t_m mixing time of overlay

V_e volume of ejecta

v_e , v_a velocity of inelastic grain capture, ricocheting escape velocity from trap, respectively

X , X_0 , X_1 overlay thickness; from small craters; averaged from large craters

Y Young's modulus

α angle of inclination to horizon

α' kinetic parameter for ejecta (cm⁻¹)

γ_t thermal inertia parameter

ϵ_i , ϵ_0 dielectric constant, of granular and of compact rock, respectively

η fraction of momentum retained after penetration of a layer

Θ_a surface temperature amplitude

κ kinetic thickness of a protective layer

λ_e wavelength

μ_1 , μ_2 , μ_3 , μ_0 total mass influx from flux components (*see* J components)

$d\mu/dt$ mass flux

ν overlapping depletion rate of craters

σ_B cumulative crater area coverage per unit time and area

σ_p contact area of grains per unit cross section

τ fractional area covered by ejecta; also total erosion relaxation time

τ_0 , τ_F , τ_t , τ_e , τ_s exponential relaxation times in various processes of crater degradation

τ_m total supply of overlay accretion time

τ_E total degradation time scale (relaxation time)

χ , χ_0 equivalent thickness of overlay annually displaced by meteorite impact

χ_s rock or boulder ablation (cm/yr)

φ_L fraction of χ ejected beyond distance L

ω_e negative gain factor around depression

REFERENCES

Alfvén, H. (1963). *Icarus* 1, 357–363.
Alfvén, H. (1965). *Science* 148, 476–477.
Alfvén, H. (1968). Personal discussion.
Allan, D. W. and Jacobs, J. A. (1956). *Geochim. Cosmochim. Acta* 9, 256–272.
Baldwin, R. B. (1949). "The Face of the Moon." Univ. of Chicago Press, Chicago, Illinois.
Baldwin, R. B. (1963). "The Measure of the Moon." Univ. of Chicago Press, Chicago, Illinois.
Baldwin, R. B. (1964a). *Ann. Rev. Astron. Astrophys.* 2, 73–94.
Baldwin, R. B. (1964b). *Astron. J.* 69, 377–392.
Bastin, J. A. (1965). *Nature* 207, 1381–1382.
Binder, A. B., Cruikshank, D. P., and Hartmann, W. K. (1965). *Icarus* 4, 415–420.
Christensen, E. M., *et al.* (1967). *J. Geophys. Res.* 72, 801–813.
Cloud, P. E. (1968). *Science* 160, 729–736.
Coffeen, D. L. (1965). *Astron. J.* 70, 403–413.
Comerford, M. F. (1966). "Comparative Erosion Rates of Stone and Iron Meteorites under Small Particle Bombardment," Preprint. Smithsonian Inst. Astrophys. Obs.
Darwin, Sir George. (1879). *Phil. Trans. Roy. Soc. London* 170, 447–530.
Dodd, R. T., Salisbury, J. W., and Smalley, V. G. (1963). *Icarus* 2, 466–480.
Dollfus, A. (1966). *In* "The Nature of the Lunar Surface" (W. N. Hess, D. H. Menzel, and J. A. O'Keefe, eds.), pp. 155–172. Johns Hopkins Press, Baltimore, Maryland.
Drake, F. (1966). *In* "The Nature of the Lunar Surface," (W. N. Hess, D. H. Menzel, and J. A. O'Keefe, eds.), pp. 277–284. Johns Hopkins, Press, Baltimore, Maryland.
Egan, W. G. and Smith, L. L. (1965). *Grumman Mem. RM-304*, 1–66.
Evans, J. V. (1962). Tech. Rept. No. 256. Lincoln Lab., M.I.T., Bedford, Massachusetts; *in* "Conference on Lunar Exploration," Paper VIII. Virginia Poly. Inst. Blacksburg, Virginia, 1962.
Evans, J. V. and Pettengill, G. H. (1963). *In* "The Moon, Meteorites and Comets" (B. M. Middlehurst and G. P. Kuiper, eds.), pp. 129–161. Univ. of Chicago Press, Chicago, Illinois.
Fedorets, V. A. (1952). *Publ. Astron. Observ. Kharkov* 2, 49–172.
Fielder, G. (1962). *J. British Astron. Assoc.* 72, 223.
Fielder, G. (1963). *Nature* 198, 1256–1260.
Fielder, G. (1965). *M.N.R.A.S.* 129, 351–361.
Fielder, G. (1966). *M.N.R.A.S.* 132, 413–422.
Gault, D. E., Quaide, W. L., Oberbeck, V. R., and Moore, H. J. (1966). *Science* 153, 985–988.
Gear, A. E. and Bastin, J. A. (1962). *Nature* 196, 1305.
Gehrels, T., Coffeen, T., and Owings, D. (1964). *Astron. J.* 69, 826–852.
Gerstenkorn, H. (1955). *Z. Astrophys.* 36, 245–274.
Gold, T. (1955). *M.N.R.A.S.* 115, 585–604.
Gold, T. (1962). *In* "The Moon" (Z. Kopal and Z. K. Mikhailov, eds.), pp. 433–439. Academic Press, New York.
Gold, T. (1966). *In* "The Nature of the Lunar Surface" (W. N. Hess, D. H. Menzel, and J. A. O'Keefe, eds.), pp. 107–121. Johns Hopkins Press, Baltimore, Maryland.
Gold, T. and Hapke, B. W. (1966). *Science* 153, 290–293.
Hagfors, T. (1966). *In* "The Nature of the Lunar Surface" (W. N. Hess, D. H. Menzel, and J. A. O'Keefe, eds.), pp. 229–239. Johns Hopkins Press, Baltimore, Maryland.
Halajian, J. D. and Spagnolo, F. A. (1966). *Grumman Mem. RM-308*, 1–89.

Hapke, B. W. and Van Horn, H. (1963). *J. Geophys. Res.* **68**, 4545–4586.

Hapke, B. W. (1964). *J. Geophys. Res.* **69**, 1147–1151.

Hapke, B. W. (1965). *Ann. N. Y. Acad. Sci.* **123**, 711–721.

Hapke, B. W. (1966a). *In* "The Nature of the Lunar Surface" (W. N. Hess, D. H. Menzel, and J. A. O'Keefe, eds.), pp. 141–154. Johns Hopkins, Press, Baltimore, Maryland.

Hapke, B. W. (1966b). *Astron. J.* **71**, 333–339.

Hapke, B. W. (1968). *Planet. Space Sci.* **16**, 101–110.

Hapke, B. and Gold, T. (1967). *Sky Telescope* **33**, 84.

Hartmann, W. K. (1965). *Icarus* **4**, 157–165 and 207–213.

Hartmann, W. K. (1966). "Nininger Meteorite Award Paper." Arizona State University, Tempe, Arizona.

Hartmann, W. K. (1967). *Icarus* **7**, 66–75.

Ingrao, H. C., Young, A. T., and Linsky, J. L. (1966). *In* "The Nature of the Lunar Surface" (W. N. Hess, D. H. Menzel, and J. A. O'Keefe, eds.), pp. 185–211. Johns Hopkins Press, Baltimore, Maryland.

Jaffe, L. D. (1965). *J. Geophys. Res.* **70**, 6129–6138.

Jaffe, L. D. (1966a). *J. Geophys. Res.* **71**, 1095–1103.

Jaffe, L. D. (1966b). *Icarus* **5**, 545–550.

Jaffe, L. D. (1967). *J. Geophys. Res.* **72**, 1727–1731.

Jaffe, L. D. and NASA Team. (1966a). *Science* **152**, 1737–1750.

Jaffe, L. D. and NASA Team. (1966b). "Surveyor I Mission Report," Part II, Tech. Rept. No. 32–1023. JPL, Pasadena, California.

Jeffreys, Sir Harold. (1947a). *M.N.R.A.S.* **107**, 260–262.

Jeffreys, Sir Harold. (1947b). *M.N.R.A.S.* **107**, 263–267.

Kopal, Z. (1965). *Boeing Sci. Res. Lab. Math. Note* No. 430, pp. 1–39.

Kopal, Z. (1966a). *In* "The Nature of the Lunar Surface" (W. N. Hess, D. H. Menzel, and J. A. O'Keefe, eds.), pp. 173–183. Johns Hopkins Press, Baltimore, Maryland.

Kopal, Z. (1966b). *Icarus* **5**, 201–213.

Kopal, Z. (1968). *Boeing Sci. Res. Lab. Math. Note* No. 566, pp. 1–43.

Krotikov, V. D. and Troitsky, V. S. (1962). *Astron. Zh.* **39** 1089–1093.

Krotikov, V. D. and Troitsky, V. S. (1963). *Usp. Fiz. Nauk* **81**, 589–639.

Kozyrev, N. A. (1959a). *Priroda*, March, 84–87.

Kozyrev, N. A. (1959b). *Sky Telescope* **18**, 561.

Kozyrev, N. A. (1962). *In* "The Moon" (Z. Kopal and Z. K. Mikhailov, eds.), pp. 265–266. Academic Press, New York.

Kuiper, G. P. (1954). *Proc. Nat. Acad. Sci. Wash.* **40**, 1096–1112.

Kuiper, G. P. (1966). *Commun. Lunar Planet. Lab.* **4.1**, 1–70.

Kuiper, G. P., Arthur, D. W. G., Moore, E., Tapscott, J. W., and Whitaker, E. A. (1960). "Photographic Lunar Atlas." Univ. of Chicago Press, Chicago, Illinois.

Levin, B. J. (1966a). *Proc. Caltech-JPL Lunar Planet. Conf.*, JPL TM 33–266, 61–76, and elsewhere.

Levin, B. J. (1966b). *In* "The Nature of the Lunar Surface" (W. N. Hess, D. H. Menzel, and J. A. O'Keefe, eds.), pp. 267–271. Johns Hopkins, Press, Baltimore, Maryland.

Linsky, J. L. (1966). *Icarus* **5**, 606–634.

Lipsky, Y. N. (1966). *Sky Telescope* **32**, 257–260.

Low, F. J. (1965). *Astrophys. J.* **142**, 806–808.

MacDonald, G. J. F. (1964). *Rev. Geophys.* **2**, 467–541.

Majeva, S. V. (1964). *Doklady Akad. Nauk SSSR* **159**, 294–297.

Marcus, A. H. (1964). *Icarus* **3**, 460–472.

Marcus, A. H. (1966a). *M.N.R.A.S.* **134**, 269–274.

Marcus, A. H. (1966b). *Icarus* **5**, 165–200 and 590–605.

Minnaert, M. (1961). *In* "Planets and Satellites" (G. P. Kuiper and B. M. Middlehurst, eds.), pp. 213–248. Univ. of Chicago Press, Chicago, Illinois.

Munk, W. S. (1968). Harold Jeffreys Lecture delivered at Burlington House, March 29, 1968.

Murphey, B. F. (1961). *Proc. Cratering Symp.* Paper G, pp. 1–13. Lawrence Radiation Lab., Livermore, California.

Murray, B. C. and Wildey, R. L. (1964). *Astrophys. J.* **139**, 734–750.

NASA. (1964). "Ranger VII Atlas. Part I—Camera A." JPL-Cal. Tech.

NASA. (1965a). "Ranger IX Atlas." JPL-Cal Tech.

NASA. (1965b). "Ranger VII Atlas. Part III—Camera P." JPL-Cal Tech.

NASA. (1967). "Surveyor III Mission Report," Tech. Rept. 32–1177, Part, I and II. JPL, Pasadena, California.

Newell, H. E. and NASA Team. (1966). "Surveyor I Preliminary Report," NASA SP-126.

Newell, H. E. and NASA Team. (1967). "Surveyor III Preliminary Report," NASA SP-146.

Nordyke, M. D. (1961). *Proc. Cratering Symp.*, Paper F, pp. 1-14. Lawrence Radiation Lab., Livermore, California.

Oetking, P. (1966). *J. Geophys. Res.* **71**, 2505–2513.

O'Keefe, J. A. (1964). *Science* **146**, 514–515.

O'Keefe, J. A. (1966a). *In* "The Nature of the Lunar Surface" (W. N. Hess, D. H. Menzel, and J. A. O'Keefe, eds.), pp. 259–266. Johns Hopkins Press, Baltimore, Maryland.

O'Keefe, J. A. (1966b). *Sci. News* **90**, 441.

O'Keefe, J. A., Lowman, P. D., and Cameron, W. S. (1967). *Science* **155**, 77–79.

Olson, S. (1966). *Am. Sci.* **54**, 458–464.

Öpik, E. (1936). *Acta Comment. Univ. Tartu* **A30**; *Tartu Observ. Publ.* **28**, No. 6 (1900).

Öpik, E. J. (1951). *Proc. Roy. Irish Acad.* **54A**, 165–199; *Armagh Observ. Contrib.* No. 6.

Öpik, E. J. (1955). *Irish Astron. J.* **3**, 245–248.

Öpik, E. J. (1956). *Irish Astron. J.* **4**, 84–135; *Armagh Observ. Contrib.* No. 19.

Öpik, E. J. (1958a). *Irish Astron. J.* **5**, 14–36; *Armagh Observ. Contrib.* No. 24.

Öpik, E. J. (1958b). *Irish Astron. J.* **5**, 79–95; *Armagh Observ. Contrib.* No. 28.

Öpik, E. J. (1958c). "Physics of Meteor Flight in the Atmosphere." Wiley (Interscience), New York.

Öpik, E. J. (1960). *M.N.R.A.S.* **120**, 404–411; *Armagh Observ. Contrib.* No. 29.

Öpik, E. J. (1961a). *Proc. Geophys. Lab.—Lawrence Radiation Lab. Cratering Symp.*, Paper S, pp. 1–28. Lawrence Radiation Lab., Livermore, California.

Öpik, E. J. (1961b). *Astron. J.* **66**, 60–67.

Öpik, E. J. (1962a). *In* "Progress in the Astronautical Sciences" (S. F. Singer, ed.), pp. 215–260. North-Holland Publ., Amsterdam.

Öpik, E. J. (1962b). *Planet. Space Sci.* **9**, 211–244.

Öpik, E. J. (1963a). *Advan. Astron. Astrophys.* **2**, 219–262.

Öpik, E. J. (1963b). *Irish Astron. J.* **6**, 39–40.

Öpik, E. J. (1963c). *Irish Astron. J.* **6**, 3–11; *Armagh Observ. Contrib.* No. 38 (1963); *Irish Astron. J.* **6**, 162–164.

Öpik, E. J. (1964). *Irish Astron. J.* **6**, 279–280.

Öpik, E. J. (1965a). *Advan. Astron. Astrophys.* **4**, 301–336.

Öpik, E. J. (1965b). *In* "Interactions of Space Vehicles with an Ionized Atmosphere," pp. 3–60. Pergamon Press, Oxford; *Armagh Observ. Contrib.* No. 52.

Öpik, E. J. (1965c). *Irish Astron. J.* **7**, 92–104; *Armagh Observ. Leafl.* No. 67.

Öpik, E. J. (1966a). *Congr. Colloq. Univ. Liege* **37**, 523–574.

Öpik, E. J. (1966b). *Congr. Colloq. Univ. Liege* **37**, 575–580.

Öpik, E. J. (1966c). *Irish Astron. J.* **7**, 141–161; *Armagh Observ. Contrib.* No. 54.

Öpik, E. J. (1966d). *In* "The Nature of the Lunar Surface" (W. N. Hess, D. H. Menzel, and J. A. O'Keefe, eds.), pp. 287–291. Johns Hopkins Press, Baltimore, Maryland.

Öpik, E. J. (1966e). *Science* **153**, 255–265.

Öpik, E. J. (1966f). *Irish Astron. J.* **7**, 201–205.

Pettengill, G. H. and Henry, J. C. (1962). *J. Geophys. Res.* **67**, 4881–4885.

Pettit, E. (1961). *In* "Planets and Satellites" (G. P. Kuiper and B. M. Middlehurst, eds.), pp. 400–428. Univ. of Chicago Press, Chicago, Illinois.

Rolsten, R. F., Hopkins, A. K., and Hunt, H. H. (1966). *Nature* **212**, 495–497.

Rosenberg, D. L. and Wehner, G. K. (1964). *J. Geophys. Res.* **69**, 3307–3308.

Ryan, J. A. (1966). *J. Geophys. Res.* **71**, 4413–4425.

Saari, J. M. (1964). *Icarus* **3**, 161–163.

Saari, J. M. and Shorthill, R. W. (1966). *Sky Telescope* **31**, 327–331.

Schmidt, O. J. (1950). *Izv. Akad. Nauk SSSR, Ser. Fiz.* **14**, No. 1, 29–45.

Senior, T. B. A. (1962). *In* "Conference on Lunar Exploration" Paper IX. Virginia Poly. Inst., Blacksburg, Virginia.

Senior, T. B. A. and Siegel, K. M. (1960). *J. Res. Nat. Bur. Std.* **D64**, 217–229.

Shoemaker, E. M. (1963). *In* "The Moon, Meteorites, and Comets" (B. M. Middlehurst and G. P. Kuiper, eds.), pp. 301–336. Univ. of Chicago Press, Chicago, Illinois.

Shoemaker, E. M. (1966). *In* "The Nature of the Lunar Surface" (W. N. Hess, D. H. Menzel, and J. A. O'Keefe, eds.), pp. 23–77. Johns Hopkins Press, Baltimore, Maryland.

Shoemaker, E. M., Hackman, R. J., and Eggleton, R. E. (1963). *Advan. Astron. Sci.* **8**, 70–89.

Shorthill, R. W. and Saari, J. M. (1965a). *Nature* **205**, 964–965.

Shorthill, R. W. and Saari, J. M. (1965b). *Science* **150**, 210–212.

Shorthill, R. W. and Saari, J. M. (1966). *In* "The Nature of the Lunar Surface" (W. N. Hess, D. H. Menzel, and J. A. O'Keefe, eds.), pp. 215–228. Johns Hopkins Press, Baltimore, Maryland.

Singer, S. F. and Walker, E. H. (1962). *Icarus* **1**, 112–120.

Slichter, L. B. (1963). *J. Geophys. Res.* **68**, 4281–4288.

Smith, B. G. (1967). *J. Geophys. Res.* **72**, 1398–1399.

Smoluchowski, R. (1966). *J. Geophys. Res.* **71**, 1569–1574.

Sorokin, N. A. (1965). *Astron. Zh.* **42**, 1070–1074.

Thompson, T. W. and Dyce, R. B. (1966). *J. Geophys. Res.* **71**, 4843–4853.

Trask, N. J. (1966). Tech. Rept. No. 32–800, pp. 252–263. JPL, Pasadena, California.

Troitsky, V. S. (1962). *Astron. Zh.* **39**, 73–78.

Turkevich, A. L., Franzgrote, E. J., and Patterson, J. H. (1967). *Science* **158**, 635–637.

Turkevich, A. L., Patterson, J. H., and Franzgrote, E. J. (1968). *Science* **160**, 1108–1110.

Twersky, V. (1962). *J. Math. Phys.* **3**, 724–734.

Urey, H. C. (1960a). *Proc. 1st. Intern. Space Sci. Symp., Nice, 1960*, pp. 1114–1122. Wiley (Interscience), New York; and elsewhere.

Urey, H. C. (1960b). *J. Geophys. Res.* **65**, 358–359; and elsewhere.

Urey, H. C. (1965). Tech. Rept. No. 32–800, pp. 339–362. JPL, Pasadena, California.

Urey, H. C. (1966). *In* "The Nature of the Lunar Surface" (W. N. Hess, D. H. Menzel, and J. A. O'Keefe, eds.), pp. 3–21. Johns Hopkins Press, Baltimore, Maryland.

Walker, E. H. (1966). *J. Geophys, Res,* **71**, 5007–5010.

Walker, E. H. (1967). *Icarus* **7**, 183–187.

Watts, C. B. (1963). *Astron. Papers Am. Ephemeris* **17**, 1–951.

Watts, R. N. (1967). *Sky Telescope* **33**, 97.

Wattson, R. B. and Danielson, R. E. (1965). *Astrophys. J.* **142**, 16–22.

Wattson, R. B. and Hapke, B. W. (1966). *Astrophys. J.* **144**, 364–368.

Wehner, G. K., Kenknight, C. E., and Rosenberg, D. (1963a). *Planet. Space Sci.* **11**, 1257–1261.

Wehner, G. K., Kenknight, C., and Rosenberg, D. L. (1963b). *Planet. Space Sci.* **11**, 885–895.

Wesselink, A. J. (1948). *Bull. Astron. Inst. Neth.* **10**, 351–363.

Whipple, F. L. (1959). *Vistas Astronautics* **2**, 267–272.

Whitaker, E. A. (1966). *In* "The Nature of the Lunar Surface" (W. N. Hess, D. H. Menzel, and J. A. O'Keefe, eds.), pp. 79–98. Johns Hopkins Press, Baltimore, Maryland.

Wildey, R. L. and Pohn, H. A. (1964). *Astron. J.* **69**, 619–634.

Author Index

Numbers in italics refer to the pages on which the complete references are listed.

A

Ahrens, T. J., 49, *103*
Alekseev, V. A., 14, 19, *26*
Aleshina, T. N., 9, 14, 19, 24, *26*, *28*
Alfvén, H., 135, 165, *333*
Allan, D. W., 158, 296, *333*
Allenby, R. J., 30, *103*
Arthur, D. W. G., 181, *334*

B

Baars, J. W., 24, *26*
Baldock, R. V., 21, *26*
Baldwin, I. E., 2, *26*
Baldwin, R. B., 109, 110, 159, 160, 161, 162, 167, 168, 169, 174, 175, 184, 185, 189, 220, 227, 300, 307, *333*
Barrick, D. E., 62, *103*
Bastin, J. A., 21, *26*, 204, 210, *333*
Batson, R. M., 84, *105*
Bay, Z., 29, *103*
Beckmann, P., 62, 67, 84, *103*
Beringer, R., 1, *26*
Bernett, E. C., 9, *26*
Binder, A. B., 203, *333*
Blevis, B. C., 34, 41, *103*
Bondar, L. N., 19, *26*
Borough, H. C., 93, *105*
Bowhill, S. A., 61, *103*
Bracewell, R. N., 3, *26*, *27*, 89, *103*
Bramley, E. N., 60, *103*
Brockelman, R. A., 73, 76, 83, 86, *104*
Broten, N. W., 19, *27*
Brouwer, D., 32, 33, *105*
Brown, W. E., 59, *103*
Browne, I. C., 37, 52, *103*

C

Cameron, W. S., 212, *335*
Campbell, M. J., 50, 51, 86, 93, *103*, *105*
Chapman, J. H., 41, *103*
Christensen, E. M., 231, 235, *333*
Clary, M. C., 42, *105*
Clegg, P. E., 21, *26*
Cloud, P. E., 164, *333*
Coates, G. T., 19, *28*
Coffeen, D. L., 201, *333*
Coffeen, T., 201, *333*
Comerford, M. F., 117, *333*
Conley, J. M., 93, *105*
Craig, K. J., 31, 32, 33, *105*
Crocker, E. A., 41, 54, *104*
Cruikshank, D. P., 203, *333*

D

Danforth, H. H., 73, 76, 83, 86, *104*
Daniels, F. B., 52, 60, 61, 62, *103*
Danielson, R. E., 203, *337*
Darwin, G. H., 151, *333*
Davidson, A. W., 19, *27*
Davies, R. D., 14, 19, 24, *26*, 84, *103*
Davis, J. R., 39, 41, 54, *103*
DeWitt, J. H., 29, 39, *103*
Dicke, R. H., 1, *26*
Dickenson, R. M., 22, *28*
Dmitrenko, D. A., 19, *27*
Dodd, R. T., 184, *333*

Brunschwig, M., 49, *103*
Bruton, R. H., 31, 32, *105*
Buettner, K. J. K., 9, *26*
Burov, A. B., 9, 24, *28*

339

Subject Index

A

Age of Moon, 216
Almuñécar beach, 122
Andesite, 49
Andesitobasalt, 25
Angular power-spectrum studies of Moon, 54, 56
Apollo 12, 290
Apollo landings, 49, 287
Apollo-type asteroids, 153, 159, 160, 183, 215, 219, 220, 222, 225, 239, 245, 247, 312, 325, 327
Arecibo Ionospheric Observatory, 30, 87, 89, 99
Arizona crater, 230
Asteroids, 159, 184
Autocorrelation function, 61
  exponential, 65, 66, 69
  Fourier transform of, 61
  Gaussian, 65
Australites, 25
Average echo intensity from Moon, 72
Axis of rotation of Moon, 35

B

Backscattered power, 76–79
Backscattering coefficients, 75, 81
Basaltic scoria, 49
Basalts, 25, 49
Boltzmann's constant, 3
Boulders on Moon, 56, 123, 197, 198, 210, 263, 290
  lifetime of, 258
Brightness temperature of Moon, 6, 10
Brightness variations of Moon, 8, 11, 12
Brown's theory (of Moon's motion), 32

C

Coesite, 117
Contour map of Moon, 159
Cosmic radiation reflected in Moon, 6
Craters on Moon, 56, 167, 184, 186, 189, 216, 217, 219, 221, 227, 269, 275, 276, 280, 281
  Alphonsus, 109, 110, 180, 184, 185, 186, 187, 188, 189, 191, 192, 194, 195, 197, 212, 213, 269, 279, 284, 286, 288, 301
    diameter of, 195
    floor of, 189–194, 301
    peak, 11, 189–191, 193, 195
    wall, 191, 194
  Archimedes, 183, 185, 220, 299
  Aristarchus, 92, 111, 209
  Aristillus, 211
  Atlas, 211
  Bullialdus, 179, 180, 181
  Campanus, 185
  Copernicus, 92, 95, 162, 181, 182, 209, 210, 299
    rays, 161, 220
    rim of, 181
    secondaries, 183, 185
  Diophantes, 211
  Eratosthenes, 181, 299
  Flammarion, 184, 185, 189
  Flamsteed's ring, 212
  Fracastorius, 185
  Kepler, 111, 209
  Kies, 185
  Langrenus, 93, 94, 193
  Le Monnier, 185
  Plato, 102, 185
  Plinius, 211
  Posidonius, 211

344